ARMED CONFLICT, WOMEN AND CLIMATE CHANGE

The gender-differentiated and more severe impacts of armed conflict upon women and girls are well recognised by the international community, as demonstrated by UN Security Council Resolution (UNSCR) 1325 on Women, Peace and Security and subsequent resolutions. Similarly, the development community has identified gender-differentiated impacts upon women and girls as a result of the effects of climate change. Current research and analysis has reached no consensus as to any causal relationship between climate change and armed conflict, but certain studies suggest an indirect linkage between climate change effects such as food insecurity and armed conflict.

Little research has been conducted on the possible compounding effects that armed conflict and climate change might have on at-risk population groups such as women and girls. *Armed Conflict, Women and Climate Change* explores the intersection of these three areas and allows the reader to better understand how military organisations across the world need to be sensitive to these relationships to be most effective in civilian-centric operations in situations of humanitarian relief, peacekeeping and even armed conflict. This book examines strategy and military doctrine from NATO, the UK, US and Australia, and explores key issues such as displacement, food and energy insecurity, and male out-migration as well as current efforts to incorporate gender considerations in military activities and operations.

This innovative book will be of great interest to students and scholars of international relations, international development, international security, sustainability, gender studies and law.

Jody M. Prescott is a lecturer at the University of Vermont, U.S.A., where he teaches environmental law, cybersecurity law and policy, and energy law.

ARMED CONFLICT, WOMEN AND CLIMATE CHANGE

Jody M. Prescott

LONDON AND NEW YORK

First published 2019
by Routledge
2 Park Square, Milton Park, Abingdon, Oxon OX14 4RN

and by Routledge
52 Vanderbilt Avenue, New York, NY 10017

Routledge is an imprint of the Taylor & Francis Group, an informa business

© 2019 Jody M. Prescott

The right of Jody M. Prescott to be identified as author of this work has been asserted by him in accordance with sections 77 and 78 of the Copyright, Designs and Patents Act 1988.

All rights reserved. No part of this book may be reprinted or reproduced or utilised in any form or by any electronic, mechanical, or other means, now known or hereafter invented, including photocopying and recording, or in any information storage or retrieval system, without permission in writing from the publishers.

Trademark notice: Product or corporate names may be trademarks or registered trademarks, and are used only for identification and explanation without intent to infringe.

British Library Cataloguing-in-Publication Data
A catalogue record for this book is available from the British Library

Library of Congress Cataloging-in-Publication Data
Names: Prescott, Jody M., author.
Title: Armed conflict, women and climate change / Jody M. Prescott.
Description: First edition. | Abingdon, Oxon ; New York, NY : Routledge, 2019. | Includes bibliographical references and index.
Identifiers: LCCN 2018038233 (print) | LCCN 2018050736 (ebook) | ISBN 9781315467214 (Master) | ISBN 9781315467207 (Adobe Reader) | ISBN 9781315467191 (ePub) | ISBN 9781315467184 (Mobipocket unencrypted) | ISBN 9781138205321 (hardback) | ISBN 9781138205352 (pbk.) | ISBN 9781315467214 (ebook)
Subjects: LCSH: Western countries—Military policy. | Women and war. | Climatic changes. | War—Environmental aspects. | War—Social aspects. | Women and the environment. | Climate change mitigation—Western countries. | Military doctrine—Western countries.
Classification: LCC UA11 (ebook) | LCC UA11 .P63 2019 (print) | DDC 355/.03351821—dc23
LC record available at https://lccn.loc.gov/2018038233

ISBN: 978-1-138-20532-1 (hbk)
ISBN: 978-1-138-20535-2 (pbk)
ISBN: 978-1-315-46721-4 (ebk)

Typeset in Bembo
by Apex CoVantage, LLC

For my wife and daughters.

CONTENTS

Acknowledgements   *viii*
List of abbreviations   *ix*

	Introduction	1
1	Armed conflict and gender	15
2	Gender and climate change	36
3	Climate change and armed conflict	58
4	NATO strategy, doctrine and gender	89
5	UK strategy, doctrine and gender	113
6	US strategy, doctrine and gender	130
7	Australian strategy, doctrine and gender	157
8	International humanitarian law and gender	173
9	Gender in military activities and operations	198
	Conclusion	235

Index   *252*

ACKNOWLEDGEMENTS

I want to thank Eric Ahlness, Jeff Farrell, Amanda Fielding, Rachel Grimes, Suzanne Levi-Sanchez, Stacey Porter, Sheila Scanlon, Jessica Sepp, Cheryl Wachenheim and Jennifer Wittwer for their generosity in allowing me to interview them. I also want to thank Clare Hutchinson, Beth Lape, Lauren Mackenzie, Jade Mali Mizutani and Stacey Porter for reviewing and providing feedback on specific parts of Chapter 9, 'Gender in military activities and operations', and Mac Squires for his guidance on the Introduction. Finally, I wish to thank the wonderful editorial staff at Routledge, Hannah Ferguson, Annabelle Harris and Matthew Shobbrook, who worked with me so patiently to burnish away so many of the rough spots in my manuscript. Any errors of course are my own responsibility.

I am also grateful to the University of Adelaide Press and the following publications for allowing me to use work I had previously published with them: *The Georgetown University Law Center Journal of Gender and the Law*, the *Asian-Pacific Law and Policy Journal*, the *Vermont Journal of Environmental Law*, *The Royal United Services Institute Journal*, *The Three Swords* and the *NATO Legal Gazette*.

ABBREVIATIONS

ACO	Allied Command Operations
ACT	Allied Command Transformation
ADDP	Australian Defence Doctrine Publication
ADFP	Australian Defence Force Publication
ADF	Australian Defence Force
ADP	Army Doctrine Publication
AJP	Allied Joint Publication
AJEPP	Allied Joint Environmental Protection Publication
ATP	Army Techniques Publication
Chairman, JCS	Chairman, Joint Chiefs of Staff
CIMIC	Civil-Military Cooperation
CVE	Countering Violent Extremism
DFID	Department for International Development
DFS	Department of Field Support (UN)
DKI APCSS	Daniel K. Inouye Asia-Pacific Center for Security Studies
DoD	Department of Defense (US)
FAO	Food and Agricultural Organisation of the UN
FCO	Foreign and Commonwealth Office
FM	Field Manual
GC I	First Geneva Convention of 1949
GC II	Second Geneva Convention of 1949
GC III	Third Geneva Convention of 1949
GC IV	Fourth Geneva Convention of 1949
GENAD	Gender Advisor
HC IV	Hague Regulations of 1907
HIV/AIDS	Human Immunodeficiency Virus/Acquired Immune Deficiency Syndrome

HN	Host Nation
HRW	Human Rights Watch
ICG	International Crisis Group
IDF	Israel Defence Forces
IDP	Internally Displaced Person
IHL	International Humanitarian Law
IPCC	UN Intergovernmental Panel on Climate Change
ISAF	International Security Assistance Force
J-1	Joint Staff Section 1 (Personnel)
J-2	Joint Staff Section 2 (Intelligence)
J-3	Joint Staff Section 3 (Operations)
J-4	Joint Staff Section 4 (Logistics)
J-5	Joint Staff Section 5 (Plans) (NATO)
J-7	Joint Staff Section 7 (Training) (NATO)
J-9	Joint Staff Section 9 (CIMIC) (NATO)
JAS	*Jama'atu Ahlus-Sunnah Lidda'awati Wal Jihad*
JDN	Joint Doctrine Note
JP	Joint Publication
JSDF	Japan Self-Defense Forces
LEGAD	Legal Advisor
LWD	Land Warfare Doctrine
MOD	Ministry of Defence
NAP	National Action Plan
NSA	NATO Standardization Agency
NSO	NATO Standardization Office
NZDF	New Zealand Defence Force
PDSS	Persons Designated with Special Status
POW	Prisoner of War
QCJWC	Quinquepartite Combined Joint Warfare Conference
ROE	Rules of Engagement
SOF	Special Operations Forces
SDDirect	Social Development Direct
UNAMA HRU	UN Assistance Mission Afghanistan Human Rights Unit
UN DPKO	UN Department of Peacekeeping Operations
UNEP	UN Environment Programme
UNIFIL	UN Interim Force In Lebanon
UNSC	UN Security Council
UNSCR	UN Security Council Resolution
USAID	US Agency for International Development
WPS	Women, Peace and Security

INTRODUCTION

Introduction

War has different and more severe impacts upon women and girls than it does on men and boys. Climate change likewise has gender-differentiated impacts, and the manner in which they register on people is strongly mediated by cultural, economic, political and religious factors. In general, women and girls experience different and more severe climate change effects than do men and boys. In areas afflicted by both armed conflict and climate change, women and girls are particularly at risk to the compounding negative effects of these forces. In turn, these risks then pose a threat to the success of military missions in these areas, especially those operations that are civilian-centric.

I spent 25 years on active duty as a military attorney in the US Army. It took a tour in Afghanistan near the end of my career for me to begin to appreciate that there were links between armed conflict, women and climate change. It has taken a few more years to understand how these links can be operationally significant, and how, from the perspective of achieving greater protection for women and girls in situations of armed conflict, military organisations might adapt their ordinary ways of doing business to leverage their strengths and resources to increase their likelihood of success in civilian-centric operations.

This book is not the product of focused field work or academic study in the areas of climate change or gender – I am an expert in neither. Instead, it is the result of my different operational, training and educational experiences in the military. These experiences have allowed me to sift through a large amount of policy, research and writing, and to identify potential paths forward for militaries to effectively incorporate an understanding of the links between armed conflict, gender and climate change into their activities and operations.

Operations and environmental damage claims

I first worked applying environmental law to operations when I was assigned to US Army Claims Service Europe in Germany in 1995. US forces in Europe were being steadily reduced after the fall of the Berlin Wall, but decades of sloppy disposal of the fuel, oil and solvents that made a modern mechanised army work were increasingly polluting the groundwater table. As the chief of the international and operational claims branch, my main job was to work with our engineer in collaboration with German state and local authorities and the engineers they had hired to conduct environmental remediation projects. We had three main goals: to ensure these projects were making progress in the clean-up, that they were being conducted consistently with German and US law, and that they were being operated with an eye towards cost-efficiency (Prescott, 1996: 234–38). Not only did I learn how difficult groundwater remediation is to accomplish, I was also introduced to the politics of water use, and grew to understand the different social, health and economic interests that different parts of a community can have regarding the availability of clean water.

I learned similar lessons working on damage claims in Bosnia-Herzegovina and Croatia for the loss of forests and crops caused by NATO and US forces (Prescott, 1998: 68–77), but from a very different cultural perspective than the one I had become familiar with in Germany. Importantly, this was also my first operational experience in a NATO headquarters. English was the common headquarters language in Sarajevo, and the commander was a US admiral. The variety and novelty of international perspectives that would come into play on any given day, however, forced me to reassess many things I had taken for granted in terms of the commonality between the US and allied military forces in terms of planning and conducting operations – a commonality that was supposedly reinforced by common doctrine and standard operating procedures.

The relationship between realistic training and doctrine

My next tour of duty was in Alaska, from 1997 to 1999, as the deputy general counsel for the US Army headquarters there. The collapse of the Soviet Union was having the same effect on Alaskan military units as it had had on forces in Germany – they were being drawn down, and repurposed. This change presented me with an unexpected opportunity. For years, the US forces under the Alaskan Command had held an annual winter exercise geared towards driving the Soviet invaders back across the Bering Straits to Siberia. By the late fall of 1998, the command had finally recognised that such an event was unlikely in the near future. The decision was made to quickly change what had been a high-intensity conflict exercise scenario into one focused on non-combatant evacuation operations. Unfortunately, not only were there no available infantry officers and soldiers to plan the evacuation exercise or to serve as role-players, there was also insufficient time according to standard training and exercise practices to train them to conduct the exercise (Headquarters, Department of the Army, 2016: 3-3–3-9).

At a staff meeting, I volunteered to organise and train the role-players and to conduct the evacuation exercise for the infantry battalion being trained. Although the chief of staff was reluctant to do so, in the end it was probably an easy choice for him to make – there simply was no one else available. So, I got the job. For the same reason, our training and exercise cadre consisted of fellow military attorneys and paralegals, and human resources officers and troops. Our raw role-players were truck mechanics from a transportation unit.

At first, we tried adapting standard training and exercise doctrine and guidance to our situation. We quickly realised it was unworkable. It was too rigid, and completely useless, for example, in helping to explain how to train a truck mechanic to understand the basics of the Vienna Convention on Diplomatic Relations so that she could accurately and with sufficient confidence play the role of a diplomat awaiting evacuation. With time at a premium, we created our own training plan that used role-playing as the primary teaching tool for the role-players. Our plan emphasised the agency of the individual soldiers as role-players in the exercise, and effectively gave them command of their specific training sites in the exercise area – not just responsibility.

Unorthodox – 'Is experiment', a Russian infantry colonel observed – but it worked. I learned useful lessons about training professional soldiers in non-traditional tasks, such as the importance of everyone knowing what the big operational picture looks like. In addition, I learned how to use a methodology to assess and mitigate operational risk – both to the soldiers and their equipment, and to the mission itself. Finally, I learned how necessary it is to have doctrine that is relevant and useful, and how flexible professional soldiers can be in adopting new ways of thinking and acting when the relevance of those new ways can be properly demonstrated to them (Prescott and Dunlap, 2000: 45–47).

Perhaps unfairly, this experience had left me with a low regard for established doctrine in general. I painfully unlearned this lesson at my next assignment, as a student and then teaching at the US Army Command and General Staff College at Fort Leavenworth, Kansas. Teaching new majors for three years at the staff college forced me to understand the larger, systemic value of doctrine to running a modern military. Our students had been successful company and battery commanders, that is, successful leading at the tactical level as captains. Their operational experiences were their own, and these strongly coloured their perspectives on the usefulness of doctrine in the field.

Like me, however, they now had to learn that doctrine serves many different functions, such as norming the experiences of the force as a whole, providing a basis for developing educational curricula and evaluating training, setting common ground for joint and multinational operations, and allowing the military to prioritise limited funding against what had been determined to be the most important mission sets. Further, being the chief legal instructor at the college now meant that I was tasked with reviewing some of the draft doctrine that was circulating through the US Army Training and Doctrine Command. It was not enough to correct typos and word choice – I had to spend a lot of time researching points I did not understand about operations to be able to make any meaningful comments.

Writing useful doctrine is hard work. Barfoed (2015) has observed, 'good doctrine must neither be based on innovative theory and beliefs alone, nor be based too much on best practice and experience from previous wars; in both cases it risks being useless at best and dangerous at worst'. Further, as Høiback (2016) has noted with no small amount of irony, and most importantly for purposes of this book, 'A doctrine can provide a steady platform for improvisation. It creates the box to think out of' (192).

An introduction to climate change and gender in operations

I returned to Alaska after finishing teaching in 2003, this time for two years as general counsel. In the wake of the 2001 Al Qaeda attacks and the US invasions of Afghanistan and Iraq, things had changed dramatically for the military in Alaska. New brigades were being created and equipped, and this meant that the new soldiers and their families needed places to live, and the units needed new and more sophisticated places to train. All of this activity required environmental analyses under US federal law – analyses that were soon challenged in litigation. For the first time, I learned that climate change was likely to have an effect on military planning and training. For example, shorter Alaskan winters meant that more variable weather events were likely to cause more rain even when the ground was still frozen, leading to severe flooding. Longer, warmer summers would increase the risk of catastrophic fires in tinder-dry black spruce forests, curtailing the use of live-fire exercises and requiring the construction of fire breaks and active fire-control measures.

At first, these environmental lessons did not seem to be particularly relevant at my next two assignments, at the main NATO training headquarters in Norfolk, Virginia, in 2005, and then at the Joint Warfare Centre, in Stavanger, Norway, in 2006. Because NATO had taken over the International Security Assistance Force (ISAF) mission in Afghanistan in late 2003, I worked as the primary legal observer/coach in the training exercises for the subordinate ISAF headquarters units rotating into Afghanistan. I also was part of the team conducting the training for the main ISAF headquarters rotations into Kabul. I relearned the lessons of Bosnia-Herzegovina and Croatia at a higher level. I began to understand how connected the actions of different staff sections needed to be with each other to work properly in an operational environment, and how hard that actually was to continuously achieve. I also learned how there is at the same time always too much information about the operational theatre in circulation within a headquarters and yet never enough to be completely helpful.

I was selected to rotate into the ISAF headquarters in Kabul in 2008 as the chief legal advisor (LEGAD). We were a small multinational legal office of British, Canadian and US attorneys. The pace of kinetic operations had begun to pick up significantly by the summer of 2008, and we found ourselves needing to learn each other's portfolios so that we could cover for each other when necessary in both kinetic and non-kinetic operations. There appears to be no official military doctrinal definition for 'kinetic' or 'non-kinetic'. In common usage, however, kinetic usually connotes the use of armed force such that enemy combatants are injured or killed or enemy property destroyed. Non-kinetic operations do not ordinarily involve the application of lethal or destructive force, and examples include information, cyber and civil-military cooperation (CIMIC) operations (Batson and Labert, 2012: 41). Stuxnet, the computer programme that apparently caused Iranian uranium enrichment centrifuges to irreparably damage themselves (Langner, 2011), is an example of an ordinarily non-kinetic means (cyber) having a kinetic effect (physical destruction or neutralisation).

In the ISAF headquarters, I might go from a dynamic targeting operation against a Taliban member on one day (often lethally kinetic), to a meeting organised by our rule of law officer with women's advocacy organisations the next (completely non-kinetic). It was at these meetings that I was first introduced to UN Security Council Resolution (UNSCR) 1325, *Women, Peace and Security*. Although, it was almost a decade old by that point, I had never read it. What members of these organisations pointed out to me, and what finally registered, was that the UN had established that the international community needed to 'fully implement humanitarian and human rights law that protects the rights of women and girls during and after conflicts' (UNSC, 2000: Preamble, para. 9), including the 1949 Geneva Conventions and the 1977 Additional Protocols. A bit sheepishly I realised that this was actually part of my job.

The purpose of these meetings, at least from my perspective at first, was not so much becoming educated on UNSCR 1325 as it was to gather data. At the strategic level, NATO was pursuing its 'Comprehensive Approach', which emphasised the importance of collaboration between civilian and military efforts to achieve better operational results more efficiently (Petersen and Binnendijk, 2007: 1–3). Within the Comprehensive Approach, the ISAF headquarters was using a holistic method of operational analysis and planning called the 'Effects-based Approach to Operations' (Beeres et al., 2012: 282–84). Our office had been given the job of assessing progress in the development of the rule of law in Afghanistan, and our rule of law officer had convinced me that metrics regarding the treatment of women in the Afghan legal system should be part of this analysis.

As a NATO LEGAD, the ideas behind the Effects-based Approach to Operations immediately appealed to me. It was a methodology that held the promise of gathering data, analysing it and then acting upon it in a way that could minimise injury or death to civilians and collateral damage to their property. It provided connection points to civilian efforts in the theatre of operations, and its use suggested that although full synchronisation with these efforts might be out of reach, complementing these civilian efforts with military initiatives to achieve overall efficiencies in operations might be possible. Further, the Effects-based Approach to Operations did not appear to require significant reorganisation or re-equipping of existing units, or new infrastructure – it was an approach, an operational mind-set, to achieving a comprehensive analysis picture of an operational theatre (Prescott, 2008: 127–28, 132–35).

I jumped at the chance to engage the rule of law area with the Effects-based Approach to Operations, in part because traditional military command and staff processes seemed poorly suited for such work. At the same time, however, I remembered my military history classes as a student back at Fort Leavenworth, especially reading and discussing Carl von Clausewitz's *On War*. I recalled enough to recognise that von Clausewitz would have found the notion of the Effects-based Approach to Operations, as I understood it, to be folly. In assessing the impact of the lack of accurate intelligence and analysis on military decision-making, he had observed that 'even if we did know all the circumstances, their implications and complexities would not permit us to take the necessary steps to deal with them' (Von Clausewitz, 1832: 153). Still, I was not convinced. Surely, the modern military headquarters infrastructure, with all of its data feeds and analysts, its computers and reach-back capabilities, and its educated, trained and experienced officers could burn through that fog of war. Surely, it could create an accurate, near real-time picture of the operational environment that would lead to more effective military efforts across the spectrum of military activity.

Should have listened to the Prussian

I was so wrong. Our analysis was both incomplete and depressing. We knew that different ISAF nations were working on initiatives to make the Afghan legal system more equal for women. Our attempts to use the large headquarters intelligence section to gather data about these projects were futile – this information simply was not being tracked. ISAF headquarters was not asking for the information, and even if the nations had provided it voluntarily, the intelligence section did not have anyone tasked to analyse it. The UN was not able to provide this information either, and the women's advocacy groups could generally only present very limited and often anecdotal data (Prescott, 2017: 12–14). It was discouraging. We were supposed to be engaging with the Afghan population to boost their support for the legitimate Afghan government, but it began to seem that we were basically neglecting half of them.

As the chief LEGAD I was on a short leash, and I spent almost all of my time in Kabul. In the spring of 2009, though, the ISAF commander sent me to Zabul Province to negotiate what were apparently a few crop-loss and land damage claims. Believing that they were using Afghan government land, US forces had begun building a large base in Suri, a very austere and rural sub-district near the Pakistani border. Unfortunately, not only was this land used by the local villagers to grow wheat, but as planned, the perimeter of the base would eventually enclose valuable vineyards and irreplaceable almond orchards. Finally, the base was being built atop a network of *karezes* (underground canals) that carried snowmelt from the mountains to the villages on the plain. In this arid land, the villagers depended on these waters (Phillips, 2009).

I spent almost three weeks negotiating with groups of village elders, landowners and tenant farmers, and charting property lines with a US Air Force engineer detachment. Villagers were generally reluctant to produce any deeds to their lands. Conceivably, landowners could have secured their documents under any of the different and mutually hostile Afghan political regimes dating back to the monarchy. Deeds, therefore, were documentary evidence of political allegiance in a land otherwise marked by general illiteracy (Prescott, 2018a: 334–35). Significantly, the most valuable deeds I did see came with 'nights of water' – irrigation rights. The different village water officials (all men) were generally involved in the discussions about the villagers' claims, and the discussions between villagers about water (at least in my presence) were only between men.

Von Clausewitz's assessment of the primacy of politics in armed conflict (Von Clausewitz, 1832: 87, 605) was completely borne out. I began to understand that the villagers' political support for the Afghan government depended on whether the base impaired their water flow, and whether the US forces turned their wheat fields into helipads and parking lots for armoured vehicles. Their political support hinged upon how we treated their environment, and I sensed that there was little room for ecological error in this high and dry place. Thinking of Alaska, I also recognised how at-risk the villagers would be to the effects of climate change if the precipitation patterns in the mountains changed, and I wondered how the families in the villages would cope with that sort of shift.

Knowing how very socially conservative rural Afghan communities were, I was not surprised that I saw no Afghan women whilst I was there. It did occur to me though, that wherever they were, they must have been very busy. Someone had to be working extra and watching the children whilst the men spent so much time with me carefully watching the progress of each other's damage claims, chatting and drinking tea.

Reassessment after a session of field tuition

A couple months later, I returned to the US and began teaching at the US Military Academy at West Point. Before going to Afghanistan, and to be fair, during much of the time I was there, gender issues had always seemed to me to be essentially human rights matters – of little relevance to the international humanitarian law (IHL) issues with which I was primarily concerned, such as dynamic targeting and civilian casualty reports. Actually, IHL had always impressed with me with its impartiality, and its fundamental treaty texts seemed to carve out an even greater protected status for women then it did men in certain respects. Further, from an IHL perspective, it did not seem necessary to distinguish the direct effects of a kinetic weapon upon a person on the basis of their sex. An injured or killed civilian was an injured or killed civilian irrespective of whether they were a woman or a man. The impact of small-arms fire or an improvised explosive device upon a person did not appear to be gender-differentiated.

However, as I began to reflect upon what I had experienced in Afghanistan, I also read more on gender. I read how armed conflict and climate change have differentiated effects on women versus men, and in particular the significance of women's ordinary gender roles of running households and serving as family caregivers, particularly in lesser developed countries. I began to question why the indirect impacts of these factors were not considered to be operational facts by militaries, because of these factors' implications for the success of stability operations and civil-military cooperation. I began to reconsider the application of IHL in kinetic engagements. Eventually, I began to realise that armed conflict, gender and climate change are not just related, but especially in the less developed countries where we seem inclined to conduct multinational operations, they might be linked in dynamic direct and indirect relationships with each other (Prescott, 2014: 768–69).

I also began to reassess the nature of the military organisations in which I had worked. I wanted to understand better how they might actually take on the task of operationalising the goals and principles of UNSCR 1325 to achieve better protection of women and girls in armed conflict situations. First, I recognised the obvious – that men are largely in charge of these organisations, that most service members are men, and that a number of these men have bad attitudes about women serving in the armed forces. I recognised that these factors could colour their understanding of the relevance of gender in operations.

Second, I understood that what I was proposing in terms of operationalising gender, particularly when I coupled it with climate change, would strike many as an abrupt break with the ordinary way of doing business, perhaps even 'revolutionary' in some senses. Revolutionary change in military organisations is not easily achieved. Their organisational missions include the application of extreme violence in circumstances of very high risk. Thus, they tend to be conservative in nature, rather than innovative (Lowery, 2017: 40), because the cost of failure is so high and the fact of failure is often irreversible. Even when these organisations do adapt, generally out of extreme necessity, there is a troubling history of backsliding to the status quo ante, as von Clausewitz had himself experienced in Prussia after the successful conclusion of the Napoleonic Wars in 1815 (Prescott, 2008: 141–42; Prescott, Iwata and Pincus, 2015: 13–16). It is worth the push to operationalise gender consistent with the requirements of the different national action plans that had been created to implement the goals and principles of UNSCR 1325, however, because our world is changing around us.

'War amongst the people' – not just the men

The international security environment of today and the near future is one increasingly characterised, in the words of General Sir Rupert Smith (2007), by 'war amongst the people' (xiii, 269–307). It is ever more civilian-centric, in which contending military and paramilitary forces wrestle more for influence over populations rather than seeking relative advantage over each other in force-on-force engagements oriented on key geographical terrain (Pfanner, 2006: 719–20, 722–26). This concept is not completely new of course – both Mao and Giap, for example, successfully operationalised it in their respective irregular/regular warfare models based in large part on rural insurgencies in the 20th century (Freedman, 2013: 183–87).

Several modern-day differences, however, have increased the likelihood of armed conflict occurring in heavily populated areas. Among these factors are mass urbanisation, the flowering of the megacity and the ever-increasing size of the global human population. Further, populations are increasingly connected and dependent upon the Internet, and the accessibility and leverage it offers to both governments and dissident movements mean that those populations will increasingly experience conflict in an informatised way (Prescott, 2014: 767–68). Climate change is tied to these trends in the human environment. Particularly in the developing world, military operations will be impacted by these new conditions to varying degrees, which will complicate efforts to promote more stable political, economic and social end-states in areas affected by armed conflict.

Location matters. The trends described previously are general, and global in nature, but they will not affect every region of a country, or every country for that matter, the same way. As to climate change, for example, Sultana (2018) has cogently and concisely observed that although 'climate change is often framed as global problem for all of humanity, the heterogeneity of its manifestations, impacts, and responses has to be carefully considered' (19). Accordingly, the effects of climate change will impact different regions and countries differently, and the severity of the potential negative impacts will likely be more pronounced in developing countries that depend on rain-fed agriculture or which are located in low-lying coastal areas. For example, different climate change models suggest that Bangladesh is particularly at risk of increased flooding, and saline infiltration of its groundwater aquifers. Sub-Saharan Africa may be at increased risk of extreme precipitation variability (Scheffran and Battaglini, 2011: S33, figure 2), as an apparent drying trend in the Mediterranean area might result in a generally wetter Sahel (Park, Bader and Matei, 2016: 943–44).

Against this backdrop, I believe that the risks posed by armed conflict and climate change to women and girls are most relevant to militaries conducting civilian-centric operations, particularly in certain parts of developing countries. In describing these operations as 'civilian-centric', I am purposefully rejecting the use of the counterinsurgency term 'population-centric' (Paul et al., 2016: 1022–23). I believe it will be more useful in evaluating the significance of different kinds of civilians to operations to consider the importance of the 'strategic civilian' (Ryan, 2016), than it is to continue with what seems to be an unspoken male-norming of the amorphous 'population' as a whole, as evidenced by the US counterinsurgency doctrine that will be discussed in detail in Chapter 6, 'US strategy, doctrine and gender'.

The relationships between armed conflict, women and climate

I am not the first to suggest that armed conflict and climate change might compound each other's negative effects upon women and girls in areas where they are likely already at risk, and that the relationships between these forces could be quite complex. In her study of the impacts of climate change on livestock and water resources in East Africa, Sundblad (2013) concluded that 'climate change, conflict and gender inequality have overlapping and sometimes mutually exacerbating impacts'. In their assessment of climate change's impacts upon women in the areas of human rights, security and economic development, Alam, Bhatia and Mawby (2015), note that 'The differential impacts of climate change on men and women are, demonstrably, more pronounced in settings that are also affected by violent conflict' (11). In its assessment of the overlapping relationship between armed conflict and climate change in the Arab region, the UN Economic and Social Commission for Western Asia found these phenomena had a reinforcing negative impact on gender equality (2016). Finally, and perhaps most importantly for purposes of this book, Fröhlich and Gioli (2015) have identified the difficulty in sorting out these complex relationships in a useful way (137–45).

Whilst valuable contributions to this under-researched area, the first three studies are not particularly well suited for use in an operational approach to these relationships. They are somewhat general in their discussions of the role of armed conflict and how these compounding effects manifest themselves. As will be discussed in Chapter 2, 'Gender and climate change', however, the conclusions that I draw from Fröhlich and Gioli's work suggest not only that an operational approach might be the best way to sort through this complexity, but that in the interests of clarity it is probably best to not try to study all three elements together at the same time.

Thus, the first three chapters of this book will work to separate out the relationships between armed conflict and women's and girls' gender roles, these gender roles and climate change, and climate change and armed conflict, to simplify the analysis of the recognised complexities. Each chapter will set out the general trends, factors and impacts related to and resulting from the relationships between these different forces. Each will then use a short case study to examine how these things might play out in any particular setting. Chapter 1, 'Armed conflict and gender', will examine the brutal Boko Haram insurgency in Nigeria. Chapter 2, 'Gender and climate change', will focus on a case study conducted of two rural South African municipalities. Finally, Chapter 3, 'Climate change and armed conflict', will look to the tragic civil war in Syria.

Because of the fundamental importance of strategy and doctrine to military education, training, planning and operations, Chapters 4 through 7 will examine the current strategic and doctrinal approaches taken by NATO, the UK, the US and Australia regarding climate change and gender, respectively. I do not ask you to join me in this examination lightly – a deep dive into doctrine is neither for the dainty nor the easily distracted. Comparative doctrinal analysis is a slog, but it is absolutely necessary to understand how NATO and these countries' military organisations believe they should be working and where their resource allocations will likely be targeted, especially given the reality of their close working relationships with each other.

Chapter 8, 'International humanitarian law and gender', will explore the protections IHL confers upon women and girls, and the history and rationales of these legal provisions. Considering IHL in this context is important, because it tends to be conservative in its application and development, much like the militaries that apply it. IHL norms exert strong but subtle influences on the ways in which militaries conduct their activities and operations, and how they develop strategy and doctrine.

Chapter 9, 'Gender in military activities and operations', will survey how military organisations have begun to incorporate perspectives on gender into military education, training, planning and operations. Although there are some obvious disconnects between strategy and policy regarding gender and actual implementation into doctrine in many military organisations, there are important efforts throughout the global community to operationalise gender effectively. Whilst no one military organisation has yet been able to fully incorporate gender throughout the complete range of its activities and operations, a survey of the actions that are being taken provides a fairly comprehensive picture of what such a fully realised programme might look like.

As to my conclusion, Gjørv and Kleppe (2016) persuasively argue that

> operations that do not have a gender analysis and perspective incorporated into their plan are in most cases not gender-neutral (as is often assumed), but gender-blind, as they lose out on understanding the full range of situations, needs and opportunities of the local population.
>
> *(255)*

They advocate for the use of an effective gender analysis that asks three basic questions of the operational theatre: how is power distributed in a host nation's society between women and men, what contributions can the different genders make to improving operational security of the mission, and how the gender mix of a deploying force can be used to advantage to engage with actors in the host nation (254–55).

This basic methodology provides a good starting point for figuring out how to make gender operationally relevant, and I wish to push it a step farther. I suggest that a gender analysis properly conducted from an operational perspective by a military organisation might conclude in the end that gender is not actually relevant from the vantage point of risk to a particular operation, at its level. To reach that point, however, a simple gender analysis is insufficient. Instead, there must also be an assessment of the risk to the service members and the operation at the strategic, operational and tactical levels if the organisation were to decide to not take an appropriate gendered approach in conducting its work. Moreover, I also believe that unless a deploying military mission is assessing a civilian-centric theatre of operations from a perspective that includes the relationships between armed conflict, women and climate change, it will be missing crucial information about the development of insecurity in that theatre and the ways its resources could be more efficiently used to contribute to the development of long-term stability (Lowery, 2017: 42–43). Examples of areas in which this might particularly play out include environmental protection in engineering activities (Prescott, 2018b), civil-military cooperation that is cognizant of the needs of different types of civilians in their environments, rebuilding resilience in the environments in which people live and farm (Prescott, 2014: 783–99) and food security (The White House, 2015). With that in mind, let's begin by examining the relationships between armed conflict and gender.

References

Alam, M; Bhatia, B and Mawby, B (2015) 'Women and Climate Change – Impact and Agency in Human Rights, Security, and Economic Development'. Georgetown Institute for Women, Peace & Security. Available at: https://giwps.georgetown.edu/wp-content/uploads/2017/09/Women-and-Climate-Change.pdf (Accessed 26 June 2016).

Barfoed, J (2015) 'Military Strategy vs Military Doctrine'. *Krigsvidenskab.dk* (17 June). Available at: www.krigsvidenskab.dk/military-strategy-vs-military-doctrine (Accessed 10 June 2018).

Batson, M and Labert, M (2012) 'Expanding the Non-Kinetic Warfare Arsenal'. *Proceedings* 138(1) (January): pp. 40–44. Available at: www.usni.org/magazines/proceedings/2012-01/expanding-non-kinetic-warfare-arsenal (Accessed 24 June 2018).

Beeres, R; Van der Meulen, J; Soeters, J and Vogelaar, A (2012) *Mission Uruzgan: Collaborating in Multiple Coalitions for Afghanistan*. Amsterdam: Amsterdam University Press.

Freedman, L (2013) *Strategy: A History*. Oxford: Oxford University Press.

Fröhlich, C and Gioli, G (2015) 'Gender, Conflict and Global Environmental Change'. *Peace Review: A Journal of Social Justice* 27(2): pp. 137–46. DOI: 10.1080/10402659.2015.1037609.

Gjørv, GH and Kleppe, TT (2016) 'Civil-Military Interaction, CIMIC and Interacting with Gender'. In: Lucius, G and Rietjens, S (eds.) *Effective Civil-Military Interaction in Peace Operations: Theory and Practice*. Cham, Switzerland: Springer International Publishing AG, pp. 249–61.

Headquarters, Department of the Army (2016) *Field Manual 7–0, Train to Win in a Complex World* (October). Washington, DC: Department of the Army. Available at: https://armypubs.army.mil/epubs/DR_pubs/DR_a/pdf/web/ARN9860_FM%207-0%20FINAL%20WEB.pdf (Accessed 24 June 2018).

Høiback, H (2016) 'The Anatomy of Doctrine and Ways to Keep It Fit'. *Journal of Strategic Studies* 39(2): pp. 185–97. DOI: 10.1080/01402390.2015.1115037.

Langner, R (2011) 'Cracking Stuxnet, a 21st-century Cyber Weapon'. TED Talks. Available at: www.ted.com/talks/ralph_langner_cracking_stuxnet_a_21st_century_cyberweapon (Accessed 14 March 2018).

Lowery, V (2017) 'Coping with Noncombatant Women in the Battlespace: Incorporating United Nations Security Council Resolution 1325 into the Operational Environment'. *Military Review* (May/June 2017): pp. 35–43.

Park, J; Bader, J and Matei, D (2016) 'Anthropogenic Mediterranean Warming Essential Driver for Present and Future Sahel Rainfall'. *Nature Climate Change* 6(10): pp. 941–45. DOI: 10.1038/NCLIMATE3065.

Paul, C; Clarke, CP; Grill, B and Dunigan, M (2016) 'Moving Beyond Population-Centric vs. Enemy-Centric Counterinsurgency'. *Small Wars and Insurgencies* 27(6): pp. 1019–42. DOI: 10.1080/09592318.2016.1233643.

Petersen, FA and Binnendijk, H (2007) 'The Comprehensive Approach Initiative: Future Options for NATO'. *Defense Horizons* 58: pp. 1–5. Available at: http://ndupress.ndu.edu/Media/News/Article/1008473/the-comprehensive-approach-initiative-future-options-for-nato/ (Accessed 3 April 2012).

Pfanner, T (2006) 'Interview with General Sir Rupert Smith'. *Review of the International Red Cross* 88 (864): pp. 216–27.

Phillips, M (2009) 'Learning a Hard History Lesson in "Talibanistan"'. *Wall Street Journal* (14 May). Available at: www.wsj.com/articles/SB124224652409516525 (Accessed 12 December 2017).

Prescott, JM (1996) 'Die Rolle des amerikanischen Schadenersatzamts bei der Umweltsanierung'. In: Brandt, A; Beudt, J and Bousonville, R (eds.) *Ruestungsaltlasten: Untersuchung, Probenahme, Sanierung.* Heidelberg: Springer-Verlag, pp. 233–38.

Prescott, JM (1998) 'Einsatzbedingte Schaeden in Bosnien-Herzegowina und Kroatien'. *Neue Zeitschrift fuer Wehrrecht* 40(2): pp. 67–78.

Prescott, JM (2008) 'The Development of NATO EBAO Doctrine: Clausewitz's Theories and the Role of Law in an Evolving Approach to Operations'. *Penn State International Law Review* 27(1): pp. 125–68.

Prescott, JM (2014) 'Climate Change, Gender and Rethinking Military Operations'. *Vermont Environmental Law Journal* 15(4) (online): pp. 766–802. Available at: http://vjel.vermontlaw.edu/files/2014/06/Prescott_ForPrint.pdf (Accessed 24 June 2018).

Prescott, JM (2017) 'Lawfare: Softwiring the Network'. *The Three Swords* 32: pp. 6–20. Available at: www.jwc.nato.int/images/stories/_news_items_/2017/LAWFARE_JWCThreeSwordsJuly17_ColPrescott.pdf (Accessed 24 June 2018).

Prescott, JM (2018a) 'Claims'. In: Fleck, D (ed.) *The Handbook of the Law of Visiting Forces*, 2nd Ed. Oxford: Oxford University Press, pp. 275–338.

Prescott, JM (2018b) 'NATO Environmental Protection Doctrine and Gender'. *NATO Legal Gazette* 40 [forthcoming in December 2018].

Prescott, JM and Dunlap, JW (2000) 'Law of War and Rules of Engagement Training for the Objective Force: A Proposed Methodology for training Role-Players'. *Army Lawyer* (September): pp. 43–47. Available at: www.loc.gov/rr/frd/Military_Law/pdf/09-2000.pdf (Accessed 24 June 2018).

Prescott, JM; Iwata, E and Pincus, R (2015) 'Gender, Law and Policy: Japan's National Action Plan on Women, Peace and Security'. *University of Hawaii Asian-Pacific Law & Policy Journal* 17(1) (online): pp. 1–45. Available at: http://blog.hawaii.edu/aplpj/files/2016/04/APLPJ_Prescott-Iwata-Pincus-FINAL.pdf (Accessed 24 June 2018).

Ryan, A (2016) 'The Strategic Civilian: Challenges for Non-Combatants in 21st Century Warfare'. *Small Wars Journal* [online]. Available at: http://smallwarsjournal.com/jrnl/art/the-strategic-civilian-challenges-for-non-combatants-in-21st-century-warfare (Accessed 20 July 2018).

Scheffran, J and Battaglini, A (2011) 'Climate and Conflicts: The ecurity Risks of Global Warming'. *Regional Environmental Change* 11: pp. S27–S39. DOI: 10.1007/s10113-010-0175-8.

Smith, R (2007) *The Utility of Force: The Art of War in the Modern World.* New York: Knopf.

Sultana, F (2018) 'Gender and Water in a Changing Climate: Challenges and Opportunities'. In: Fröhlich, C; Gioli, G; Cremades, R and Myrttinen, H (eds.) *Water Security Across the Gender Divide.* Cham, Switzerland: Springer International Publishing AG, pp. 17–33.

Sundblad, L (2013) 'Climate Change, Conflict, and Gender Inequality'. *Afrikan Sarvi, Horn of Africa Journal* 2/3013. Finnish Somalia Network. Available at: https://afrikansarvi.fi/issue6/62-artikkeli/177-climate-change-conflict-and-gender-inequality (Accessed 26 June 2018).

The White House (2015) *National Security Strategy.* Washington, DC: The White House. Available at: http://nssarchive.us/wp-content/uploads/2015/02/2015.pdf (Accessed 21 October 2016).

UN Economic and Social Commission for Western Asia (2016) *Conflict, Climate Change and Their Mutually Reinforcing Impact on Gender Imbalances in the Arab Region, Technical Paper*

2 (10 June). Beirut: UN. Available at: www.unescwa.org/sites/www.unescwa.org/files/publications/files/gender-climate-change-conflict-english.pdf (Accessed 26 June 2018).

UN Security Council (2000) Resolution 1325. *On Women, Peace and Security*. UN Doc. S/RES/1325 (31 October).

Von Clausewitz, C (1832) *On War*. Paret, P and Howard, ME (trans.) 1976. Princeton, NJ: Princeton University Press.

1
ARMED CONFLICT AND GENDER

Beginning with UNSCR 1325 on women, peace and security, the last two decades have witnessed a sustained and growing world-wide effort to establish a normative platform upon which to base national and international policies and initiatives to achieve greater protection of women and girls in situations involving armed conflict. UNSCR 1325 has been amplified by a number of additional UN Security Council resolutions, which for the most part have focused on measures to address sexual and gender-based violence against civilians, and particularly against women and girls. Examples include UNSCR 1820, noting the need to prevent sexual and gender-based violence in armed conflict (UN, 2008), and UNSCR 1960 (UN, 2010), which defined institutional tools for combatting impunity of perpetrators of sexual and gender-based violence, and setting out steps to prevent violence and to protect women from it.

To implement the requirements of UNSCR 1325, nations have been encouraged to develop their own national action plans (UN Security Council, 2005: 2–13). As of July 2018, 74 countries had established their own plans. Some of these plans are very recent, such as Mozambique's, which was established in June 2018 (PeaceWomen, 2018). Other countries, however, began establishing their national action plans shortly after UNSCR 1325 was promulgated. The UK, for example, was already working under its fourth national action plan in 2018, as will be discussed in Chapter 5, 'UK strategy, doctrine and gender'. Before exploring the gendered effects of armed conflict that UNSCR 1325 was intended to address, it is important to note that neither the resolution itself nor the international and national efforts to implement its goals and principles are without blemish or critics. To best assess the impact of UNSCR 1325, it is therefore useful to first briefly review the general context of armed conflict with regard to women and girls today, and to specifically examine what textual language the UN Security Council agreed to when it passed UNSCR 1325.

War's impacts and UNSCR 1325's main provisions

In general, armed conflict today tends to happen more often in areas which are less economically developed. Martin (2009) estimates that between 70 and 75 percent of people dislocated by war as either refugees or displaced persons are women and their minor children, and that children account for about 50 percent of all refugees (1). Kim and Conceição (2010) estimate that between 1990 and 2005, over 3 million people died in armed conflicts occurring in developing countries (31–38). Women in these areas are more likely to have a relatively inferior social and economic status as compared to men, and this gender discrimination and its effects make women more at risk to the impacts of armed conflict (Gardam and Jarvis, 2001: 8–9). These negative effects occur irrespective of whether women are in the roles of civilians caught up in the fighting, refugees or even combatants (Prescott, 2013: 87–90).

With this general picture in mind, it is now useful to examine what the UN Security Council committed to in approving UNSCR 1325 (UN, 2000). In its Preamble, the resolution set out four main conditions that needed to be accomplished to achieve greater protection for women and girls. First, there needed to be 'equal participation and full involvement' by women 'in all efforts for the maintenance of peace and security', particularly with regard to decision-making. Second, gender perspectives needed to be 'mainstreamed' into peacekeeping operations. Third, specialised training needed to be developed 'for all peacekeeping personnel on the protection, special needs and human rights of women and children in conflict situations'. Fourth, and as noted earlier in the Introduction, the UN Security Council recognised the need 'to implement fully international humanitarian and human rights law that protects the rights of women and girls during and after conflict'.

'Full' implementation of IHL and international human rights law does not mean implementation of only those areas that IHL and international human rights law have in common, such as their respective prohibitions of sexual and gender-based violence. This is not what the resolution states. In this sense, the significant number of UN Security Council resolutions subsequent to UNSCR 1325 that focus on sexual and gender-based violence are undoubtedly necessary and beneficial, but they might have obscured the full scope of UNSCR 1325's requirements (Prescott, 2013: 103–04). The potential implications of 'full' implementation to the other areas of IHL such as the use of force in kinetic engagements must be explored to comply with the resolution's requirements. Although other criticisms of UNSCR 1325 or efforts to implement it are not so focused on IHL and are therefore largely outside the scope of this book, they are worth briefly reviewing to better understand the interests and positions of other stakeholders in the resolution's implementation at the national and international levels.

Critique of UNSCR 1325

Some of the criticisms of UNSCR 1325 and national implementation efforts are political or philosophical, but others are more directly tied to operational concerns. For example, some argue that the steps taken by different nations under their national action plans to incorporate gender perspectives in their military activities and operations are inconsistent with the intent of the drafters of the resolution. The drafters' intent, these writers argue, was to eliminate war, not to 'make war safe for women', and the 'militarisation' of UNSCR 1325 is therefore counterproductive to achieving the resolution's goals (Weiss, 2011; Shepherd, 2016: 332–33). Whilst this was undoubtedly part of the historical context of the resolution's creation, the subjective intent of civil society actors cannot control the meaning of its plain text as approved by the UN Security Council. 'Full' implementation of IHL suggests that the council recognised that it would still be quite some time before war became safe for women.

Winslow (2009) notes that certain feminists are critical of UNSCR 1325's failure to address root problems because of its focus upon women and girls, rather than identifying the relationship between masculine identities during armed conflict and violent behaviour patterns that are accepted during those times because they are seen as 'male' (534). This criticism could be operationally relevant, although perhaps not in the way these writers were thinking. For example, some studies have attributed armed conflict between different pastoral peoples in Kenya to the impact of climate change (Pearson, 2017: 2–3). Other studies of conflict between two of these peoples, the Turkana and the Pokot, suggest instead that one significant component of the raiding of each other's cattle are sets of 'masculinities' – young men striving to meet societal expectations of masculine strength and to fulfil the traditional male role of providing these valuable assets and food for their families (Schilling, Froese and Naujoks, 2018: 181).

Winslow (2009) notes as well that others have criticised UNSCR 1325 because it does not include ethnicity or class relations (543). This omission might also have operational consequences. One review of the UN mission in Eritrea and Ethiopia suggests that UN women soldiers from Ghana and Kenya interacted better with local women than did their ethnic European counterparts (Valenius, 2007: 36). Finally, in the wake of revelations of widespread sexual abuse against women and girls in the Democratic Republic of Congo by UN peacekeeping troops shortly after the passage of UNSCR 1325 (UN Secretary General, 2005), the UN's commitment to UNSCR 1325 was seriously questioned. Although there have been subsequent cases of similar misconduct by UN troops (Oladipo, 2017) and the UN has been criticised for the handling of these cases as well (Anderlini, 2017), it is important to note that the UN has undertaken important steps to help educate and train personnel regarding their responsibilities to protect women and children, a number of which will be discussed in Chapter 9, 'Gender in military activities and operations'. With these criticisms in mind, let's now turn to the more specific ways in which women and girls are impacted by armed conflict.

Women and girls in armed conflict

Impacts on women and girls as civilians caught up in the fighting

There has been significant progress in recent years, but historically analyses and reports on armed conflicts' effects on civilians have not tended to differentiate experiences of civilians based on their gender. Indeed, Gardam and Charlesworth (2000) note that the particular concerns of women have been marginalised in such assessments (150). Although women's and girls' gender-differentiated experiences of war occur within specific contexts, it is useful first to assess the general conditions under which many women in the world live today. In terms of relative affluence, women constitute the majority of the more than one billion people who live in poverty. Because of their traditional family roles in many cultures, women are generally less mobile than men, less likely to be employed outside the home (Gardam and Jarvis, 2001: 8–9), and generally less educated (Gardam and Charlesworth, 2000: 151, 153). Armed conflict tends to exacerbate these factors to the further detriment of women. Importantly, however, these descriptions are general – they do not apply everywhere all the time, and care must be taken in any specific operational gender analysis to determine whether and how they do.

Although modern armed conflicts involve women as military leaders (Davis, 2010; Biank, 2014) and combatants (Wright, 2011; Lemmon, 2015), the majority of women who are impacted by armed conflict are in fact civilians (Gardam and Charlesworth, 2000: 152), and they ordinarily do not have much control over the manner in which war affects them. In terms of physical insecurity, Jones (2010) notes that women are very often victims of rape, sexual assault, virtual enslavement and torture during the course and often in the aftermath of armed conflict (19–20, 58, 84, 101, 134–35, 143, 150–51, 161–62, 172). As described by Valenius (2007), other forms of sexual violence against women in these circumstances include 'forced prostitution, . . . forced impregnation, forced maternity, forced termination of pregnancy, forced sterilisation, . . . strip searches and inappropriate medical examinations' (20).

The brutal sexual assaults that occur during the course of armed conflict can have serious and long-lasting negative health effects for victimised women and girls. One study of rape survivors from the Democratic Republic of Congo (Leaning, Bartels and Mowafi, 2009) noted that 85 percent of the women reported vaginal discharge afterwards, and that 10 percent reported having been impregnated. Particularly serious were the large number of women who reported symptoms consistent with having suffered fistulas (tears in their vaginal, urinary and anal tracts causing unnatural connections between the two) after their rapes – 41 percent reported discharging urine or faeces from the vagina. And 77 percent of these women also reported experiencing insomnia and nightmares, and 91 percent reported suffering from feelings of shame and fear (188).

The ripple effects of these injuries upon women degrade their levels of social and economic security, as well as those of their children (Jones, 2010: 58, 161–62). Dharmapuri (2011) explains that rape and sexual violence against women can destabilise the economic productivity of communities, because they are unable to safely leave their homes and villages to go the fields to tend their crops, or to markets to sell and obtain food. This diminishes their ability to feed their families and earn currency (64). Women and girls can also find themselves at higher risk of contracting sexually transmitted diseases, because they are at greater risk of sexual violence in the lawless and chaotic situations that often arise in humanitarian disaster and armed conflict situations, and might even need to trade sex for food and other essential supplies (Decker et al., 2009: 72).

Further, Gardam and Jarvis (2001) observe that 'Armed conflict is invariably accompanied by the disintegration of societal structures, which will leave women more disadvantaged because of their inferior position than men' (20). Because women in general have access to fewer economic resources to begin with in many societies, they are both less able to resist the hardships caused by armed conflict and less likely to have the resilience to rebuild after the conflict has ended. In addition, women may be dependent upon the men of their families to earn a living, and armed conflict will often take men away from their homes and livelihoods as either combatants or refugees (Gardam and Charlesworth, 2000: 151–53). In terms of the gender-differentiated impact of armed conflict, Buscher (2009) notes that in non-international armed conflicts, whilst young men find themselves forming a recruiting pool for combatant groups, often because of their limited economic choices, young women and girls will often find themselves dropping out of school to take on chores additional to their ordinary tasks such as drawing water, collecting biomass for energy and caring for family members (88).

Women and teenaged girls therefore often become the providers and protectors for their families, in addition to the time-consuming domestic duties they already perform (Gardam and Charlesworth, 2000: 153; Valenius, 2007: 23). Coupled with their often lower levels of education and training, these great responsibilities are not matched by opportunities for women and girls to obtain meaningful and remunerative work. The UN High Commissioner for Refugees (UNHCR; 2008) finds this is particularly the case with the diversion of resources that occurs in conflict situations that might otherwise be used for capacity-building purposes (306–10), and Gardam and Jarvis (2001) note that it occurs as well in the generally austere conditions that characterise post-conflict situations (40).

Unfortunately, women civilians experience armed conflict's negative effects even if they are not caught up directly in the fighting. Buscher (2009) details how armed conflict disrupts livelihoods, by decreasing access to markets and financial assets, reducing access to land for growing crops or herding animals, and damaging vital infrastructure (89). The insecurity caused by armed conflict often restricts women's freedom of movement, often forcing them to lose their livelihoods (88–89). In addition, because women in these conflict zones are not likely to be able to equitably access economic resources even during times of relative peace, their resilience to withstand the hardships caused by armed conflict or to rebuild quickly once the fighting is done might be severely impaired (Gardam and Charlesworth, 2000: 151, 153).

Impacts on women and girls as refugees

Refugees in general are 'exposed to foreign – and often inadequate – living conditions, and consequently tend to be more prone to accidents, injuries and disease'

(Gardam and Charlesworth, 2011: 124). The risks of such events, however, are greater for women than they are for men. Women who find themselves as refugees from armed conflict must survive under particularly trying circumstances. Refugee camp living conditions are often substandard, and women therefore often lack appropriate hygienic facilities or basic medical care. Shortages of contraceptive supplies mean that women are at risk of increased pregnancy during these times of intense stress (Gardam and Charlesworth, 2011: 125). Refugee camps also present an increased risk of sexual violence to women, but the specialised medical services necessary to treat victims of sexual assault might be in very short supply (UN News Centre, 2011).

This is not to say that sexual assault and its effects are suffered only by female refugees. Storr (2011) points out that men are likewise subject to sexual violence, and they too are at greater risk of its occurrence when they are uprooted from their homes during armed conflict. Sexual violence against men tends to be underreported, notes Valenius (2007), and reported by victims as 'torture' rather than sexual assault because of masculinity concerns and cultural taboos in different societies (18). For these reasons, the UN High Commissioner for Refugees (2012) has drawn up a list of indicators to help identify men who have been victims of sexual violence (8–9). The international community has also been paying more attention to this violence experienced by male refugees as a result of war, as demonstrated by the recent UN High Commissioner for Refugees' report on the sexual abuse of men and boys in the Syrian Civil War (Chenoweth, 2017). Women and girl refugees, however, still constitute the majority of the victims (UNHCR, 2003: 6).

In refugee camps, when food is scarce, cultural attitudes will often mean that women and girls eat what is left over after the men and boys are fed – or they might simply not eat at all (Gardam and Charlesworth, 2000: 155). Women, girls and boys must often exchange sexual favours for food in refugee camps (Women's Commission for Refugee Women and Children, 2002: 27). Jones (2010) has identified the special challenges that face women who are permanently displaced from their homes in making new lives as refugees (172). She also notes that because of their generally inferior status, women often find themselves without marketable skills as they try to adapt to life as refugees (172, 202–28).

Buscher (2009) notes that for those female refugees who flee to urban areas rather than camps, there might be more employment opportunities, but often they will be required to seek permission to work there, which can be difficult for them to obtain (94). Often, those who had engaged in land-based work raising their own crops or tending livestock are only able to find work in menial and marginalised trades, such as maids or servants, which not only likely pay little but can also be socially humiliating for them as well (90). Even if they do return to their homes, these women often find themselves marginalised in peace negotiations and in the rebuilding and reconciliation process. Further, Valenius (2007) reports that although the 'official' conflict might be over, women will still face the threat of continued violence, discrimination and untreated medical problems in these often austere environments (16, 22–23).

Impacts on women and girls as combatants

There can be very real differences between the experiences of a professional woman soldier in the military of a developed country and those of a woman who finds herself in the role of a combatant in a less developed country, often on behalf of a faction or an insurgency rather than her country's formal military. As gender barriers continue to fall in the militaries of developed countries, women now find that many combat positions that were once closed to them solely on the basis of their sex are available to them. Canada and New Zealand opened all of their military positions to women in 2001 (Cawkill et al., 2009: 17, 27), Australia in 2011 (Department of Defence, n.d.) and the UK (Ministry of Defence and Prime Minister's Office, 2016) and the US (Pellerin, 2015) in 2016. This potentially increases opportunities for promotion and positions that will further develop their leadership and technical skills.

Even Japan, which by many different measures is very male-normative and male-dominated in its economic and political spheres, with defined and limited social and economic gender roles for women, has begun making a deliberate effort to increase the number of women in its Self-Defense Forces, and to open up military positions to them (Prescott, Iwata and Pincus, 2016: 1–2), as will be discussed in detail in Chapter 9, 'Gender in military activities and operations'. Frühstück's (2007) research has shown that historically, Japanese female service members have been concentrated in communications, intelligence, medical and accounting specialties, and in general duties units (91). As an example of increased gender equality, the Japan Air Self-Defense Force recently opened up the role of fighter pilot to women (NBC News, 2017). Prior to that, women had not been allowed to fly fighters because of concern that it would impact their ability to bear children (Ministry of Defense, 2014: 351).

In many developed countries today, female service members' employment and operational experiences will be increasingly more like their male comrades'. As they work in these new roles they have chosen, these women will still be receiving the same benefits, health care and otherwise, to which they would be entitled under their respective national systems if they were serving in what were considered more traditional positions for women service members. These women are likely to be recognised for their leadership and actions in combat (Wright, 2011). Returning home, although there continue to be strong concerns that the needs of female veterans are neither fully understood nor properly provided for (Neuhaus and Crompvoets, 2013: 530–32; Medill Reports Chicago, 2016), they will likely be reintegrating into societies that accept and value their career choices. They will have developed skills and resources that would allow them a high degree of economic agency in civilian life. The position of women combatants in less developed countries, particularly in non-international armed conflict situations (Dharmapuri, 2011: 62), can be quite disadvantaged by comparison.

Young women and girls in these conflicts might find themselves fighting on behalf of factions or insurgencies, often unwillingly, without the social and economic safety net their countries' regular military organisations have available to them. In some cases, women combatants might have been members of their countries' armed forces but decided to switch sides. In many others, they may be fighting for revenge because of losses inflicted upon their families. Although these women and girls perform many tasks that are same as the men they fight alongside (Korge, 2013), their roles, statuses and experiences are not completely the same (Dharmapuri, 2011: 62). For example, Valenius (2007) notes that these women fighters also function in many logistical roles, such as cooking, setting up camp and transporting equipment and materiel; often as sexual partners for the men; and as spies (23).

These women and girls sometimes achieve leadership roles in their irregular military units, and become well-known for their martial skills, such as 'Black Diamond', a Liberian rebel soldier who became prominent for her ferocity in the Liberian civil war (Huffman, 2013: 7–8, 20–22). Another example might be Colonel Umuraza, a rebel officer in the M23 insurgent group in the Democratic Republic of Congo, whose story is among the many of female fighters documented in Francesca Tosarelli's photo documentary 'Ms. Kalashnikov' (2013). The women who fight in the Iraqi Kurdish *Peshmerga* are perhaps one of the better-known examples of female frontline combatants serving in an army that has in many respects the characteristics of a national military force (Nilsson, 2018: 265–71). Similarly, the daring raid into Iraq in 2014 by Syrian Kurdish women soldiers and police belonging to the People's Protection Units and Women's Protection Units to rescue tens of thousands of Yazidi refugees on Mount Sinjar from Islamic State forces (Tax, 2016: 41–42) stands out as an example of small unit leadership and tactical competence. As one rescued Yazidi man noted, 'Women fighters – they saved us. My society, Yazidi society, is more, let's say, traditional. I'd never thought of women as leaders, as heroes, before' (Enzinna, 2015).

Although not the case with the Kurdish female combatants (Tax, 2016: 143–46), in many other non-state forces women will often bear and care for children as the result of rape or relationships with the men, including forced marriage, and their reintegration into post-war societies presents them with very significant challenges. Disarmament, demobilisation and reintegration provisions in post-war settlements often fail to recognise their special needs (Valenius, 2007: 23). Wessells (2011) has pointed out that official programmes are often heavily military and security oriented, and therefore ill-suited to address returning women fighters' needs in civilian society. Conversely, unofficial programmes conducted by humanitarian organisations might be better crafted to help the women fit back in to their communities, but they tend to be under-resourced and unevenly available to impacted populations (189–90).

Wessells (2011) also notes that when these women and girls return, the behaviour patterns that they developed as fighters might seem unruly and disrespectful to their civilian communities. They might also experience sexual harassment (192–93, 200) and other forms of gender discrimination (Worthen et al., 2012: 26). They must often endure the social stigma that may attach to them as former female combatants, such as being called 'rebel girls' (Wessells, 2011: 192). Further, while participating in the fighting, they were at risk from sexually transmitted diseases and damage to their reproductive systems, because they often were required to have sex with male fighters on demand, and the medical care they received while fighting was likely inadequate. These factors will often lead to these women not being seen as desirable marriage prospects (Wessells, 2011: 189, 192–93), which will also likely act to compromise their reintegration into their communities (IRIN, 2011). Women who were married when they were abducted into the fighting may find that their husbands reject them upon their return (Wessells, 2011: 193).

Because of this, when these young women and girls seek to reintegrate into communities, they often experience psycho-social distress and social isolation (Worthen et al., 2012: 25). Decker and her co-authors (2009) point out that the 'restoration of physical, sexual, reproductive, and mental health is an important part of the reintegration process, and is typically a key priority of women and girls exiting situations of sexual exploitation' (81). In one study of a reintegration programme conducted in Sierra Leone between 2001 and 2005 after the disastrous civil war there, the returning women actually described the stresses of trying to reintegrate as worse than what they had experienced as fighters (Wessells, 2011: 192). Significant numbers might not be able to return to their home communities, and those that do often find that it is difficult to resume relationships with their parents (Worthen et al., 2012: 26).

These returning young women and girls do not necessarily have economically valuable skills, and might find themselves relegated to occupations such as collecting firewood or farming other people's land (Worthen et al., 2012: 25–26). Worse yet, the combination of stigmatisation, rejection by their communities and their lack of livelihood skills may force them to enter into sex work (Wessells, 2011: 189), which could unfortunately confirm negative opinions by community members as to their promiscuity. Not only will those who are young mothers of children fathered by fighters often suffer from stigmatisation, but their children will too (Worthen et al., 2012: 25–26).

Further, for girls who were fighters, reintegration programmes often try to fit them into vocational training that emphasises traditional gender roles, and therefore fail to accommodate the very non-traditional experiences the girls had as combatants (IRIN, 2013). On the other hand, the Sierra Leone programme noted earlier focused on four main lines of effort to smooth the reintegration process: increased access to health care, development of livelihood skills and income, increased educational opportunities and the strengthening of 'community mechanisms for protection and reconciliation' (Wessells, 2011: 193). Wessels also reports that evaluation of the programme after its completion noted significant increases in measures of reintegration progress by individual participants, often by proxy, such as increases in calorie intake, micro-financing loan repayment timeliness and the number of marriages (193–202).

All of these factors and experiences of women and girls as civilians, as refugees and as combatants are of course generalised when extracted from their specific contexts and categorised this way. In any given operational theatre, these effects and problems might not all manifest amongst affected women and girls, and these same effects and problems might emerge in different times and different places. Therefore, it is important to consider how these factors and experiences might play out in a specific context so as to understand the complexities they present to conducting useful and actionable gender analysis in a theatre of operations. Let's turn now to a case study of the impact of an armed conflict upon women and girls that has gained widespread international attention – the civil war between the government of Nigeria and the forces of a jihadi organisation known often by its abbreviation 'JAS' in Nigeria, and in the West as Boko Haram.[1]

A case study in armed conflict's impacts on women – Boko Haram's operationalisation of gender

Background

The Boko Haram insurgency in Nigeria provides an unfortunately useful example of how these different, gendered factors affecting women and girls as civilians and as combatants play out in in a non-international armed conflict. Boko Haram began as a very conservative Muslim religious movement around 2002 in Borno state, calling for the rejection of Western culture and Nigerian institutions that were not consistent with Islam (Comolli, 2015: 45–50; Matfess, 2017b: 10–11). The International Crisis Group (ICG) notes that this part of Nigeria is characterised by the practice of a conservative form of Islam, buttressed by the enactment of a strict form of Sharia law in many of the states that make up the region (2016: 2). Christians are a minority group in this region (Kriesch, 2017).

Economically, Nigeria's northeast is impoverished. In 2012, 75 percent of the population of Nigeria's northern states were living in poverty (Comolli, 2015: 2). Women may own household goods and have access to farm and pastoral land, but they are rarely land holders; in fact only about 4 percent of the landowners in this area are women. Women are marginalised from political life (Matfess, 2017b: 2), and the International Crisis Group (2016) notes that the female illiteracy rate was 72 percent in 2013 (2). Matfess (2017b) reports that only about 4 percent of girls manage to complete their secondary education (2017b: 47). Women and girls tend to marry fairly young, and females in this area have a high fertility rate, which is sadly coupled with a high maternal mortality rate (ICG, 2016: 3).

Led by the charismatic cleric Mohammed Yusuf, Boko Haram steadily gathered increasing numbers of supporters, and increasing political and financial influence in the area (Comolli, 2015: 52, 79–80; Matfess, 2017b: 17). Boko Haram's message of Islamic order, Islamic education for women and a recognised status for women also attracted a significant number of female followers (ICG, 2016: 4, 5). For a significant number of women, life under Boko Haram held the promise of being more predictable and less onerous than in the more secular northern communities (Matfess, 2017a: 6, 57–58). As Boko Haram grew in strength, friction developed between it and the Nigerian government, culminating in violence between Boko Haram members and police forces, and attacks on police stations and government buildings (Comolli, 2015: 53).

The Nigerian government violently cracked down on Boko Haram in July 2009, sending in military units to support the police. Comolli (2015) recounts how Yusuf surrendered to the military, and was then turned over to the police, who apparently then executed him (54–56). Surviving members of the movement dispersed, and began conducting revenge and guerrilla attacks. Men suspected of having assisted the Nigerian security forces were attacked and killed. Kidnappings of women and children, initially of Christians, began in 2013, ostensibly to secure hostages to exchange for detained family members of Boko Haram leaders. From this point on, the kidnapping of women became a common Boko Haram tactic. As civilian vigilante groups from Christian and Muslim communities emerged to fight Boko Haram later in 2013 (the so-called 'Civilian Joint Task Force'), Boko Haram fighters began capturing Muslim women as well (ICG, 2016: 6–8). Although there were documented instances of killings of women, particularly those who sought to escape or who refused to convert to Boko Haram's teachings, as the insurgency continued Boko Haram tended to spare them, but kill the men it captured outright (Bauer, 2016: 62; ICG, 2016: 7; Matfess, 2017b: 25, 89).

Ironically, Boko Haram's attacks on women have resulted in women assuming combatant and leadership roles that probably would not have occurred otherwise. Nigerian women have now joined the different vigilante groups protecting their communities (ICG, 2016: 11), performing searches of women suspected of smuggling weapons for Boko Haram or being suicide bombers (Comolli, 2015: 128). Collyer (2017) reports how one Muslim woman, Aisha Bakari Gombi, who grew up hunting with her grandfather, now works as a hunter of Boko Haram, assisting the Nigerian military and local people by tracking and recovering abductees by leading a team of men. Another Muslim woman, Hamsat Hassan, is a fellow insurgent hunter likewise leading a team of men, who joined the fight because Boko Haram fighters kidnapped her sister.

Treatment of abducted women and girls by Boko Haram

Over time, as the Nigerian military and the vigilante groups have clawed back territory from Boko Haram control, kidnapped women and girls have managed to escape or have been freed. A 2014 Human Rights Watch (HRW) report based on interviews with returning Christian women and girls abducted by Boko Haram documented the abuses they suffered at the hands of their jihadi captors. Some were threatened with death unless they renounced their religion and accepted Islam, whilst others were threatened with whippings or beatings if they did not. Those who refused to convert were often forced to become ammunition bearers in military operations, or forced to work as porters carrying plunder from villages that had been attacked by Boko Haram. Unfortunately, Matfess (2017b) reports that a number of women also suffered sexual abuse at the hands of their Nigerian military liberators (30–31).

A number of female victims stated that they had been subjected to rape, generally after they had been forced to marry Boko Haram fighters. According to social workers who have worked with some of the victims, the rape of abductees has been underreported due to the conservative nature of the communities in north-eastern Nigeria, which has led to stigmatisation of victims of sexual abuse (HRW, 2014). Other reports based on interviews with women who escaped suggest that hundreds of women were raped, many repeatedly (Bloom and Matfess, 2016: 110). Fearing rejection from her husband, one married woman interviewed by Human Rights Watch who had been abducted and raped chose to not inform him of what had happened to her, which complicated her receiving the medical and psychological care she required (Braunschweiger, 2014).

Braunschweiger (2014) also detailed another young woman being used as a decoy to lure Christian men into Boko Haram control, where they would be given the choice of either converting to Islam or being killed. When the captured men refused, their Boko Haram captors slit their throats. Although this woman managed to escape, she was later informed that because Boko Haram fighters believed she was pregnant with their leader's child, they were seeking to recapture her. In this sense, Boko Haram's physical and psychological impacts on the returning women can continue long after they escape its grasp. Those displaced by fighting involving Boko Haram suffer negative impacts as well, even if they were not directly attacked.

The camps for internally displaced persons

Perhaps as many as 2.8 million people have been displaced by the conflict, and in 2017, Matfess reported that more than 4.4 million people were experiencing severe food insecurity, the majority of both sets being women and children (2017b: 146–47). Women and girls who escape from Boko Haram have often found themselves having to live in government-run camps for internally displaced persons (IDPs), or sometimes informal camps operating with little or no government assistance. Women and girls in these camps, which are generally staffed and guarded by men, have been victims of sexual violence, or have had to prostitute themselves to obtain food, money or permission to leave the camps (ICG, 2016: 14).

The camps have often been lacking in adequate food, clean water and proper medical facilities and services for pregnant women (Omole, Welye and Abimbola, 2015: 941), particularly the informal camps (Matfess, 2017b: 174–75). Within the camps, women who were abducted by Boko Haram are often viewed with disfavour and socially isolated. In research conducted by the UN Children's Fund (UNICEF) and International Alert (2016) among women in IDP camps, the services and items provided to the women and girl survivors of sexual abuse that were particularly appreciated were 'psycho-social support, maternal health care and distribution of hygiene kits' (12). They also noted that 'The lack of livelihood opportunities for women inside IDP camps, especially for those who have been rejected by their husbands and separated from families, has reportedly created destitution. Some women and girls have exchanged sexual activities for money' (13).

Difficult homecomings

When women returning from areas controlled by Boko Haram sought to reintegrate back into their communities, they often found themselves the subject of suspicion as to whether their loyalties still held with the group. People in the receiving communities feared that they might become suicide bombers (ICG, 2016: 15) or even recruit others to join Boko Haram (UNICEF, 2016: 15). Even when young women and girls are accepted by their families, the psychological damage they have suffered can make it difficult to actually reintegrate with them (Baker, 2017). Women with children fathered by Boko Haram fighters have also often been regarded unfavourably by their families, even in instances where the pregnancy was forced, given the conservative mores of the north-eastern communities regarding extra-marital sex. The children brought back by these young mothers have often been stigmatised themselves, as being of 'bad blood' (ICG, 2016: 15), and that they will grow up to be the next generation of fighters (UNICEF, 2016: 9). Community members sometimes threatened to kill these children (Bauer, 2016: 47).

In some cases, the social ostracism is so intense that women will move to different locations to avoid it (Amusan and Ejoke, 2017: 58), further delaying their reintegration into civil society. The UN Children's Fund and International Alert (2016) report that there are often concerns among family members and the young women and girls whether they will be able to marry later (17). In many cases, because of the physical and psychological trauma experienced by the kidnapped women, they are not healthy enough to fully engage with efforts to provide them vocational training or employment (Amusan and Ejoke, 2017: 58). By way of comparison, Yazidi women abducted by Islamic State of Iraq and the Levant fighters in Iraq suffered many of the same conditions and attacks experienced by these Nigerian women (Tax, 2016: 40–41). Yazidi refugees too have struggled to recover after the trauma of the Islamic State attacks, and still evidence psychological scars despite being in relatively safe locations (Erdener, 2017: 63–69). Clearly, these sorts of impacts are not limited to victims of Boko Haram.

Rehabilitation and 'de-radicalisation'

International attention focused on this civil war in 2014 after Boko Haram fighters raided a Christian girls' school in Chibok and kidnapped 276 girls. When 82 of them were released in May 2017, they were placed in a specially designed residential facility where they stayed for nine months, receiving health services, psychological counselling, trauma therapy and remedial education courses (Baker, 2017). As the Nigerian government regained control over areas of the country's northeast, the international community also became aware that it was not just the female victims of Boko Haram who required attention, but its female supporters as well. In one case, a number of wives of Boko Haram members and their children were placed in a 'de-radicalisation' centre. These women were under armed guard, and not allowed to leave the centre, but women social workers were actually responsible for them. It is unclear whether participating in this programme would actually assist these women and their families reintegrate back into their communities (ICG, 2016: 16).

In another case, the director of the Nigerian National Security Agency's Behavioural Analysis and Strategic Communications office, Dr Fatima Akilu, announced in 2015 that 20 women and girls who had been recruited by Boko Haram were undergoing a rehabilitation and de-radicalisation programme (The NEWS, 2015). This programme provided medical care, psycho-social support, counselling for post-traumatic stress disorder, education and livelihood training, but it was no longer in operation by the end of 2015 (UNICEF, 2016: 10). The programme was funded by the EU and the UK (Freeman, 2015). The UN Children's Fund and International Alert (2016) reported that 'Communities believe that the government-led "de-radicalisation" process for those who spent time with JAS in captivity is a necessary precondition for their reintegration' (18). After she left government in 2015, Dr Akilu founded the Neem Foundation, a non-governmental organisation that specialises in research and providing services in the field of preventing violent extremism (Neem Foundation, 2016), and which builds on the work she and her team performed with their earlier Countering Violent Extremism Programme for the National Security Agency. In a later interview, Dr Akilu explained that although women might successfully complete the de-radicalisation programme, they might still experience stigmatisation in their communities (Matfess, 2017a).

Matfess (2017b) notes that there are reports of young women who have been through this sort of intensive programme who then decide to re-join their husbands in Boko Haram. The reasons vary, but for many it seems that they have assessed that their quality of life was actually better in the insurgency – for example, if they were married to more senior Boko Haram commanders and enjoyed a higher social status. The payment of the bride price by many suitors directly to women in areas under Boko Haram control rather than to the women's families appears to have been a significant factor in many women's perceptions of their treatment under the insurgency. Perhaps related to this calculation, the de-radicalisation programmes for women have been critiqued as not focusing on allowing the women to acquire skills or providing training that could help these young mothers, now single, support themselves and their families (60–61).

The complexity of gendered effects

To focus only on the grave sexual abuse of the women and girls Boko Haram fighters kidnapped is to fail to appreciate the full scope of the gender complexity of this armed conflict. Tragically, many young women have become suicide bombers (Bloom and Matfess, 2016: 105). This is in contrast to Al Shabaab in Somalia, another African jihadi terrorist group, which rarely appears to use female suicide bombers (Davis, 2017: 113–14). Although Davis states that 'For Boko Haram, the only operational role for women is suicide-bombers' (2017: 107), Boko Haram might also have found places for women in its ranks as actual leaders and fighters. Interviews with former Boko Haram fighters suggest that it had between 500 and 1,000 female members, who mainly served as foot soldiers. Others served as recruiters for the organisation, as spies and as explosives experts (Botha and Abdile,

2016: 3). This report of Boko Haram's use of women as actual combatants is perhaps corroborated to a degree by certain other reports of female insurgents taking a direct part in hostilities (ICG, 2016: 10), and Matfess's recent field work in the area of Boko Haram's insurgency (2017b: 130–34). Matfess did note reports of women with higher-ranking social status in Boko Haram becoming 'enforcers' of Boko Haram theology when the men fighters are away from their communities (122–23).

Summary

The gendered nature of armed conflict's impacts upon women and girls poses significant problems for countries and communities trying to stabilise their security situations in times of armed conflict. The impacts have profound negative effects on community cohesion, economic development and individual women's and girls' physical and mental health. From an operational perspective, it is important for militaries to realise that regardless of how successful their actions against insurgents might be in terms of destroying the insurgents or neutralising their influence in civilian-centric environments, these efforts alone might not be sufficient to generate the durable stability in these areas that the militaries are seeking to achieve. Unless complementary efforts provide some degree of employment and health security to female victims of armed conflict, as well as fostering environments where they are less subject to social, political and economic discrimination, lasting stability could be elusive.

To be operationally useful, gender analysis must be granular, that is, accurate, detailed and specific to the operational theatre. For example, from the perspective of understanding the strengths and nature of one's adversary, as noted previously regarding areas formerly controlled by Boko Haram, marriage to Boko Haram fighters was often viewed positively by women and their families in both a social and an economic sense. The International Crisis Group (2016) reported that officially, wives appear to have enjoyed a more favourable status than mere captives, and that Boko Haram fighters would often offer relatively attractive dowries for brides (8–9). Further, as noted earlier, the dowries were often paid directly to the women rather than their families, incentivising young women to voluntarily agree to marriage to Boko Haram insurgents (Matfess, 2017a). Also, women might find themselves, willingly or not, in the role of enforcers of insurgent ideology or religious practices, and this further complicates their reintegration into post-conflict societies (Moaveni, 2015).

If Mao was correct that the people are like the sea in which the insurgent swims (1937: Chapter 6), these gender-related factors cannot be discounted in terms of Boko Haram's continuing ability to wage its war amongst the people and retain a degree of resiliency in the face of concentrated government military action (*Nigeria*, 2016: 21262–63). This presents a dilemma potentially for deploying forces – insurgencies might actually enjoy a sense of heightened legitimacy among local women as compared to the status quo society ostensibly protected by the host nation the deploying forces are there to support. Challenging deficiencies in the governments of host nation allies is delicately done. If, however, such positions are supported by a rigorous gender analysis that helps focus on specific action items rather than broad cultural judgements, this conversation could be conducted properly and respectfully, and deployed force units should not therefore shy away from these tasks for fear of giving unintended offence. Finally, from an environmental perspective, these issues also factor into community, family and individual resilience to the effects of climate change in those areas where climate change is having a gendered impact upon women and girls. With that in mind, let's examine the relationship between gender and climate change.

Note

1 The full name of the organisation is *Jama'atu Ahlus-Sunnah Lidda'awati Wal Jihad*, sometimes also referred to by the abbreviation of this name, JAS.

References

Amusan, L and Ejoke, UP (2017) 'The Psychological Trauma Inflicted by Boko Haram Insurgency in the North Eastern Nigeria'. *Aggression and Violent Behavior* 36: pp. 52–59. DOI: 10.1016/j.avb.2017.07.001.

Anderlini, SN (2017) 'UN Peacekeepers' Sexual Assault Problem'. *Foreign Affairs* (9 June). Available at: www.foreignaffairs.com/articles/world/2017-06-09/un-peacekeepers-sexual-assault-problem (Accessed 11 April 2018).

Baker, A (2017) 'Boko Haram's Other Victims'. *Time* (27 June). Available at: http://time.com/boko-harams-other-victims/ (Accessed 11 April 2018).

Bauer, W (2016) *Stolen Girls – Survivors of Boko Haram Tell Their Story*. New York: New Press.

Biank, T (2014) *Undaunted: The Real Story of America's Servicewomen in Today's Military*. New York: Dutton Caliber.

Bloom, M and Matfess, H (2016) 'Women as Symbols and Swords in Boko Haram's Terror'. *Prism* 6(1): pp. 105–21. Available at: http://cco.ndu.edu/Portals/96/Documents/prism/prism_6-1/Women%20as%20Symbols%20and%20Swords.pdf (Accessed 24 June 2018).

Botha, A and Abdile, M (2016) 'Summary: Getting Behind the Profiles of Boko Haram Members and Factors Contributing to Radicalisation versus Working for Peace'. Finn Church Aid, the Network for Religious and Traditional Peacemakers, and the International Dialogue Centre. Available at: https://frantic.s3-eu-west-1.amazonaws.com/kua-peacemakers/2016/10/Boko_Haram_Study_Botha_Abdile_SUMMARY_disclaimer_included_2016.pdf (Accessed 13 June 2018).

Braunschweiger, A (2014) 'Interview: Life After Escaping Boko Haram's Clutches'. *Human Rights Watch* (27 October). Available at: http://features.hrw.org/features/Interview_2014/Life_After_Escaping_Boko_Harams_Clutches/index.html (Accessed 3 March 2018).

Buscher, D (2009) 'Women, Work, and War'. In: Martin, SF and Tirman, J (eds.) *Women, Migration and Conflict*. Dordrecht: Springer Science and Business Media, pp. 87–106.

Cawkill, P; Rogers, A; Knight, S and Spear, L (2009) *Women in Ground Close Combat Roles: The Experiences of other Nations and a Review of the Academic Literature, Report No. DSTL/CR37770 V3-0*. Hants, UK: UK Defence Science and Technology Laboratory. Available at: https://assets.publishing.service.gov.uk/government/uploads/system/uploads/attachment_data/file/27406/women_combat_experiences_literature.pdf (Accessed 13 June 2018).

Chenoweth, S (2017) '"We Keep It In Our Heart" – Sexual Violence Against Men and Boys In The Syria Crisis'. UN High Commissioner for Refugees. Available at: https://data2.unhcr.org/es/documents/download/60864#_ga=2.121844236.641198675.1512507880-2095888809.1417795315 (Accessed 6 June 2018).

Collyer, R (2017) 'Meet Aisha, a Former Antelope Hunter Who Now Tracks Boko Haram'. *The Guardian* (8 February). Available at: www.theguardian.com/world/2017/feb/08/antelope-hunter-boko-haram-nigeria (Accessed 17 January 2018).

Comolli, V (2015) *Boko Haram – Nigeria's Islamist Insurgency*. London: Hurst & Company.

Davis, D (2010) 'Female Engagement Team Commander Speaks about Mission in Afghanistan'. *DVIDS* (16 October). Available at: www.dvidshub.net/video/97875/female-engagement-team-commander-speaks-about-mission-afghanistan (Accessed 7 August 2012).

Davis, J (2017) *Women in Modern Terrorism – From Liberation Wars to Global Jihad and the Islamic State*. Lanham, MD: Rowman & Littlefield.

Decker, MR; Oram, S; Gupta, J and Silverman, JG (2009) 'Forced Prostitution and Trafficking for Sexual Exploitation Among Women and Girls in Situations of Migration and Conflict: Review and Recommendations for Reproductive Health Care Personnel'. In: Martin, SF and Tirman, J (eds.) *Women, Migration, and Conflict: Breaking a Deadly Cycle*. Dordrecht: Springer Science +Business Media BV, pp. 63–86.

Department of Defence (n.d.) 'Removal of Gender Restrictions from ADF Combat Role Employment Categories'. Available at: www.defence.gov.au/women/ (Accessed 29 May 2018).

Dharmapuri, S (2011) 'Just Add Women and Stir?' *Parameters* 41(1): pp. 56–70. Available at: www.peacewomen.org/assets/file/Resources/Academic/1325-1820-partpp_justaddwomenandstir_gender_peaceprocesses__parametersquarterly_2011.pdf (Accessed 24 June 2018).

Enzinna, W (2015) The Rojava Experiment. *New York Times Magazine* (29 November). Available at: www.nytimes.com/2015/11/29/magazine/a-dream-of-utopia-in-hell.html (Accessed 4 December 2017).

Erdener, E (2017) 'The Ways of Coping with Post-war Trauma of Yezidi Refugee Women in Turkey'. *Women's Studies International Forum* 65: pp. 60–70. DOI: 10.1016.j.wsif.2017.10.003.

Freeman, C (2015) 'Fears for Nigeria's Counter-radicalisation Programme as British-Trained Head is Ousted.' *The Telegraph* (7 October). Available at: www.telegraph.co.uk/news/worldnews/africaandindianocean/nigeria/11915708/Fears-for-Nigerias-counter-radicalisation-programme-as-British-trained-head-is-ousted.html (Accessed 27 December 2017).

Frühstück, S (2007) *Uneasy Warriors – Gender, Memory, and Popular Culture in the Japanese Army*. Oakland, CA: University of California Press.

Gardam, J and Charlesworth, H (2000) 'The Need for New Directions in the Protection of Women in Armed Conflict'. *Human Rights Quarterly* 22(1): pp. 148–66.

Gardam, J and Charlesworth, H (2002) 'Protection of Women in Armed Conflict' (Workbook Readings, Section 2). In: Department for International Development and Department of Foreign Affairs and International Trade (eds.) *Gender and Peacekeeping Online Training Course*. Available at: www.genderandpeacekeeping.org/ (Accessed 27 May 2011).

Gardam, JG and Jarvis, MJ (2001) *Women, Armed Conflict and International Law*. The Hague: Kluwer Law International.

Huffman, A (2013) *Here I Am – The Story of Tim Hetherington, War Photographer*. New York: Grove Press/Atlantic Monthly Press.

Human Rights Watch (2014) '"Those Terrible Weeks in their Camp" – Boko Haram Violence against Women and Girls in Northeast Nigeria'. *Human Rights Watch* (27 October).

Available at: http://features.hrw.org/features/HRW_2014_report/Those_Terrible_Weeks_in_Their_Camp/index.html (Accessed 29 December 2017).

ICG (International Crisis Group) (2016) *Nigeria: Women and the Boko Haram Insurgency, Africa Report no. 242*. Available at: https://d2071andvip0wj.cloudfront.net/242-nigeria-women-and-the-boko-haram%20Insurgency.pdf (Accessed 13 June 2018).

IRIN Staff (2011) 'Fewer "I do's" for Former Female Rebels'. *IRIN News* (23 February). Available at: www.irinnews.org/report/92017/sri-lanka-fewer-i-dos-former-female-rebels (Accessed 13 June 2018).

IRIN Staff (2013) 'Girl Child Soldiers Face New Battle in Civilian Life'. *IRIN News* (12 February). Available at: www.irinnews.org/analysis/2013/02/12/girl-child-soldiers-face-new-battles-civilian-life (Accessed 13 June 2018).

Jones, A (2010) *War Is Not Over When It Is Over – Women Speak Out From The Ruins Of War*. New York: Macmillan Publishers.

Kim, N and Conceição, P (2010) 'The Economic Crisis, Violent Conflict, and Human Development'. *International Journal of Peace Studies* 15(1) (Spring/Summer): pp. 29–43. Available at: www.gmu.edu/programs/icar/ijps/vol15_1/KimConceicao15n1.pdf (Accessed 24 June 2018).

Korge, J (2013) 'Ms. Kalashnikov: The Women Rebels of Congo'. *Der Spiegel Online* (30 September). Available at: www.spiegel.de/international/world/francesca-tosarelli-photographs-women-rebel (Accessed 23 October 2017).

Kriesch, A (2017) 'Boko Haram Insurgency Weighs on Minority Christians Three Years After Chibok'. *Deutsche Welle*. Available at: www.dw.com/en/boko-haram-insurgency-weighs-on-minority-christians-three-years-after-chibok/a-38415778 (13 April 2017) (Accessed 9 February 2018).

Leaning, J; Bartels, S and Mowafi, H (2009) 'Sexual Violence during War and Forced Migration'. In: Martin, SF and Tirman, J (eds.) *Women, Migration, and Conflict: Breaking a Deadly Cycle*. Dordrecht: Springer Science +Business Media BV, pp. 173–99.

Lemmon, GT (2015) *Ashley's War: The Untold Story of a Team of Women Soldiers on the Special Ops Battlefield*. New York: HarperCollins Publishers.

Mao, T-t (1937) *On Guerrilla Warfare*. Available at: www.marxists.org/reference/archive/mao/works/1937/guerrilla-warfare/ch06.htm (Accessed 24 June 2018).

Martin, SF (2009) Introduction. In: Martin, SF and Tirman, J (eds.) *Women, Migration, and Conflict: Breaking a Deadly Cycle*. Dordrecht: Springer Science +Business Media BV, pp. 1–22.

Matfess, H (2017a) 'Rescued and Deradicalised Women Are Returning to Boko Haram. Why?' *African Arguments* (1 November 2017a). Available at: http://africanarguments.org/2017/11/01/rescued-and-deradicalised-women-are-returning-to-boko-haram-why/ (Accessed 29 December 2017).

Matfess, H (2017b) *Women and the War on Boko Haram: Wives, Weapons, and Witnesses* London: Zed Books.

Medill Reports Chicago (2016) 'Women Warriors Have Special Health Needs, Researchers Report'. *Medill News Service* (28 January). Available at: http://news.medill.northwestern.edu/chicago/women-warriors-have-special-health-needs-researchers-report/ (Accessed 23 December 2017).

Ministry of Defence and Prime Minister's Office (UK) (2016) 'Ban on Women in Ground Close Combat Roles Lifted'. *GOV.UK* (8 July). Available at: www.gov.uk/government/news/ban-on-women-in-ground-close-combat-roles-lifted (Accessed 18 December 2017).

Ministry of Defense (Japan) (2014) *Defense of Japan 2014*. Tokyo: Ministry of Defense. Available at: www.mod.go.jp/e/publ/w_paper/2014.html (Accessed 4 February 2015).

Moaveni, A (2015) 'ISIS Women and Enforcers in Syria Recount Collaboration, Anguish and Escape'. *New York Times* (21 November). Available at: www.nytimes.com/2015/11/22/world/middleeast/isis-wives-and-enforcers-in-syria-recount-collaboration-anguish-and-escape.html (Accessed 20 December 2017).

NBC News (2017) 'Japan Approves Female Fighter Pilots'. *NBC* (11 November). Available at: www.nbcnews.com/video/japan-approves-female-fighter-pilots-564477507848 (Accessed 3 December 2017).

Neem Foundation (2016) 'About NEEM – What We Do'. *Neem Foundation*. Available at: www.neemfoundation.org.ng/ (Accessed 5 January 2018).

Neuhaus, SJ and Crompvoets, SL (2013) 'Australia's Servicewomen and Female Veterans: Do We Understand Their Health Needs?' *The Medical Journal of Australia* 199(8): pp. 530–32. DOI: 10.5694/mja13.10370.

'Nigeria: War Against Boko Haram Not Over' (2016) *Africa Research Bulletin: Political, Social and Cultural Series* (December 12): pp. 21262–63. DOI: 10.1111/j.1467-825X.2017.07424.x.

Nilsson, M (2018) 'Muslim Mothers in Ground Combat Against the Islamic State: Women's Identities and Social Change in Iraqi Kurdistan'. *Armed Forces and Society* 44(2): pp. 261–79. DOI 10.1177/0095327X17699568.

Oladipo, T (2017) 'The UN's Peacekeeping Nightmare in Africa'. *BBC* (5 January). Available at: www.bbc.com/news/world-africa-38372614 (Accessed 19 February 2018).

Omole, O; Welye, H and Abimbola, S (2015) 'Boko Haram Insurgency: Implications for Public Health'. *The Lancet* 385: p. 941. DOI: 10.1016/S0140-6736(15)60207-0.

PeaceWomen (2018) 'Member States – National Action Plans for the Implementation of UNSCR 1325 on Women, Peace and Security'. PeaceWomen.org. Available at: www.peacewomen.org/member-states (Accessed 20 July 2018).

Pearson, D (2017) 'Climate Change and Violent Conflict in Kenyan Pastoral Lands: UN Sustainable Development Goals Present a Way Forward'. Future Directions International. Available at: www.futuredirections.org.au/wp-content/uploads/2017/12/Climate-Change-and-Violent-Conflict-in-Kenyan-Pastoral-Lands.pdf (Accessed 10 June 2018).

Pellerin, C (2015) 'Carter Opens All Military Occupations, Positions to Women'. *US Department of Defense News* (3 December). Available at: www.defense.gov/News/Article/Article/632536/carter-opens-all-military-occupations-positions-to-women/ (Accessed 22 January 2018).

Prescott, JM (2013) 'NATO Gender Mainstreaming and the Feminist Critique of the Law of Armed Conflict'. *Georgetown Journal of Gender and the Law* 15(1): pp. 83–131.

Prescott, JM; Iwata, E and Pincus, B (2015) 'Gender, Law and Policy: Japan's National Action Plan on Women, Peace and Security'. *Asian-Pacific Law & Policy Journal* 17(1) [online]. Available at: http://blog.hawaii.edu/aplpj/files/2016/04/APLPJ_Prescott-Iwata-Pincus-FINAL.pdf (Accessed 24 June 2018).

Schilling, J; Froese, R and Naujoks, J (2018) '"Just Women" Is Not Enough: Towards a Gender-Relational Approach to Water and Peace-Building'. In Fröhlich, C; Gioli, G; Cremades, R and Myrttinen, H (eds.) *Water Security Across the Gender Divide*. Cham, Switzerland: Springer International Publishing AG, pp. 173–96.

Shepherd, LJ (2016) 'Making War Safe for Women? National Action Plans and the Militarization of the Women, Peace and Security Agenda'. *International Political Science Review* 37(3): pp. 324–35. DOI: 10.1177/0192512116629820.

Storr, W (2011) 'The Rape of Men: The Darkest Secret of War'. *The Guardian* (16 July). Available at: www.theguardian.com/society/2011/jul/17/the-rape-of-men (Accessed 9 December 2017).

Tax, M (2016) *A Road Unforeseen – Women Fight the Islamic State*. New York: Bellevue Literary Press.

The NEWS Staff (2015) 'Nigerian Agency Saves 22 Female Suicide Bombers'. *The NEWS* (30 June). Available at: http://thenewsnigeria.com.ng/2015/06/nigerian-agency-saves-22-female-suicide-bombers/ (Accessed 6 December 2017).

Tosarelli, F (2013) 'Ms. Kalashnikov'. Available at: www.francescatosarelli.com/ms-kalashnikov/e4wemgonbikmyn37kvjetdpn5bcw86 (Accessed 3 December 2017).

UN High Commissioner for Refugees (2003) 'Sexual and Gender-Based Violence Against Refugees, Returnees and Internally Displaced Persons – Guidelines for Prevention and Response'. UN High Commissioner for Refugees. Available at: www.unhcr.org/3f696bcc4.pdf (Accessed 3 September 2016).

UN High Commissioner for Refugees (2008) *UNHCR Handbook for the Protection of Women and Girls*. Geneva: UNICEF.

UN High Commissioner for Refugees (2012) 'Working With Men and Boy Survivors of Sexual and Gender-Based Violence in Forced Displacement'. Refworld.org. Available at: www.refworld.org/pdfid/5006aa262.pdf (Accessed 3 September 2016).

UNICEF (UN Children's Fund) and International Alert (2016) '"Bad Blood": Perceptions of Children Born of Conflict-Related Sexual Violence and Women and Girls Associated with Boko Haram in Northeast Nigeria'. Available at: www.unicef.org/nigeria/Nigeria_BadBlood_EN_2016.pdf (Accessed 13 June 2018).

UN News Centre Staff (2011) 'UN Official Voices Concern Over Reports of Rape of Somali Women Fleeing Famine'. *UN News Centre* (11 August). Available at: www.un.org/apps/news/story.asp?NewsID=39282#.WlEQVd-nGUk (Accessed 13 July 2013).

UN Secretary-General (2005) 'Comprehensive Review of the Whole Question of Peacekeeping Operations in all their Aspects'. A/59/710 (24 March). Available at: www.un.org/ga/search/view_doc.asp?symbol=a/59/710 (Accessed 13 June 2018).

UN Security Council (2000) 'Women, Peace and Security'. S/RES/1325 (31 October). Available at: www.un-documents.net/sr1325.htm (Accessed 13 June 2018).

UN Security Council (2005) 'Presidential Statement, Women and Peace and Security' (S/PRST/2005/52 (27 October). Available at: www.un.org/en/ga/search/view_doc.asp?symbol=S/PRST/2005/52 (Accessed 6 September 2017).

UN Security Council (2008) 'Women and Sexual Violence'. S/RES/1820 (19 June). Available at: www.securitycouncilreport.org/atf/cf/%7B65BFCF9B-6D27-4E9C-8CD3-CF6E4FF96FF9%7D/CAC%20S%20RES%201820.pdf (Accessed 13 June 2018).

UN Security Council (2010) 'Women and Peace and Security'. S/RES/1960 (16 December). Available at: https://undocs.org/S/RES/1960 (2010) (Accessed 13 June 2018).

Valenius, J (2007) 'Gender Mainstreaming in ESDP Missions'. *Chaillot Paper No. 101*. European Institute for Security Studies. Available at: www.nato.int/ims/2007/win/opinions/Gender%20mainstreaming%20in%20ESDP%20missions.pdf (May 2007) (Accessed 23 February 2018).

Weiss, C (2011) 'We Must Not Make War Safe for Women'. *OpenDemocracy 50.50* [blog]. Available at: www.opendemocracy.net/5050/cora-weiss/we-must-npt-make-war-safe-for-women (Accessed 13 February 2018).

Wessells, M (2011) 'The Reintegration of Formerly Recruited Girls: A Resilience Approach'. In: Cook, DT and Wall, J (eds.) *Children and Armed Conflict: Cross-Disciplinary Investigations*. New York: Palgrave Macmillan, pp. 189–204.

Winslow, D (2009) 'Gender Mainstreaming: Lessons for Diversity'. *Commonwealth & Comparative Politics* 47(4): pp. 538–39. DOI: 10.1080/14662040903381453.

Women's Commission for Refugee Women and Children (2002) 'UNHCR Policy on Refugee Women and Guidelines on Their Protection: An Assessment of Ten Years of

Implementation'. Refworld.org. Available at: www.refworld.org/pdfid/48aa83220.pdf (Accessed 7 November 2017).

Worthen, M; McKay, S; Veale, A and Wessells, M (2012) 'Reintegration of Young Mothers'. *Forced Migration Review* 40: pp. 25–26. Available at: www.fmreview.org/young-and-out-of-place/worthen-et-al.html (Accessed 24 June 2018).

Wright, W (2011) 'Two Members of Cultural Support Team Receive Combat Action Badges'. *DVIDS* (12 September 2011) Available at: www.dvidshub.net/news/76916/two-members-cultural-support-team-receive-combat-action-badges (Accessed 24 June 2018).

2
GENDER AND CLIMATE CHANGE

Before looking into the gendered impacts of climate change, it is important to set the global backdrop against which these impacts should be viewed and understood with regard to the status of women and girls in general. As to women's political participation, for example, by 2012, the number of women in national parliamentary seats had increased 6 percent since 1990, which was progress, but they still only made up 18 percent of national legislatures (Farhar, Osnes and Lowry, 2014: 154). By the end of 2017, this percentage had only increased to 23.5 percent (Inter-Parliamentary Union, n.d.). Perhaps another example of modest progress is that in developing countries between 1990 and 2010, the percentage of women employed in non-agricultural work increased from 29 to 34 percent (Farhar, Osnes and Lowry, 2014: 160). Whether this actually resulted in an improvement in their working conditions is unknown, because many rural women and girls migrate to urban centres in search of non-agricultural work, where the conditions under which they labour can be grim (Westerman, 2017).

Importantly (and this does not always appear to be appreciated in media accounts of climate change's effects), the impacts of climate change are not uniform across a population in a given area. Even in developed countries, it is not just the poor who are disproportionately affected, but it is poor women who bear the brunt of the impact (Aguilar, 2009: 79). For example, after Hurricane Katrina in the US in 2005, Chindarkar (2012) reports that households headed by poor women in New Orleans were less likely to return to the city and rebuild than those headed by married couples, because these women had been unable to afford adequate home or renter's insurance prior to the storm (5). Denton (2002) has noted, however, that in developing countries, gender and sex-related differences in climate changes' effects appear to be much more pronounced (18).

Verner (2012) has assessed that at least for rural women in developing countries, the three primary factors of gender-based risk to climate change's effects are unequal access to resources, unequal opportunities to change or improve their livelihoods, and exclusion from decision-making (279). In this regard, a more encouraging statistic is that between 1991 and 2010, the percentage of girls in developing countries enrolled in primary and secondary school increased from about 75 to 89 percent (Farhar, Osnes and Lowry, 2014: 163). This increase is important for many reasons, not the least of which is that studies indicate that poorly educated women and girls are less likely to employ sustainable climate change adaptation and mitigation techniques (Aguilar, 2009: 59). Education, and its attendant empowerment in economic, social and perhaps even political terms, is therefore also an enabler in developing skills useful in fighting the negative effects of climate change.

From an operational perspective, evaluating a mission area solely on the basis of gendered risk is not likely to lead to actionable intelligence and analysis for commanders, planners and operators. Thus, this chapter will use an impacts-mapping methodology suggested by researchers at the Institute for Development Studies, University of Sussex, to track the linkages between climate change and gendered effects (Brody, Demetriades and Esplen, 2008: 3–10). This impacts-categorisation of the climate change-related effects on the civilian population lends itself well to operational analysis because it could provide the basis for a focused data-collection plan and its format is one that would likely be more easily understood by a military audience. Further, it could provide a simplified bridge for data and analysis exchange between military forces, civilian agencies and international and non-governmental organisations operating in the same theatre.

Convincing military organisations to invest in such an approach requires making the case that these effects potentially exist in operational theatres and that they are worth measuring and acting upon. One drawback in making such a case is that the relevant evidence in this area tends to be more qualitative than quantitative, and sometimes even anecdotal. In lieu of an analysis more grounded in factual data, the best option remaining is perhaps testing a qualitative meta-assessment of the relevant literature by comparing it with later largely qualitative studies to determine whether there are any inconsistencies between them that suggest caution should be used in relying upon this literature in an operational gender analysis.

Fortuitously, in 2012 Goh conducted a very thorough review of the linkages between gender and climate change using Brody et al.'s impacts-categorisation, from the perspectives of well-being and economic assets. Her assessment addressed not just whether published research identified gender-differentiated impacts of climate change, but also whether these impacts were more severe for women and girls. This chapter will describe her largely positive findings in this regard, and then assess whether more recently published research corroborates her conclusions. Next, the value of identifying an operational approach that could prove useful not just in assessing the relationships between gender and climate change but also in setting the stage for Chapter 4's discussion of climate change and armed conflict will be considered. Finally, to place this discussion in context, a case study assessing climate change and gender impacts from South Africa will be examined in detail.

Climate change's impacts on women and girls

In their study on mapping climate change impacts and gender, Brody and her co-authors (2008) identified several categories of climate change impacts that could benefit from further research and analysis. Of these, six in particular appear promising categories for military intelligence collection and analysis in civilian-centric operations: agricultural production; the closely related area of food security; health; water and energy resource availability and use; and migration induced by climate change and natural disasters related to climate change (3–10). Although the specific impacts in these categories will of course be dependent on the specific circumstances of an operational theatre, it is useful to describe in general the sorts of gendered effects that climate change might have under each of these categories. This description would be useful in helping set the scope of any sex and gender-disaggregated information collection plan in the field.

Agricultural production

Women's roles in agriculture demonstrate a high degree of variation across the developing world, but 'women farmers share a common set of gender-based disadvantages' (Speranza and Bikketi, 2018: 127). Payne (2016) notes that in many developing countries, women are more likely to be engaged in small-scale, low-technology farming and have less secure forms of access rights to land than men, which can place them particularly at risk to climate change's effects (53). The lack of title to land in their own names in many areas complicates women's participation in decision-making processes as to resource use and allocation. For example, Verner (2012) points out that women who do not own land might be unable to secure credit for seeds and improvements on the land they work, and might find that their access to water for use on the land is insecure. Further, unless they are recognised as landowners, and therefore clear stakeholders, women might also find themselves unable to become full members in rural organisations that make decisions which affect resource use, and could therefore be excluded from decision-making (287).

Although they constitute a small minority of global societies, it is important to note that there are exceptions in developing countries to male-dominated systems of agricultural production. For example, the Minangkabau ethnic group that lives in the highlands of West Sumatra is matrilineal – and daughters inherit rice paddies and houses rather than sons (Sankari, 2016). In contrast, and describing a much more typical situation, Hillenbrand (2010) notes that in parts of Bangladesh, gendered norms regarding asset control lead to 'an assumption that women in agriculture are concerned with subsistence only', which reinforces institutional and policy biases that worsen 'women's disadvantage in accessing markets, credit, technology and services, and perpetuates the lack of recognition surrounding women's role in farming' (413).

In addition to unequal access to resources and exclusion from decision-making, Verner (2012) finds that rural women ordinarily perform time-consuming, 'non-mechanized, labor-intensive, non-capital intensive activities', such as the many tasks associated with animal husbandry (281). For example, a typical woman farmer in Sub-Saharan Africa is estimated to put in a 16-hour workday on average, counting time spent growing food and performing domestic chores (Farhar, Osnes and Lowry, 2014: 159). For the most part, Verner (2012) reports, this labour is unpaid (282).

Food security

Food security is directly related to agricultural production, but it is not the same thing. Food security has been defined by the UN's Food and Agriculture Organisation (FAO) as existing 'when all people, at all times, have physical, social and economic access to sufficient, safe and nutritious food which meets their dietary needs and food preferences for an active and healthy life' (FAO, n.d.). In this regard, it must be acknowledged that although the general trend of climate change's impacts on food supplies is expected to result in lower crop yields globally over time, in certain places at certain times there will likely be localised increases in agricultural productivity as the climate shifts (Payne, 2016: 51). Whilst these temporary upswings would likely promote food security in these areas (assuming the local population had access to the harvested crops), climate change's negative impacts on neighbouring areas could lead to a greater regional or national level of food insecurity at the same time.

In addition, Aguilar (2009) points out that the opportunities to grow certain crops can be gender-differentiated. In developing Oceania island states, for example, men and women will often work different crops (22). In Yemen, crops such as groundnuts, pumpkins and leafy vegetables are considered 'women's crops', and the role women play in selecting seeds for the next growing season has a direct impact on the biodiversity of these food supplies (58). Research has shown that women in developing countries often rely upon crop biodiversity to deal with climatic variability (82). In Zanzibar, seaweed has become a cash crop for the women who cultivate it – men are generally not involved (Ash, 2018). These gendered aspects of agricultural production can lead to direct impacts on families' food security in a time of climate change.

Food security is also directly related to health, the next category identified by Brody and her co-authors. Insufficient nutrition is a major factor in the increased mortality that results from humanitarian crisis situations. Martin (2009) notes that pregnant or nursing women who receive insufficient nutrition are not only unable to fully nourish their developing foetuses or babies, they are also more susceptible to disease themselves (9). Further, as Chindarkar (2012) has observed, in many societies women and girls are likely to eat last in their families after the men and boys have been fed (1–2), meaning that instances of food insecurity may result in them becoming more quickly malnourished (Aguilar, 2009: 59).

Health

Payne (2016) notes that there are two primary mechanisms of negative health impacts due to climate change, first, the increase in disease that can be directly associated with extreme weather events, and second, the longer-term trends in disease that are related to the broader effects of climate change, such as the spread of insect disease vectors with increased average temperature, decreased agricultural yields and the problems of increased urbanisation (47–48, 50). Continuing with the example of pregnant women, they might be at increased risk of contracting malaria because of changes in climate. Changes in their hormone levels might make them more attractive to mosquitos, and even if they are using anti-mosquito netting, their increased frequency of urination may expose them to greater risk of infection if they leave their dwellings to visit latrines. Further, if they are primarily responsible for drawing water, they are exposed to active mosquitos for greater periods of time (50).

Women's and girls' gender roles within the family in many societies in developing countries also increase risks to their health. As the primary caregivers in families, they are more exposed to others who are ill and might have contagious diseases (Aguilar, 2009: 59). They are often solely or primarily responsible for gathering biomass for cooking, and when climate shifts diminish local supplies, they must venture farther afield to find sufficient fuel. Collecting and carrying it long distances is demanding work, and leads to an increase in physical injuries, such as bone fractures. Cooking with solid fuels in enclosed spaces without clean cook-stoves, as many women and girls in developing countries do, has significant adverse health effects on their respiratory health, leading to increases in deaths from smoke-related ailments (Farhar, Osnes and Lowry, 2014: 157–58). Aguilar (2009) has identified that these gendered health effects take a toll on the time available for girls to devote to their studies, complicating their ability to obtain educations that could help lift them out of poverty (59).

Water and energy resource availability and use

As noted earlier, women in developing countries are often responsible for drawing water for their families' household use (Verner, 2012: 281), and gathering energy-producing biomass (Muchiri, 2008: 5). For example, drawing water is not just laborious, it can also require a significant amount of time particularly when water resources are scarce. One study has calculated the amount of time spent by women and children (primarily girls) in Africa drawing water to be as much as 40 billion hours a year. As Goetz et al. (2009) have noted, when water is scarce, women must go farther to draw it, and their many other chores are still waiting for them upon their return (37).

Gathering biomass for fuel is also time-consuming, and Gardam (2014) has pointed out that unsustainable rates of consumption in certain areas mean that women must often range ever farther to secure necessary supplies of firewood to be burned directly or converted to charcoal (45, 51–52) irrespective of climate change. Since women are often responsible for gathering firewood and water, two resources which climate change might render in scarcer supply, the longer distances women will travel completing these chores would expose them to greater risk of violence in zones of armed conflict (Prescott, 2014: 775).

Because of the time spent in gathering and using water and energy resources, as well as other household chores, not only do women and girls generally have less time available to become educated, they also lack the time to undertake other economic activity that could generate cash earnings (Goetz et al., 2009: 37). Tellingly, in communities where mechanical energy is available to draw water, work the soil and move crops to market, girls demonstrate improved school attendance and academic achievement increases by the equivalent of as many as two grades (Farhar, Osnes and Lowry, 2014: 155). Bolton (2015) reminds us, however, that in some developing countries, biomass is currently abundant and can therefore be used more cheaply and sustainably for cooking than other fuels (14–15).

Despite the fact that women might be the primary users of water and energy resources as they go about their daily livelihoods, their access to these resources might be completely dependent upon men in their families being successful in securing these opportunities. For example, as noted in the Introduction, in certain rural parts of Afghanistan, the title to land may include the right to draw a certain number of 'nights' of limited irrigation water. The actual distribution of water is often determined at the village level in meetings between male village members and the local water official – meetings at which women are not allowed (Prescott, 2014: 776). Similarly, Van Aken and De Donato (2018) conducted a case study of a rural village on the West Bank of the Jordan River, near Bethlehem, where spring water had traditionally been divided between villagers who owned land near the springs. The water was allocated under a system that provided a certain number of hours of irrigation every eight days. Negotiations regarding the rotations of water use were essentially conducted by the men, and the water flows through irrigation systems were generally controlled by men (77–78).

Interestingly, in certain situations, there may be an intra-gender component to the expenditure of women's time in gathering and using water resources and other chores. For instance, Clement and Karki (2018) have assessed the creation of 'multiple-use water systems', that is, 'drinking water systems that are specifically designed to provide additional water for small-scale commercial horticulture in homestead land', in rural western Nepal (154). This programme led to significant amounts of time saved in drawing water for many women. For daughters-in-law living in their husbands' family homes, however, this time savings was transferred to other chores that might have been done later, apparently at the direction of their husbands' more senior female family members (160). Similarly, in communities in Khatlon Province, Tajikistan, mothers-in-law often hold sway over the daily work activities of their sons' wives (Levi-Sanchez, 2018: 131–48), creating in effect a matrilineal line of authority within a largely male-dominated society.

Migration induced by climate change

Chindarkar (2012) reports that when more variable climates or extreme weather events impact rural peoples' ability to sustain themselves on their land, out-migration of men to seek paying employment is fairly common (3). In some cases, men may even choose to migrate in search of work that they consider gender-appropriate, rather than remaining at home and taking up livelihood patterns associated with women (Clement and Karki, 2018: 165–66). In the event that climate change-induced factors lead to male members of the family migrating to cities in search of work, women will often find themselves picking up additional chores that the men had previously performed. They may risk greater chances of being expelled from their families or suffering sexual violence (Chindarkar, 2012: 3). Further, Aguilar (2009) has identified that certain adaptation measures, such as those related to reversing desertification, are often labour-intensive, and might further reduce the amount of time women have available to perform their ordinary chores (86).

When conditions are so bad that entire families migrate away from their communities, there are a number of challenges that they face irrespective of gender, and there are certain highly gendered consequences. For example, Martin (2009) reports that between 1993 and 2003, the average time refugees spent in displaced circumstances almost doubled from nine years to 17 years (6). This means that refugees might be unable to transfer their practical skills in growing crops or herding animals to their children, thus losing these valuable skill sets (Buscher, 2009: 93). Further, Martin (2009) cautions that if refugees and displaced persons find themselves unable to conduct their ordinary livelihoods, they might find themselves resorting to measures such as illegal harvesting of biomass and stealing crops or livestock, as well as prostitution and the marketing of contraband (8).

Those women who are willing and able to migrate might not have marketable skills they can use to help reduce the stresses of even local migration (Alber, 2011: 33). From a gendered-impact perspective, programmes to help refugees and displaced persons develop new livelihoods that are sustainable can face significant challenges. For example, these programmes might be conducted along traditional gender lines, such as sewing and handicrafts for women, which might not yield sufficient income to make up for the loss of their ordinary economic activities, or for which there might not be sufficient demand in the local market (Martin, 2009: 8).

Further, lower levels of education among women and girls also complicate the delivery of vocational training programmes that could add sustainable economic value to their skill sets (Buscher, 2009: 95). Women migrants might also be disadvantaged by their lack of opportunities to learn different languages because of their ordinary livelihood patterns, which can make it harder for them to find meaningful work in host countries (Núñez Carrasco, 2016: 223–34, 227). Regarding employment, Buscher (2009) has identified that another aspect of dislocation that can disproportionately affect women and girls is the loss of community and social arrangements for taking care of children (89).

Natural disasters related to climate change

Severe weather events associated with climate change include flooding, storms and heat waves. The impact of these events is greatest upon less-wealthy countries, and less-wealthy populations in general. These events have gendered effects. In both rural and urban areas, severe weather events pose different challenges for women than they might for men. For example, in societies in which women's mobility is constrained and gender norms tend to keep them in their homes, women are less likely to evacuate, and thus more likely to perish when natural disasters strike their homes (Chindarkar, 2012: 2). Thus, Alber (2011) reports that in the 1991 cyclone in Bangladesh, mortality rates for women were 7.1 percent, but for men, only 1.5 percent. Similarly, in the 2004 tsunami, there were 3,972 fatalities among women in Amapura, Sri Lanka, but only 2,124 men are recorded as having died (33). In urban environments, the effects of heat stress are likely exacerbated, particularly in those cities that suffer from overcrowding and insufficient sanitary infrastructure (Payne, 2016: 49).

Sultana (2018) has noted that the mobility constraints experienced by women who are pregnant or nursing might also tend to keep them from moving away from danger. Also, women might be less likely to leave their homes in disaster situations, to avoid having to associate with non-familial men and run the risk of sexual harassment and abuse in shelters, and to try to save their belongings. These belongings are not just mere things with sentimental attachment, they can also be the home gardens and small-animal livestock that provide direct subsistence for the women's families. Damages to even unimproved infrastructure, such as unpaved roads washed out by flooding, not only restrict women's movement to markets, but if the men in their families have migrated seeking employment elsewhere, it may be difficult to find the labour to effect the necessary repairs (24–25).

Although women generally appear to suffer these negative impacts more severely than men, it is important to note that depending on the situation, these experiences might be reversed. For example, as a result of Hurricane Mitch in Central America in 1998, more men than women died, perhaps as a result of the men working outside more and participating in rescue efforts (Payne, 2016: 48). Further, in the aftermath of the storm, Honduran men found themselves having to perform significant extra work to repair damaged roads and water pipes, and this was perceived by them as being the third most significant impact of the storm after lower coffee production and less income. Women too had additional work, but theirs appears to have been accommodated in the ordinary daily routine of domestic chores with less disruption, and this additional work did not register in the top five perceived negative impacts upon them of the storm (Paolisso, Ritchie and Ramirez, 2002: 187).

Women who are dislocated by natural disasters face particular risks if they must shelter in camps established for refugees and internally displaced people. These camps often have substandard sanitary conditions, and lack in quality gynaecological and obstetric personnel and supplies (Martin, 2009: 9; Alber, 2011: 33). Alber (2011) explains that

> Reproductive health issues include for instance, the need for sanitation during menstruation and after giving birth, constrained mobility during pregnancy and higher nutritional needs during lactation. During menstruation, women need adequate sanitation in privacy and personal safety which is often not ensured after a disaster.

(16–17)

For example, after the 2005 earthquake in Pakistan, relief workers learned that local customs regarding privacy as to latrine use and menstrual sanitation needs required that the latrines be screened so that men and women could not see each other entering them, and that special menstrual units needed to be constructed to allow women to clean themselves and their non-disposable sanitation supplies (Nawaz et al., 2010: 83).

Further, the way in which camps are constructed can lead to greater risks for women and girls of sexual or gender-based violence. Often, they are poorly lit and overcrowded. Camp latrines might be located some distance from women and girls' living spaces, and sources of biomass for cooking might also be distant, increasing their risks of attack when they venture out to forage (Zetter and Boano, 2009: 210–13). These relief shelters might also not provide sufficient security against sexual abuse or exploitation.

Some studies have indicated that women suffer more than men in general from the psycho-social impacts of natural disasters, perhaps because they are responsible for looking after other family members and dealing with the effects of broken social ties and separated families. Further, depending on the situation, women may find it difficult to access relief aid directly because they are not heads of households (Chindarkar, 2012: 3–5). Because of the way certain skills have been gendered in their societies, women who are the heads of households might find themselves unable to build shelters from provided relief supplies, and might have to exchange sex in return for men assembling their shelters (Benelli, Mazurana and Walker, 2012: 228).

Finally, designing vocational programmes for women in camps must effectively deal with another gendered aspect of dislocation – the availability of time for trainees. Buscher (2009) observes that whilst men and boys may have lost their original occupations, such as farming and herding, in camp situations they may now have an abundance of time to learn new trades. Women and girls, on the other hand, generally still find themselves with time-consuming domestic chores that must be completed under more difficult conditions (90). It is important to note that certain studies indicate that that a majority of refugees and displaced persons might not even live in formal camps, but may instead seek shelter in urban areas in informal settlements. Because they are dispersed, they might not be able to access the aid and relief services made available to people in established camps (Zetter and Boano, 2009: 214–16).

Assessing the literature from the perspectives of well-being and assets

Mindful of this range of possible gendered effects of climate change upon women and girls, and that much of the literature describing these effects is qualitative, it is useful to compare meta-analysis of it with later literature to see whether there are any inconsistencies that would suggest that these effects are perhaps too variable to be factored into an analysis of gender-based risk in a given operational area. Goh's 2012 assessment of the literature dealing with the gendered impacts of climate change, although focused on the different impacts to women's and men's assets and well-being in developing countries, is marked by her carefulness and her candour in describing the state of published research and analysis on the topic. It is particularly well suited to this task.

Within six of the impact areas identified for mapping the relationships between climate change and gender by Brody, Demetriades and Esplen, Goh tested the literature to determine first whether it actually supports the hypothesis that 'Climate-related shocks affect men's and women's well-being and assets differently', and second whether it supports the hypothesis that 'Climate-related shocks affect women more negatively than men' (4). Goh noted that much of the information available 'comes from self-published literature by international organizations, nongovernmental organizations, and private foundations, as well as aid and disaster relief organizations' (5). Further, a majority of these publications 'are case studies that are specific to a certain area due to the highly contextual nature of the subject' (5). Accordingly, Goh focused her review on case studies that provided evidence that was relevant to her two hypotheses.

As to her review of agricultural production case studies, Goh concluded that increasing climate variability tended to lower agricultural production, and that although this increased the workload for both women and men, women experienced a heavier workload because of the domestic chores they were ordinarily expected to perform. This greater workload and women's lesser access to agricultural tools and information could result in a more severe impact of climate change upon women (8–9). Similarly, in the related area of food security, Goh found that climate change's impacts were both gender-differentiated and more severe upon women. She observed that in climate-changed impacted communities,

> women in particular are more hard-pressed to provide meals for their families since they are primarily responsible for household food security and ... reduce their food intake so others may eat more, or part with assets such as jewellery and small livestock or take on additional work' to mitigate the deficit in food supplies.
>
> *(10)*

Regarding health impacts, Goh assessed that 'the differential impacts from climate change on women's and men's physical health are not very distinct in the literature reviewed' (11), but that the indirect effects of malnutrition and the psychological and emotional costs of having to provide care for their families under very challenging circumstances appeared to impact women more severely than men (11–12). Goh found the negative and more severe impacts of climate change in the areas of water and energy use upon women and girls to be more obvious because of the identifiable time and labour burdens placed upon them to obtain adequate water and biomass supplies (12–13), as well as in natural disasters. Given women's ordinary 'roles as caregivers, social expectations of what is acceptable for women in different societies, and lack of access to income generating activities and assets', Goh found that women and children were more at-risk to both the immediate impacts as well as the long-term effects of natural disasters (17).

She was not convinced, however, that the impacts of climate-changed induced migration and conflict on women's assets had a more severe effect upon them than on men (15). Whilst her review of migration-related case studies is extensive, this is perhaps the one section of her literature review that is not as thorough as the rest of her study, specifically with regard to the impact of conflict. Here, Goh cites only one case study, Omolo's 2011 work on conflict between pastoral peoples in northern Kenya. In terms of her methodology, however, it is important to remember that she focused her review on case studies that met a particular standard in terms of relevant evidence to test her hypotheses regarding climate change's gendered effects on assets. The conflicts described by Omolo did precisely that, because the armed conflict between the Turkana and their Pokot neighbours manifested itself in raids and counter-raids to steal cattle (2011: 88). It is not clear that there were any other case studies or research available that met Goh's criteria.

Comparison with more recent literature

There are limitations in testing Goh's conclusions using the same methodology focused on case studies with more recent literature, because there has not been an abundance of new case studies or a large amount of new research that meet her criteria that have been published since late 2012. Further, those which have been published are potentially subject to the same concerns regarding hard data and bias that Goh identified in her meta-analysis. Given this however, there is still sufficient later-published work to allow a fairly well-developed qualitative comparison.

Agricultural production and food security

In a recent work, Huyer (2016) focuses on the need to couple new technologies and applications with increased resources and control of agricultural assets for women and equitable decision-making between women and men, to effectively reduce gendered gaps in agricultural production and aspects of food security (106–12). Her analysis confirms the continuing disadvantaged state of many women working in agriculture in developing countries identified in Goh's literature review, and provides new examples of how this gender-differentiation more severely impacts women. For instance, in one section of Uganda, 80 percent of the men were able to listen to daily weather forecasts on the radio, whilst only 20 percent of the women were able to do so (107). In a time of increasing extreme weather events, this difference in the availability of climate change-related information might have significant impacts on these women's ability to mitigate climate change's effects upon their crops (109).

With regard to food security, Quisumbing and her co-authors (2015) assessed the results of four different agricultural interventions focusing on high value food products from a gender perspective (707–22). Two of the projects, one in

Mozambique and the other in Bangladesh, were aimed at the local dairy industries, and sought to increase capability and production among smallholders. The third project provided women in Burkina Faso training and supplies such as chickens, seeds and gardening supplies so that they would be able to create higher-yielding home gardens. The forth project was conducted in Uganda, and involved the introduction of bio-fortified orange sweet potato to groups of largely women farmers to increase the uptake of vitamin A in their families' diets (709). Their assessment identified a number of gender-related issues also considered by Goh that lead to women occupying a disadvantaged place in important respects vis-à-vis men in the area of food production, and found that nutrition-oriented projects tended to do a better job of benefitting women because they sidestep men's control of high value assets (722–23).

Health

Already cited extensively in this chapter, Payne's 2016 assessment of the gender-differentiated impacts of climate change on health, and the more severe effects registered by women as they likely experience more disease as a result of more direct contact with insect vectors and indirectly because of malnourishment and their roles as family caregivers (45–59) is consistent with Goh's analysis.

Water and energy availability and use

In focusing on water and the climate change-gender relationship in Southeast Asia, Sultana (2014) highlights the simultaneous phenomena of both too much water, in the form of 'floods, storm surges, cyclones, riverbank erosion, waterlogging', and too little usable water, because of 'pollution, drought, salinity, [and] desertification' (375). She finds, unfortunately, that the 'Differentiated vulnerabilities based on gender are often obscured in discussions of specific locations (e.g., floodplains)' (375), and that 'Few scholars have focused on the ways that gender is a key factor in impacts, adaptation, or mitigation in the voluminous literature on climate change' (373). Sultana is careful to note that climate change impacts on water availability and use will have impacts on people irrespective of their gender (375), but with the case studies and evidence she cites she makes it clear that there are gendered impacts, and that in many cases women and girls suffer from these impacts disproportionately (376–79).

As to energy, in 2016 the World Health Organization (WHO) stated that approximately 3 billion people currently cook and heat their dwellings with biomass, generally using open fires and simple stoves. It estimates that 4 million people die prematurely each year from illnesses caused by cooking with solid fuels. Women and children are at higher risk then men because of the greater amount of time they spend around domestic hearths. Further, the gathering of biomass fuel requires significant expenditures of time, and exposes women and children to injury and violence, as well as taking them away from their studies in school (WHO, 2016). This later information is consistent with Goh's assessment.

Migration induced by climate change

In a fairly recent article reviewing the literature regarding climate change and migration, Brzoska and Fröhlich (2016) were sceptical of estimates that climate-change induced migration of hundreds of millions of people will occur within the next 30 years, because these forecasts were derived from models based on scant empirical evidence and were insufficiently complex to account for the role of mediating factors, such as 'the availability of external humanitarian assistance and or existing social networks' (194). These are valid points, and they highlight the necessity to use a more contextualised approach to looking at the relationships between climate change and migration. To better assess Goh's analysis in this regard, it is useful to consider a study of climate change and migration conducted of Bangladesh.

This study, conducted in 2015, noted the risk posed to Bangladesh by extreme weather events such as cyclones, storm surges and flooding, and the increasing frequency of these events as climate change has progressed, accompanied by droughts and the salinization of groundwater in coastal areas (Rabbani et al., 2015: 2–3). The study noted generally that 'While research on climate change-induced migration in itself is scarce, its impact on women is under-explored' (5). In the storm, drought and flood prone areas surveyed in the study, climate-change induced migration consisted primarily of men seeking employment in cities and commercial areas, and women tended to remain behind with the families in their home villages (15).

The study also noted that in many cases, the men are not able to remit enough money back to their families to cover domestic expenses, and the women must take on jobs to earn money in addition to their regular domestic chores and responsibilities. Women left behind must still deal with the natural hazards and conditions that prompted the men to leave, and in 60 percent of communities surveyed, women experienced harassment from non-family men in their husbands' absence. In 40 percent of the communities, younger women avoided leaving their homes at all, meaning that they were no longer able to attend school. Some 60 percent of women were fearful about their homes being burgled and their possessions being stolen, which led to significant psychological stress. In the coastal areas, for example, 40 percent of women whose husbands had migrated reported having suffered physical assault, sometimes sexually (Rabbani et al., 2015: 23–24). The findings of this study are consistent with Goh's assessment.

Natural disasters related to climate change

Fothergill (1996) wrote one of the earliest comprehensive assessments of the relationships between gendered risks and disasters over 20 years ago. Taking a historical approach that is notable for its holism, Fothergill marshalled a wide collection of data and analyses that supported the idea that women, particularly poor women, suffer disproportionately in instances of natural and human-caused disasters (35–48). Almost 15 years later, based on their field work and review of the results of more recent scholarship and analysis on disasters, Richter and Flowers (2010) came to the same conclusions (211–19). Almost 20 years later, in their article encouraging the greater involvement of women in disaster preparedness and capacity building as a means to mitigate the impact of natural disasters upon women, Ashraf and Azad (2015) surveyed the literature addressing disasters' gendered impacts, and identified the differentiated and more severe impacts of disasters upon women as compared to men (137–41). This well-documented and continuing thread of disaster-related scholarship identifying the disproportionate effect of natural disasters upon women is consistent with Goh's assessment.

An operational approach?

In an article that described the evolution of thinking and research on the relationships between climate change, gender and conflict, Fröhlich and Gioli (2015) noted that 'Gender has long been identified as an important variable in both conflict (de-)escalation processes and vulnerability or adaptive capacity toward global environmental change' (137). However, they found that whilst 'there is wide agreement on the importance of gender within environmental issues, large discrepancies and debates still prevail on how exactly it is relevant . . . and how it is to be incorporated and operationalized in research frameworks' (139). One of the challenges they specifically identified in resolving these gaps is the importance of collecting and correlating sex-disaggregated data across different scales of analysis so as to develop a better empirical basis for understanding gender-related impacts (142). Further, they argued that in understanding how conflict, gender and climate change relate to each other it was important that researchers avoid the 'victimization fallacy', that is, that 'women are mainly or even exclusively the victims of global environmental change, resource scarcity and conflict' (145). They were concerned that this led to both a 'victimization bias in research and analysis' and a denial of agency to women in bringing about social change (145).

Their scepticism is healthy – faulty analysis based on incomplete data might be actionable in a military operation, but it is generally not useful in the long run. These concerns are directly related to a point Fröhlich and Gioli made regarding the relationships between armed conflict, women and climate change that is particularly important. They observed that although the most current UN Intergovernmental Panel on Climate Change (IPCC) report at the time of their writing noted gendered impacts resulting from climate change, it addressed neither the 'conflictive repercussions' of gender inequalities nor 'the impact of conflict on gender differences in resilience, vulnerability, and adaptive capacity toward climate change' (142).

This lack of analysis did not surprise them, because it squared with their own apparently incompletely successful efforts to address these issues in numerous workshops they had conducted on the topic. They attributed this in part to the tendency to either generalise 'to the point of banal platitudes' or drown 'in the complexity of local specificities' as they wrestled with understanding these relationships (142). Their experience, although understandably unsatisfactory for them, is very useful for this book. Perhaps the best analytical framework to address such a complex set of relationships involving armed conflict, women and climate change is to actually look at them from a military operational perspective, and to think of them in terms of risks posed to particular civilian groups such as women and girls, and the risks posed to mission success if the threats to such groups were to play out in any given theatre of operations.

Also from an operational perspective, it is likely too complex to consider the relationships between armed conflict, gender and climate change all together at one time. Instead it is more useful to isolate these relationships and consider them in pairs first, as we have done with armed conflict and gender, and will shortly do with climate change and armed conflict. To get a better idea of how the relationships between gender and climate change could play out in an area of civilian-centric operations, let's turn now to a short case study of a climate-change impacted area in South Africa – which, for purposes of clarity, is not marked by armed conflict.

Two municipalities in South Africa: a case study

About 20 years ago, researchers began to model and forecast the effects of global climate change upon South Africa. Although in general the different estimates of climate change impacts concurred in predicting longer summer-like conditions as a result of significant atmospheric warming and reduced precipitation, they varied significantly in how these effects would actually play out. Some models predicted fairly modest decreases in precipitation. A team from the University of Natal, however, predicted a large decrease in water catchment in the western half of the country by 2015 with a more gradual drying out in the eastern regions by 2050 (Van Jaarsveld and Chown, 2001: 13–14) Unfortunately, as evidenced by the recent water crisis experienced by Cape Town as a result of a multi-year drought (Raphelson, 2018), even this prediction was perhaps optimistic.

Babugura (2010) conducted a detailed case study of the gendered effects of climate change in the villages of two South African municipalities located in KwaZulu-Natal Province, in eastern South Africa. Agriculture is the predominant economic sector in both municipalities, and both have widespread poverty and high unemployment levels in their rural areas (11). Approximately half of all South Africans live in rural areas, and the majority of these rural dwellers are women (19). HIV/AIDS is a very significant threat to villagers' well-being, and caring for those infected strains household resources. Traditionally, in these communities women's roles were centred on the house, taking care of the family, and tending the home garden. Men generally made all the important decisions, looked after the family's livestock and were expected to protect family assets (36).

The villagers participating in the study recognised that they were experiencing climatic change, as evidenced by increased variability in the rains and warmer temperatures, and that these changes were having negative effects on their crops, water and health (37–39). As climatic conditions became harsher, most men eventually lost their livestock. This had forced both men and women to seek alternative livelihood activities outside the agricultural sector (36, 43). Women found that the changes in climate made their work both longer and more arduous, for example, many were now walking longer distances to get water, in times of greater heat. Men, on the other hand, were generally not able to find other employment, and this resulted in stress as a result of being unable to fulfil their traditional gender roles as providers for their families. Importantly, although a number of men had migrated to urban areas in search of work, some men had begun to shift in their gender roles and assist the women in their families by gathering biomass and drawing water (44). Women who were left as the heads of their households, however, had to do all the work in caring for their families (53).

As a result of land redistribution programmes put in place after the end of the apartheid government and changes in customary law, it was easier for women to acquire, inherit and hold land in their own names than it had been in the past. Unfortunately, due to the high levels of poverty that afflicted both women and men in these municipalities, it was difficult for them to marshal the resources and obtain the credit necessary to invest in improvements that would have made the land truly usable, such as irrigation or machinery. Even if they did, it appears that they had no meaningful access to the markets necessary to support commercial farming as compared to the subsistence farming in which most of them engaged (47–48, 59). This situation was likely exacerbated by the low levels of educational attainment among both men and women in these communities (60).

In terms of women's inclusion in decision-making, they were not excluded from community councils in these municipalities, and in female-headed households, which formed the majority of households in these areas, women made all the decisions, rather than deferring to a male relative. In male-headed households, decision-making varied, but in families where the women were the primary earners, they often made all the decisions. Further, women who earned their own income felt no need to turn it over to their male partners (50). Women study participants reported being motivated to find livelihood activities that generated income because of the high levels of unemployment among men, and because men tended to spend their earnings on alcohol and prostitutes (55).

Babugura's assessment of the gender-differentiation of climate change's impacts in these communities conforms to the general pattern seen in the literature concerning gender and climate change and in more specific case studies – women were taking on additional work to cope with the effects of climate variability. This extra work was physically and emotionally draining as they sought to make ends meet and take care of their children. The effects on men, who were generally unemployed, were more psychological than physical, but this led them to engage in negative coping behaviours (such as alcohol use and engaging prostitutes) that had long-term impacts on their families' well-being (65). Importantly though, these gendered impacts were also accompanied by shifts in traditional gender roles, showing that these social relationships were dynamic and evolving as these communities struggled to cope with climate change (66).

Summary

Gradual climate change and civilian-centric operations

As discussed at the beginning of this chapter, climate change will impact military operations in two interrelated ways, first as a general trend towards a particular type of weather, and second, as a generator of more frequent, more extreme weather events. Both of these components of climate change will have gendered-differentiated impacts upon women and girls, and generally speaking, these impacts will tend to be more severe for females than for males. An overall trend towards a particular type of weather over time may result in identifiable impacts upon women and girls in the areas of food production and security, health, water and energy, and migration that are similarly gender-differentiated. These effects might be gradual, and not lend themselves to convenient quantitative assessment. In these instances, as shown by Babugura's qualitative study, even in very poor communities the extended impact timeline of climate change provides a chance to adapt to and to mitigate its effects. Importantly, one of these coping measures might actually be a shift in gender roles in the community.

From an operational perspective, the extended impact timeline of the gradual weather change component offers opportunities and risks in a civilian-centric mission. Diligent staff work at a headquarters' home station might craft a very thorough operational area analysis of the population of the country into which the unit would be deploying. Unless this analysis is based on recent data that looks at the host country from a gender perspective, that is, on a sex-disaggregated basis, trends in gender roles evolving under the threat of climate change might be missed. Conceivably, this could lead to mistaken assumptions as to where the deploying military forces should devote resources in promoting stability amongst the civilian population.

For example, looking at the villages studied by Babugura, the military might decide that the real problem in these areas is that men no longer had the cattle they once herded, and that this loss of livelihood was leading to migration in search of employment and negative behaviours fostered by idleness. Depending on the situation, these behaviours could include joining insurgencies, which would likely be identified as a significant threat to both the mission and to its personnel. However, unless the military is also planning on bringing back the rains on a regular basis, well-meaning efforts such as increased extension service training on caring for livestock or providing materials to build stronger pens for cattle would be counterproductive in terms of building resilience in the communities. A gender analysis would show instead that the people themselves are moving in a different direction as they seek to provide for their families, and that the most useful single thing might instead be the installation of water taps near homes to free up more time for women to undertake alternative economic activities and keep their daughters in school.

Extreme weather events and civilian-centric operations

In his influential study of fluctuations among different populations of animals and plants in their specific environments, CS Holling (1973) defined resilience as 'the persistence of relationships within a system; a measure of the ability of systems to absorb changes of state variables, driving variables, and parameters, and still persist' (17). For example, damage from an earthquake could be reduced through the application of stricter building codes in a city, thereby making the infrastructure more resilient to the earthquake's impacts, as shown by Japanese efforts (Glanz and Onishi, 2011). This physical resilience leads to more of the city remaining usable after the earthquake, and likely works to reduce victims' need for both short-term relief and long-term rebuilding, which could itself promote social resilience, and likely enhance societal stability. Cross (2015) reports that where this resilience is lacking, for example in low-cost and so-called 'informal' buildings damaged by the 2015 earthquake in Nepal, the poorest in a community can suffer disproportionate damages. Discrimination in the distribution of aid, perhaps itself a sign of a lack of resilience in the Nepalese governmental system, compounded this suffering for many (Lord and Moktan, 2017: 128–30), and the effects of this lack of system resilience were still being felt by people years later (Amnesty International, 2017).

Although it is understood that armed conflict and natural disasters have gender-differentiated impacts upon populations that are not identical (Shteir, 2013: 4–5), extreme weather events could have a gendered shock effect similar to armed conflict's impacts on women and girls in many respects. The impact timeline is dramatically foreshortened, and particularly in poor communities, resources that provide resilience capabilities among the civilian population in times of gradual climate change, such as harvested and portable food or jewellery to sell so that such food could be bought, are quickly consumed. This heightens women's and girls' risks to the impacts in the other areas such as food security and health. From an operational perspective, however, it could be argued that as to any military response, disasters such as droughts and floods are not really that different from earthquakes or tsunamis. Sure, climate change might have been a causal factor, but dealing with the immediate exigencies of the disaster is the priority. Gender considerations, including those related to climate change, might not be taken into account in the rush to save lives and provide shelter during the course of the expedited planning that precedes the military relief operation.

But prudent planning principles suggest that they should be. The promotion and execution of appropriate climate change mitigation and adaptation measures could help build resilience in a society's food, water and energy supplies. This resilience could manifest itself as a lesser degree of need in the aftermath of an extreme event, perhaps making it a proxy for increased stability. For example, local food supplies might only need relief supplementation, rather than replacement. Farming families could remain on their land more often and longer, minimising the trauma of displacement to other, often less safe, locations. The greater the initial resilience within communities afflicted by extreme weather events, the smaller the scale of the military intervention required to try to bring the host nation back to a stable state.

Military organisations should understand that the investment in resilience, in addressing climate change as a process rather than just acknowledging it as a fact, is likely crucial to long-term stability in many theatres of operation. Unless this is acknowledged, their efforts to boost physical security in a host nation will be quietly abraded by the creeping friction of environmental degradation upon the societies they are seeking to support. Unless they recognise the gendered aspects of these issues, scarce military resources might be misdirected and fail to develop the degree of resilience possible in civilian-centric operations through inclusion of women and girls. Potentially, misdirected efforts hasten the day of their return when the next climate-change induced disaster strikes. With this in mind, let's turn to the relationship between climate change and armed conflict.

References

Aguilar, L (2009) *Training Manual on Gender and Climate Change*. San José, Costa Rica: Global Gender and Climate Change Alliance.

Alber, G (2011) 'Gender, Cities and Climate Change (Thematic report prepared for Cities and Climate Change – Global Report on Human Settlements 2011)'. UN Habitat.org. Available at: https://unhabitat.org/wp-content/uploads/2012/06/GRHS2011ThematicStudyGender.pdf (Accessed 24 June 2018).

Amnesty International (2017) 'Nepal: Two Years On, the Government Continues to Fail Marginalised Earthquake Survivors'. Amnesty International.org. Available at: www.amnesty.org/en/latest/news/2017/04/nepal-two-years-on-the-government-continues-to-fail-marginalised-earthquake-survivors/ (Accessed 30 March 2017).

Ash, Lucy (2018) 'The Crop that Put Women on Top in Zanzibar'. *BBC* (3 July). Available at: www.bbc.com/news/stories-44688104.

Ashraf, MA and Azad, MAK (2015) 'Gender Issues in Disaster: Understanding the Relationships of Vulnerability, Preparedness and Capacity'. *Environment and Ecology Research* 3(5): pp. 136–42. DOI: 10.13189/eer.2015.030504.

Babugura, A (2010) *Gender and Climate Change: South Africa Case Study*. Capetown: Heinrich Böll Foundation South Africa. Available at: www.boell.de/sites/default/files/assets/boell.de/images/download_de/ecology/south_africa.pdf (Accessed 24 June 2018).

Benelli, P; Mazurana, D and Walker, P (2012) 'Using Sex and Age Disaggregated Data to Improve Humanitarian Response in Emergencies'. *Gender & Development* 20(2): pp. 219–32. DOI: 10.1080/13552074.2012.687219.

Bolton, S (2015) 'Establishing a Women-Driven Clean Cookstove Distribution Network in Eastern Indonesia'. *Boiling Point: A Practitioner's Journal on Household Energy, Stoves and Poverty Reduction* 66: pp. 14–17. Available at: www.esmap.org/sites/default/files/resources-document/GENDERWomen%20Energy%20and%20Economic%20Empowerment.pdf (Accessed 13 March 2018).

Brody, A; Demetriades, J and Esplen, E (2008) *Gender and Climate Change: Mapping the Linkages*. Institute of Development Studies, University of Sussex. Available at: www.bridge.ids.ac.uk/sites/bridge.ids.ac.uk/files/reports/Climate_Change_DFID.pdf (Accessed 22 December 2017).

Brzoska, M and Fröhlich, C (2016) 'Climate Change, Migration and Violent Conflict: Vulnerabilities, Pathways and Adaptation Strategies'. *Migration and Development* 5(2): pp. 190–210. DOI: 10.1080/21632324.2015.1022973.

Buscher, D (2009) 'Women, Work, and War'. In: Martin, SF and Tirman, J (eds.) *Women, Migration and Conflict*. Dordrecht: Springer Science +Business Media BV, pp. 87–106.

Chindarkar, N (2012) 'Gender and Climate Change Induced Migration: Proposing a Framework for Analysis'. *Environmental Research Letters* 7: pp. 1–7. DOI: 10.1088/1748-9326/7/2/025601.

Clement, F and Karki, E (2018) 'When Water Security Programmes Seek to Empower Women – a Case Study from Western Nepal'. In: Fröhlich, C; Gioli, G; Cremades, R and

Myrttinen, H (eds.) *Water Security Across the Gender Divide*. Cham, Switzerland: Springer International Publishing AG, pp. 151–69.

Cross, R (2015) 'Nepal Earthquake: A Disaster that Shows Quakes Don't Kill People, Buildings Do'. *The Guardian* (30 April). Available at: www.theguardian.com/cities/2015/apr/30/nepal-earthquake-disaster-building-collapse-resilience-kathmandu (Accessed 3 January 2017).

Denton, F (2002) 'Climate Change Vulnerability, Impacts, and Adaptation: Why Does Gender Matter?' *Gender & Development* 10(2): pp. 10–20. DOI: 10.1080/13552070215903.

FAO (Food and Agriculture Organization of the UN) (n.d.) 'Food Security Statistics'. FAO.org. Available at: www.fao.org/economic/ess/ess-fs/en/ (Accessed 16 June 2018).

Farhar, BC; Osnes, B and Lowry, EA (2014) 'Energy and Gender'. In: Halff, A; Sovacool, BK and Rozhon, J (eds.) *Energy Poverty: Global Challenges and Local Solutions*, Oxford, UK: Oxford University Press, pp. 152–79.

Fothergill, A (1996) 'Gender, Risk and Disaster'. *International Journal of Mass Emergencies and Disasters* 14(1): pp. 35–56. Available at: www.ijmed.org/articles/96/download/ (Accessed 24 June 2018).

Fröhlich, C and Gioli, G (2015) 'Gender, Conflict and Global Environmental Change'. *Peace Review: A Journal of Social Justice* 27(2): pp. 137–46. DOI: 10.1080/10402659.2015.1037609.

Gardam, J (2014) 'A Role for International Law in Achieving a Gender Aware Energy Policy'. In: Babie, P and Leadbetter, P (eds.) *Law as Change: Engaging with the Life and Scholarship of Adrian Bradbrook*. Adelaide: University of Adelaide Press, pp. 43–58.

Glanz, J and Onishi, N (2011) 'Japan's Strict Building Codes Saved Lives'. *New York Times* (11 March). Available at: www.nytimes.com/2011/03/12/world/asia/12codes.html (Accessed 17 December 2017).

Goetz, AM; Cueva-B, H; Eddon, R; Sandler, J and Doraid, M (2009) 'Who Answers to Women? Gender & Accountability'. UNIFEM.org. Available at: www.unifem.org/progress/2008/media/POWW08_Report_Full_Text.pdf (Accessed 24 June 2018).

Goh, AHX (2012) *A Literature Review of the Gender-Differentiated Impacts of Climate Change on Women's and Men's Assets and Well-Being in Developing Countries. Collective Action and Property Rights, Working Paper No. 106*. International Food Policy Research Institute.org. Available at: http://ebrary.ifpri.org/cdm/ref/collection/p15738coll2/id/127247 (Accessed 20 July 2018). DOI: 10.2499/CAPRiWP106.

Hillenbrand, E (2010) 'Transforming Gender in Homestead Food Production'. *Gender & Development* 18(3): pp. 411–25. DOI: 10.1080/13552074.2010.521987.

Holling, CS (1973) 'Resilience and Stability of Ecological Systems'. *Annual Review of Ecology and Systematics* 4: pp. 1–23. DOI: 10.1146/annurev.es.04.110173.000245.

Huyer, S (2016) 'Closing the Gender Gap in Agriculture'. *Gender, Technology and Development* 20(2): pp. 105–16. DOI: 10.1177/0971852416643872.

Inter-Parliamentary Union (n.d.) 'Women in National Parliaments'. Inter-Parliamentary Union.org. Available at: http://archive.ipu.org/wmn-e/world.htm (Accessed 3 February 2018).

Levi-Sanchez, S (2018) 'Local Governance in Khatlon, Tajikistan'. In: Laruelle, M (ed.) *Tajikistan on the Move: Approaches, Fieldworks, Topics*. Lanham, MD: Rowman & Littlefield, pp. 131–48.

Lord, A and Moktan, S (2017) 'Uncertain Aftermath: Political Implications of the 2015 Earthquakes in Nepal'. *Accord* 26: pp. 128–32. Available at: www.c-r.org/accord/nepal/uncertain-aftermath-political-impacts-2015-earthquakes-nepal (Accessed 17 February 2018).

Martin, SF (2009) 'Introduction'. In: Martin, SF and Tirman, J (eds.) *Women, Migration, and Conflict: Breaking a Deadly Cycle*. Dordrecht: Springer Science +Business Media BV, pp. 1–22.

Muchiri, L (2008) 'Gender and Equity in Bioenergy Access and Delivery in Kenya'. PISCES. Available at: https://assets.publishing.service.gov.uk/media/57a08b9540f0b64974000c40/Kenya-Gender.pdf (Accessed 24 June 2018).

Nawaz, J; Lal, S; Raza, S and House, S (2010) 'Oxfam Experience of Providing Screened Toilet, Bathing and Menstruation Units in its Earthquake Response to Pakistan'. *Gender & Development* 18(1): pp. 81–86. DOI: 10.1080/13552071003600067.

Núñez Carrasco, L (2016) 'Migration, Gender and Health'. In: Gideon, J (ed.) *Handbook on Gender and Health*. Cheltenham, UK: Edward Elgar, pp. 223–34.

Omolo, NA (2011) 'Gender and Climate Change-induced Conflict in Pastoral Communities: Case study of Turkana in Northwestern Kenya'. *Foresight for Development* pp. 81–102. Available at: www.foresightfordevelopment.org/sobipro/55/314-gender-and-climate-change-induced-conflict-in-pastoral-communities-case-study-of-turkana-in-northwestern-kenya (Accessed 5 March 2018).

Paolisso, M; Ritchie, A and Ramirez, A (2002) 'The Significance of Gender Division of Labor in Assessing Disaster Impacts: A Case Study of Hurricane Mitch and Hillside Farmers in Honduras'. *International Journal of Mass Emergencies and Disasters* 20(2): pp. 171–95. Available at: www.ijmed.org/articles/573/download/ (Accessed 24 June 2018).

Payne, S (2016) 'Gender, Health and Climate Change'. In: Gideon, J (ed.) *Handbook on Gender and Health*. Cheltenham, UK: Edward Elgar, pp. 45–59.

Prescott, JM (2014) 'Climate Change, Gender, and Rethinking Military Operations'. *Vermont Journal of Environmental Law* 15(4): pp. 766–802. Available at: http://vjel.vermontlaw.edu/files/2014/06/Prescott_ForPrint.pdf (Accessed 24 June 2018).

Quisumbing, AR; Rubin, D; Manfre, C; Waithanji, E; Van den Bold, M and Olney, D (2015) 'Gender, Assets, and Market-oriented Agriculture: Learning from High Value Crop and Livestock Projects in Africa and Asia'. *Agriculture and Human Values* 32(4): pp. 705–25. DOI: 10.1007/s10460-015-9587-x.

Rabbani, MG; Khan, ZM; Tuhin, MH; Naznin, Z; Emran, DA and Karim, S (2015) 'Climate Change and Migration in Bangladesh: A Gender Perspective'. Bangladesh Centre for Advanced Studies for UN Women. Available at: http://www2.unwomen.org/-/media/field%20office%20eseasia/docs/publications/2016/01/climate%20change%20%20migration%20in%20bangladesh%20a%20gender%20perspective.pdf?la=en&vs=4849 (Accessed 13 February 2018).

Raphelson, S (2018) 'Drought-Stricken Cape Town Braces for Water to Run Out in April'. *National Public Radio* (23 January). Available at: www.npr.org/2018/01/23/579784235/drought-stricken-cape-town-braces-for-water-to-run-out-in-april (Accessed 24 June 2018).

Richter, R and Flowers, T (2010) 'Gender-Aware Disaster Care: Issues and Interventions in Supplies, Services Triage and Treatment'. *International Journal of Mass Emergencies and Disasters* 28(2): pp. 207–25. Available at: www.ijmed.org/articles/573/download/ (Accessed 24 June 2018).

Sankari, R (2016) 'World's Largest Matrilineal Society'. *BBC* (22 September). Available at: www.bbc.com/travel/story/20160916-worlds-largest-matrilineal-society (Accessed 12 January 2018).

Shteir, S (2013) 'Gendered Crises, Gendered Reponses. Civil-Military Occasional Paper 01/2013'. Australian Civil-Military Centre. Available at: http://acmc.gov.au/wp-content/uploads/2013/09/Gendered-Crises-Gendered-Responses.pdf (Accessed 2 February 2018).

Speranza, CI and Bikketi, E (2018) 'Engaging with Gender in Water Governance and Practice in Kenya'. In: Fröhlich, C; Gioli, G; Cremades, R and Myrttinen, H (eds.) *Water*

Security Across the Gender Divide. Cham, Switzerland: Springer International Publishing AG, pp. 125–50.
Sultana, F (2014) 'Gendering Climate Change: Geographical Insights'. *The Professional Geographer* 66(3): pp. 372–81. DOI: 10.1080/00330124.2013.821730.
Sultana, F (2018) 'Gender and Water in a Changing Climate: Challenges and Opportunities'. In: Fröhlich, C; Gioli, G; Cremades, R and Myrttinen, H (eds.) *Water Security Across the Gender Divide*. Cham, Switzerland: Springer International Publishing AG, pp. 17–33.
Van Aken, M and De Donato, A (2018) 'Gender and Water in the Middle East. Local and Global Realities'. In: Fröhlich, C; Gioli, G; Cremades, R and Myrttinen, H (eds.) *Water Security Across the Gender Divide*. Cham, Switzerland: Springer International Publishing AG, pp. 61–82.
Van Jaarsveld, AS and Chown, SL (2001) 'Climate Change and its Impacts in South Africa'. *Trends in Ecology and Evolution* 16(1): pp. 13–14. DOI: 10.1016/S0169-5347(00)02037-1.
Verner, D (2012) *Adaptation to a Changing Climate in the Arab Countries: A Case for Adaptation Governance and Leadership in Building Climate Resilience*. Washington, DC: The World Bank. Available at: file:///C:/Users/jmprescott/Downloads/734820PUB0REVI000PUBDATE01101602012%20(1).pdf (Accessed 24 June 2018).
Westerman, A (2017) '4 Years After Rana Plaza Tragedy, What's Changed For Bangladeshi Garment Workers?' *National Public Radio* (30 April). Available at: www.npr.org/sections/parallels/2017/04/30/525858799/4-years-after-rana-plaza-tragedy-whats-changed-for-bangladeshi-garment-workers (Accessed 17 January 2018).
WHO (World Health Organization) (2016) 'Fact Sheet No. 292, Household Air Pollution and Health'. World Health Organization. Available at: www.who.int/mediacentre/factsheets/fs292/en/ (Accessed 17 December 2018).
Zetter, R and Boano, C (2009) 'Gendering Space for Forcibly Displaced Women and Children: Concepts, Policies and Guidelines'. In: Martin, SF and Tirman, J (eds.) *Women, Migration and Conflict*. Dordrecht: Science +Business Media B.V, pp. 201–27.

3
CLIMATE CHANGE AND ARMED CONFLICT

To help set the context for the examination of the potential relationships between climate change and armed conflict as set out in the scientific literature, this chapter will first briefly discuss the relationship between armed conflict and environmental degradation, and then consider arguments that interest in climate change by militaries is self-serving from a security institution perspective, and does nothing really to address the underlying problems giving rise to changes in the global climate. A short discussion of popular and political perceptions and characterisations of the armed conflict/climate change relationship will follow, to be followed in turn by a review of the approach taken by the IPCC in assessing the state of the scientific literature on this topic. The last area to be considered before reviewing the scientific literature will be the national security and policy documents of two NATO partners, the UK and the US, with a comparison of the different approaches of these two close allies.

After examining these two national approaches, this chapter will survey the progression in research and analysis of the data regarding climate change and armed conflict, a fair amount of which was considered by the IPCC in its 2014 report. Where potentially useful from an operational perspective, this chapter will point out challenges with certain methodologies and conclusions. To help place the results of these studies in context, this chapter will close with a detailed examination of climate change and the civil war in Syria, and weigh the information and analysis relevant to that disastrous armed conflict after the punishing drought that occurred in most parts of the country between late 2006 and either 2008 or early 2009. The complexities in identifying the relationships between the drought and the subsequent outbreak of civil war strongly suggest that causality will never be established with scientific certainty for Syria, and perhaps for other places as well. However, the reasonable possibility of a positive correlation between the two (although indirect) means that in the course of prudent planning for civilian-centric operations, analysis specific to different localities within a theatre of operations should be developed with an understanding of the effects of climate change and armed conflict on particular populations, including the gendered effects.

Armed conflict, environmental degradation and climate change risks

The direct and indirect causal relationships between armed conflict and subsequent environmental degradation are well described in historical and scientific literature (Austin and Bruch, 2000: 1–4, 303–08). Plagued by warfare since the Soviet invasion in 1979, Afghanistan is perhaps one of the better-known examples of this. Although there were significant instances of agricultural infrastructure being destroyed or abandoned directly because of the conflict (Formoli, 1995: 66), the UN Environment Programme (UNEP) (2003) assessed that the most significant problem was probably the mismanagement of water resources because of the collapse of effective governance at the national and local levels, which led to increasing desertification in many areas (10, 14, 95). Further, it also observed that in areas where water was more plentiful, illegal logging for construction materials and cooking fuel had depleted Afghanistan's forests, and along with unregulated grazing, had exposed soils to greater erosion (11). The poor security situation in Afghanistan has not just resulted in continued illegal forest harvesting (Trofimov, 2010; Anwari and Siddique, 2017), it has also delayed and compromised reforestation efforts after the trees have been cut down (Groninger, 2012: 834).

Causality between climate change and armed conflict, on the other hand, has only become the subject of focused research and analysis in the last 15 years or so, and this possible relationship is not yet as well understood. Interestingly, some of the military organisations that have seemed reluctant to innovate in terms of incorporating gender perspectives into their activities were relatively quick to recognise some of the potential threats posed by climate change's effects upon their activities and operations – at least in certain respects. Before examining any possible causal relationships between climate change and armed conflict, it is useful to consider some critical perspectives on certain national efforts in this regard.

'Militarisation' of climate change?

A number of writers have criticised what they see as the 'militarisation' of climate change policy. Nagel (2015), for example, has noted how in the later Bush Administration US military experts (men) were beginning to assess climate change from the perspective of national security. In keeping with the male-dominated nature of the military, the US Department of Defense's research and action agenda in her view thereafter reflected 'classical military/masculine tactics for attacking climate change: large-scale geo-engineering approaches to manage solar radiation or sequester carbon, energy research focused on military missions, conflict scenario planning, climate forecasting and modelling, and global resource assessments' (205). Missing was any real interest on the US military's part in exploring '"softer" (feminine) research topics on the health, social, racial, gender or age disparities in climate change impacts, factors that promote community resilience in the face of disasters, or strategies for reducing the carbon emissions causing climate change' (205).

Similarly, Bonds (2015) argued that instead of tackling the root causes of climate change, including carbon emissions, US military officials appear to be focused on a two-track approach that treats climate change as a fact rather than a process. First, he saw the US military seeking technical solutions to the problems posed by climate change, such as the impacts upon naval bases of rising sea levels. Second, Bonds noted strategy documents that proposed increased cooperation with developing countries so that they are better able to handle humanitarian disasters that might result from climate change. He was concerned that these efforts would lead to 'undemocratic and sometimes extremely repressive outcomes' for nations of the Global South (210), and that they would provide a plausible explanation to the US public to justify the sustainment of US global militarism (212).

Both Nagel and Bonds are correct in certain important respects. Although the situation might change because of the implementation of the *Women, Peace, and Security Act of 2017*, as will be discussed in detail in Chapter 7, 'US strategy, doctrine and gender', the US Department of Defense does not yet appear to have made a concentrated effort to understand and operationalise disparities in climate change impacts among at-risk population cohorts such as women and girls. Further, the US defence establishment does indeed appear to be treating climate change as just another fact to be dealt with in terms of hardening infrastructure and conducting humanitarian assistance operations (Prescott, 2014: 768), rather than as a process with gendered effects. Where then is space carved out in political and military thinking to deal with climate change's actual relationship to armed conflict?

Climate change's role in conflict: perceptions vs. analysis

At times, reporting in the media (Vidal, 2011; Fountain, 2015) and the statements of political officials (Borger, 2007; Denyer, 2014) seem to have jumped ahead of what research in the area was actually showing in terms of the probability of a positive correlation between climate change and armed conflict, to say nothing of actually establishing a causal link between the two. Although these reports and statements are not necessarily accurate, they are perhaps understandable. Seter (2016) argues that regardless of whether scholarly analysis establishes a definitive link, the current global focus on issues of climate change might cause politicians to feel 'compelled to generate policies to address this issue' (2). For example, in his address to the graduating class at the US Coast Guard Academy in 2015, President Obama stated that although climate change was not the cause of conflicts occurring around the world

> we also know . . . that severe drought helped to create the instability in Nigeria that was exploited by the terrorist group Boko Haram. It's now believed that drought and crop failures and high food prices helped fuel the early unrest in Syria, which descended into civil war in the heart of the Middle East.
>
> *(Obama, 2015)*

President Obama would likely have had access to quality intelligence throughout the course of his administrations – perhaps scientific analysis by the sprawling US intelligence community was the basis for his knowledge and belief. In a recent article, Adams et al. (2018) suggest that there might be a research bias in the scientific literature on the subject, such that it has been more convenient for climate-change researchers to focus on certain areas in which conflict has occurred or is ongoing. They suggest that because research has focused on areas at risk for conflict, researchers are not analysing regions and countries that are at risk for negative climate change impacts but which are not experiencing armed conflict (200–01).

In their analysis of English-language climate change/conflict studies, Adams and his co-authors found that the countries which presumably English-speaking researchers focused on were mainly African countries that had been British colonies, with Asia being less studied and South America and Oceania hardly at all (200, 202). They do note that had they analysed French or Spanish language journals their results 'would probably yield a different picture of countries and regions most frequently mentioned' in climate change/conflict studies (202). Nonetheless, they conclude that English-language studies which identify a positive link between climate change and conflict suffer from the taint of convenience bias because conflict was already occurring in places where it was easier to gather data (202).

However, they did not explain why English-language studies that found no positive or no strong positive relationship between climate change and conflict did not suffer from this bias problem. Further, they provided no convincing explanation why researchers seeking to determine whether there are links between climate change and conflict would choose to go to a place where there was no conflict. This is a curious study from an operational perspective in certain respects, but it presents important questions about data reliability, analytical bias, perceptions and conclusions that this chapter will seek to explore.

The Intergovernmental Panel on climate change's security assessment

The IPCC has taken a conservative approach in assessing what scientific research says about the relationship between armed conflict and climate change. In its first three assessment reports, the IPCC did not address a linkage between climate change and armed conflict at all. In its *4th Assessment Report* in 2007, the IPCC explored the possibility of a causal link between armed conflict and climate change for the first time. With regard to Sub-Saharan Africa, the IPCC noted that conflict was one possible factor that in combination with a climate stressor such as drought or flooding could result in a complex disaster (IPCC, 2007a: 442). Although armed conflicts had recently occurred in different states in the area, climate change itself was only seen as a possible contributing factor in conflicts related to resource scarcity, such as a lack of water. Structural inequalities in these states, resource mismanagement, land distribution, environmental degradation and ethnic clashes were seen as more likely causes of conflict than climate change itself (443). Accordingly, in its

Synthesis Report, the IPCC noted only that 'vulnerability to climate change can be exacerbated by other stresses', including conflict (IPCC, 2007b: 14).

In its *5th Assessment Report* in 2014, the IPCC again analysed the research and literature dealing with the relationship between armed conflict and climate change. For purposes of its assessment, the IPCC defined an armed conflict as one which resulted in 'more than 25 battle-related deaths in a year' (IPCC, 2014: 771). The IPCC noted that most of the research examining this relationship focused 'on the connections between climate variability and intrastate conflicts', and that although some studies found a weak relationship between the two, others had found no such relationship. The IPCC assessed that 'collectively the research does not conclude that there is a strong positive relationship between warming and armed conflict' (771).

The IPCC did assess that there was high agreement among researchers as to the factors that increase the likelihood of civil war in countries, such as low per capita income and inconsistent political institutions, and that a number of these factors were sensitive to climate change (771–72). Similarly, the IPCC also noted that in situations where 'property rights and conflict management institutions are ineffective or illegitimate', mitigation and adaptation measures that modified people's 'access to resources have the potential to create and aggravate conflict' (773). In contrast, the IPCC noted that research was in high agreement that in circumstances where these risk factors were 'extremely low' (771), climate change impacts on armed conflict were negligible. In conclusion, the IPCC noted that it was not possible to make reliable statements about the relationship between climate change and armed conflict 'given the absence of generally supported theories and evidence about causality' (771).

In light of this analysis, it will be no surprise that the strategy, policy and doctrine of both the UK and the US take a very cautious approach to the operational significance of climate change. In fact, their approach is so careful that it does not register in their operational doctrine at all, and its treatment in their respective national-level strategies is at best uneven. UK strategy at least views climate change as a potential driver of instability, whilst current US strategy simply ignores it.

The UK security assessment of climate change

UK national security strategy

In its 2010 national security strategy, the UK government acknowledged that 'the physical effects of climate change are likely to become increasingly significant as a "risk multiplier" exacerbating existing tensions around the world' (Her Majesty's Government, 2010: 17) Published in 2015, the latest UK national security strategy sets out the government's intent and path to implement national security measures through 2020 (Her Majesty's Government, 2015: 9). It is very detailed in terms of its recognition of the potential impacts of climate change in the international security environment, describing it as one of the 'drivers of instability' along with social inequality, rapid and unplanned urbanisation, and global economic shocks (17). Accordingly, the government decided that the focus of UK development efforts would be 'on building stability overseas in fragile states' (17). In particular, the UK would help finance increased access to clean energy and efforts to fight deforestation to help build resilience among the poorest and most at-risk people (66).

As to how the impacts of climate change would play out, the 2015 strategy noted differentiated impacts depending upon location. For example, heightened risk to the Middle East and North Africa because of water stress and high population growth rates was identified as a possible impact, whilst rising sea levels would 'threaten coastal cities and small islands' (Her Majesty's Government: 21). The strategy also acknowledged the possibility that disruptions in populations, food supplies and economies could result from more extreme weather events, and could lead to 'political instability, conflict and migration' (21). Pragmatically, it assessed that it would cost less and save more lives to invest in building resilience capacity than it would simply to respond to such disasters after they happened. Also from a perspective of pragmatism, the 2015 national security strategy did not see climate change itself as a priority threat to the UK at least until 2020, but instead assessed that its full effects were more likely to be seen sometime after 2035 (65, 86).

There has been some concern that despite the recognition of climate change as a serious security threat in the text of the 2010 and 2015 national security strategies, the implementing national security risk assessments have not accorded it the priority it might deserve. Lunn and Scarnell (2015) note that neither the 2010 nor the 2015 National Security Risk Assessments used the phrase 'climate change' (5). The 2015 national security strategy listed climate change as one of five risks 'which remain important and need to be addressed' (10). However, it is not accorded a prominent status despite it likely being a factor in at least two of the top-ranked risks in the 2015 National Security Risk Assessment, namely 'Major Natural Hazards' and 'Instability Overseas', and a possible factor in at least two others, namely 'International Military Conflict' and 'Public Health' (32).

UK Ministry of Defence forecasting studies

The Ministry of Defence (2017) has established an internal strategy for ensuring that its installations and activities become more resilient to the effects of climate change, but the most comprehensive treatment of the impacts of climate change on the international security environment is found in two documents that seek to forecast likely trends and their future effects. The first document looks out to 2035, and the second sets its horizon a decade later, in 2045. *Future Operating Environment 2035 (FOE 2035)* sees climate change resulting in rising sea levels and an increase in the intensity, frequency and duration of extreme weather events by the 2035 forecast horizon, leading to direct losses 'of life, physical destruction, disease and famine' (Development, Concepts and Doctrine Centre, 2014b: 3). Secondary results might include increased 'migration, social unrest, instability and conflict that could affect the UK's interests' and water scarcity (3).

From an operational perspective, *FOE 2035* sees these effects as likely driving an increased need for the Ministry of Defence to conduct humanitarian operations along with non-governmental organisations – but in a supporting role (9). Consistent with General Sir Rupert Smith's idea of 'war amongst the people', it also recognises that as these trends manifest themselves, 'it is amongst the human terrain where the most dynamic and radical change can be expected' (25). In this sense, urbanisation presents significant challenges, because not only are cities expected to grow and become more physically complex and densely populated, but those located in coastal areas will also likely be among the first to suffer the effects of rising sea levels (25).

The second forecasting document, *Global Strategic Trends – Out to 2045 (GST 2045)*, provides a more detailed picture of both the operational assessment of climate change's effects and the implications of those changes for the operations of British forces. *GST 2045* focuses on three main areas where the impacts of climate change are likely to be most noticeable: food production and water scarcity, rising sea levels and human health (Ministry of Defence, 2014a: 21–23, 42). As to food production and water scarcity, it assesses that because of changing weather patterns, the likelihood of loss or contamination of water supplies is almost certain to lead to crop destruction in the short term, and significant decreases in farmland fertility in the long run (42).

Although climate change might open up new areas for cultivation, and in certain areas particular crops might experience some benefits, *GST 2045* assesses that on balance the effects upon agriculture will be negative on a global scale. Importantly, despite the likelihood of significant human population growth by 2045, it assesses that the amount of arable land available for cultivation will remain roughly the same, and that any productivity increases in the high latitudes will be offset by decreases in the tropical regions. At the same time, maritime food productivity might be reduced because of ocean warming, acidification and unsustainable harvesting practices (23, 42).

Regarding rising sea levels, not only does *GST 2045* note that they will potentially threaten the ever-growing coastal cities, particularly in areas affected by tropical cyclones, it specifically identifies those in Asia which are perhaps most threatened, including Kolkata, Mumbai, Shanghai, Ho Chi Minh City, and Bangkok. The size of the populations of these cities will likely increase the scale of any potential humanitarian crises (xvi, 17, 152). It does not include Singapore on this list because the city-state is assessed as having 'sufficient resources to protect itself from catastrophic effects' (152). *GST 2045* also notes that although no major urban areas in either Australia or New Zealand have been identified so far as being at high risk of flooding, significant portions of both countries' populations do live near the coast. Finally, it highlights the concern that low-lying island countries, such as the Maldives or Vanuatu, might experience large-scale out-migration – but also that in many instances, affected people will simply be without sufficient capital to resource their movement away from threatened areas (8, 152).

Finally, from the perspective of human health, *GST 2045* notes that climate change's impacts could include malnourishment because of decreased agricultural production, a decline in safe drinking water and a lack of secure shelter from more frequent extreme weather events. It also identifies the mental health issues that could result from dealing with the stresses of these impacts in addition to the physical problems that they pose. *GST 2045* also identifies the general threat to human respiratory health posed by the interaction between higher temperatures and pollution, such as higher ozone levels, and between higher temperatures and favourable conditions for the increased transmission of disease through invertebrate disease vectors such as insects and snails (6–10).

Most usefully from an international security perspective, *GST 2045* then forecasts the ripple effects of these impacts across the globe. As to Europe, it expects the effects of climate change to be less severe in general than in other regions of the world, but it that even within Europe increased water scarcity in the south could result in decreased agricultural productivity there. It assesses that shocks to the global trading system are possible because of economic interdependencies between countries more likely to be severely impacted by extreme weather events and their global trading partners, and that there is a heightened chance for the outbreak of conflict due to food and water scarcity in highly threatened areas such as those in Africa (xvi, xix, 138). In sum, it finds that it 'seems likely that developing countries will feel the economic impact of climate change particularly sharply, as they are unlikely to have the resources to mitigate its effects and as successfully as more developed countries' (33).

The US security assessment of climate change

A gradual awakening

Interestingly, when the US Central Intelligence Agency first examined the threats posed by climate change in 1974, the concern at that time was not that global warming would be the driver of climatic variability. Rather, based on analysis of data available at that time, there appeared to be a significant likelihood that global temperatures would drop by the end of the second millennium and have pronounced negative effects on agriculture across the world (Directorate of Intelligence, 1974: 26–33). A decade and a half later, global cooling was no longer seen as the potential driver of climate change that it was once feared to be. The challenge posed by global warming instead was perhaps first officially recognised as a security issue by the US government in 1990, when the Chairman of the US Senate Armed Services Committee, Senator Nunn, stated that global warming, as an aspect of environmental destruction, was a 'growing national security threat' (Shabecoff, 1990).

In a 2003 study on the impacts of global warming commissioned by the US Department of Defense, researchers posited a worst case scenario in which the climate could change abruptly as a result of a slowdown in the ocean's heat-bearing currents (Schwartz and Randall, 2003: 1). The study's authors envisioned that 'an abrupt climate change scenario could potentially de-stabilize the geo-political environment, leading to skirmishes, battles, and even war due to resource constraints such as' shortages of food and water, and 'Disrupted access to energy supplies due to extensive sea ice and storminess' (2). This study was widely reported in the US and international media at the time, in part because of the Bush Administration's reticence to accept that climate change was occurring (Davidson, 2004), and in part because of the very dire picture it portrayed of a potential near-term climate-change impacted world – 'Secret report warns of rioting and nuclear war', 'Britain will be "Siberian" in less than 20 years', and 'Threat to world is greater than terrorism' (Townsend and Harris, 2004). It is not obvious that the study's sensationalised treatment in the press advanced thoughtful consideration of its extrapolations.

The reports of the National Intelligence Council, collective products of the US intelligence community, have been more measured in their analysis. In its 2008 statement to the US House of Representatives' Committee on Intelligence and the then-Committee on Energy Independence and Global Warming, the council assessed that not only was the US less likely to be affected by climate change and better resourced than most nations to deal with it, but that 'the most significant impact for the United States will be indirect and result from climate-driven effects on many other countries and their potential to seriously affect U.S. national security interests' (National Intelligence Council, 2008: 4). These impacts could negatively impact domestic stability in certain states by exacerbating 'poverty, social tensions, environmental degradation, ineffectual leadership, and weak political institutions' (4–5).

In its 2016 report, the council assessed that these changes were already happening, and that in the near-term (five years) security risks linked to climate change were likely to arise from distinct extreme weather events and from increasing shortages of resources, such as water. In the mid-term (20 years), not only would these sorts of events remain threats according to the council, but 'the projected effects of climate change will play out in the combination of multiple weather disturbances with broader, systemic changes, including the effects of sea level rise' (National Intelligence Council, 2016: 3). Multiple events, 'when several extreme weather events occur within a small region or short time', were assessed as being able to 'cause more damage than very powerful single ones' because of the cumulative impact they would have on community resilience (5). Although the extensive damage suffered by the US islands in the Caribbean during the 2017 hurricane season was likely caused at least in part by factors other than climate change (Fawkes, 2017), it provides an illustration of how debilitating the impacts of multiple events might be.

US national security strategy

In the last version of the US national security strategy published by the Obama Administration (The White House, 2015), the 'accelerating impacts of climate change' (Preface) were recognised as being one of the main threats to global security and US interests, along with terrorism, the proliferation of weapons of mass destruction, global economic crises, severe outbreaks of infectious disease on a global scale, major energy market disruptions and the problems associated with weak or failing states (2, 7). Whilst this strategy placed the US as a leader in efforts to confront climate change impacts both at home and abroad (12), this focus on climate change in national-level security strategy did not survive the change in US administrations in 2017. Under the Trump Administration, the most recent version of the US national security strategy (The White House, 2017) does not mention climate change at all – the few mentions of 'climate' are mainly in the context of fostering favourable business and investment climates (20–22). The one use of 'climate' in the document that is related to climate change is not really positive in the sense of taking action to address climate impacts. Instead, it notes that 'Climate policies will continue to shape the global energy system', and that 'U.S. leadership is indispensable to countering an anti-growth energy agenda that is detrimental to U.S. economic and energy security interests' (22).

US Department of Defense strategy and policy

In 2010, under the Obama Administration, the US Department of Defense formally addressed climate change for the first time in its *Quadrennial Review*. The Department of Defense assessed that climate change had the potential to 'spark or exacerbate future conflicts' (Department of Defense, 2010: iv). In particular, it noted that 'climate change could have significant geopolitical impacts around the world, contributing to poverty, environmental degradation, and the further weakening of fragile governments' (85).

The 2014 *Quadrennial Review* further amplified the Department of Defense's approach to climate change, and provided a more nuanced appraisal of the relationship between armed conflict and climate change. First, it recognised that 'The impacts of climate change may increase the frequency, scale, and complexity of future missions, including defense support to civil authorities' (Department of Defense, 2014: vi). Specifically, the Department of Defense recognised that 'Climate change may exacerbate water scarcity and lead to sharp increases in food costs', and that the impacts of climate change 'will influence resource competition while placing additional burdens on economies, societies, and governance institutions around the world' (8). It also assessed these impacts as 'threat multipliers that will aggravate stressors abroad such as poverty, environmental degradation, political instability, and social tensions – conditions that can enable terrorist activity and other forms of violence' (8). The 2014 *Quadrennial Review* also reiterated the department's support for 'building humanitarian assistance and disaster response capabilities . . . with our allies and partners' (8).

The US Department of Defense's approach under the Obama Administration, while positive in terms of coming to grips with the realities of climate change, was incomplete because it did not explicitly address the importance of the environment and gender in the US military's interaction with local civilian populations in foreign countries as a cost-effective and efficient means to address climate change and promote stability. As required by the US Congress, in 2018 the quadrennial reviews were replaced by the new US *National Defense Strategy*, which is a classified document. Issued under the Trump Administration, the first non-classified summary of the strategy makes no mention of climate change (US Secretary of Defense, 2018), but perhaps it is addressed in some fashion in the classified version.

In sum, the IPCC's approach to evaluating the research on the relationships between climate change and armed conflict is conservative but not unjustified given of the lack of consensus on the topic, and it instead points to populations' risks of experiencing negative impacts from the effects of climate change as being more clearly related to other causal factors leading to armed conflict than climate change itself. In terms of their treatment of the international security aspects of climate change, the UK and particularly the US strategies and policies, both at the national and the defence ministry levels, show an uneven evolution in terms of their treatment of the operational significance of climate change. It can be fairly argued that the most recent documents under the Trump Administration actually show a regression in this regard as compared to the positions taken earlier by the Obama Administration, or at least a negative reassessment. Mindful of this, it is now useful to turn to the research and analysis on the relationship between climate change and armed conflict, and to assess whether and to what degree these national positions are consistent with these findings.

A review of the analysis of the relationships between climate change and armed conflict

To begin this evaluation, it is helpful to first review two fairly recent assessments of the development of that literature, the first by Salehyan in 2014, and the second by Buhaug in 2016. For purposes of this chapter, comparing the findings of these two assessments will be sufficient to outline the factors and conclusions that would be most useful to a military planner or operator. Thorough review of the entire body of this research would be valuable, but it is beyond the scope of this chapter's purpose.

Climate change affects conflict behaviour

In his evaluation of the disparity between findings in scientific research on the link between climate change and armed conflict, Salehyan (2014) found that these studies 'have adopted a broad array of methodological approaches, units of analysis, temporal scales, indicators of climate/weather, and definitions of conflict' (2). Accordingly, he noted that whilst 'this pluralism can sometimes be productive, it can also lead to what seems like a cacophony of different findings' (2). Salehyan cautioned that 'A finding, or lack thereof, at one level of analysis does not necessarily translate well to other levels of analysis', and that determining whether climate change will lead to conflict 'is rarely an inference that can be made from research designs that can only capture particular aspects of the climate-conflict nexus' (4). From a perspective of methodology, however, he assessed that the more recent research uses data that is more directly related to the relationship, rather than having somewhat more distant proxies stand in for the lack of data (3–5).

For example, Salehyan noted the analysis of sub-national violence in India by Wischnath and Buhaug (2014: 6–15) and their finding that there was a significant increase in armed violence in the year immediately following a year with bad harvests. Importantly, rather than use rainfall and temperature as proxies for food and water availability, they used data on actual changes in grain production (Salehyan, 2014: 4). Similarly, Salehyan favourably described the methodology used by von Uexkull (2014: 16–26) in her study of the connections in Africa between rain-fed agriculture, drought and a particular form of violence, organised armed conflict. Von Uexkull looked at both single-year droughts and sustained drought, and concluded that both types resulted in a significantly increased risk of violence in particular localities, more so in areas of rain-fed agriculture (Salehyan, 2014: 4). Salehyan also noted studies with rigorous methodologies that provide mixed results on the climate change-conflict nexus, such as that of Ide and his co-authors (Ide et al., 2014: 68–81). In sum, Salehyan concluded that although 'most scholars now agree there is a climatic influence on conflict behavior', there remains a lack of consensus 'about how or why the climatic variable – alone or in conjunction with other factors – matter for violence' (Salehyan, 2014: 1).

Conflict as a result of the interaction between the risks posed by climate change, local land use and government response

Writing two years later, Buhaug (2016) also saw empirical climate change and conflict research and analysis as having occurred in two phases. The first set tended to rely on simple meteorological proxies for climate change, such as temperature and rainfall, in seeking to determine whether there was a connection in between climate change and armed conflict. Most of these analyses did not consider 'interactive or conditional climate effects' (334). In general, the work from this stage of research and analysis did not show 'a general and robust climate-conflict connection' (334).

For Buhaug, an important distinction between the first and second phases of this research was that the latter examined possible indirect links between climate change and conflict, and took a multi-stage approach in 'considering the conflict impact of adverse socioeconomic changes assumed to be affected by rapid climate shifts' (334). These indirect links included food insecurity, production shocks and migration – human-driven phenomena more directly related to human behaviour. Importantly, he noted that whilst the correlation between climate change and resulting conflict is now perhaps more positive than earlier studies suggested, the reverse correlation between armed conflict and resulting environmental change is much stronger. Also, as compared to stable and wealthy societies that have the capability and resources to handle climate change situations in a peaceful manner, Buhaug found that 'Conflict affected societies, . . . many of which struggle with endemic political chaos, corruption, and poor economic growth, are unable to adapt to and cope with these challenges on their own' (335).

The value in following Buhaug's holistic approach to incorporating interactive socioeconomic factors into the analysis of whether an indirect result of climate change has the potential to lead to conflict is demonstrated by the article by Brzoska and Fröhlich (2016) noted earlier in Chapter 2, 'Gender and climate change'. In it, Brzoska and Fröhlich were sceptical of the potential for climate-change-induced migration to cause conflict. They noted that estimates of hundreds of millions of people being forced by climate change to migrate from their homes were based on little empirical evidence, and that the estimates did not properly include the role that adaptation measures could play in allowing people to remain in place (193, 204). Further, they were not convinced of 'the conflictive power of environmentally induced population movements' (204) because there was scant empirical proof that such migrations result in armed conflict.

Brzoska and Fröhlich allowed that certain case studies, such as those examining the resettlement of victims of Bengali floods and storms into other Indian states, do show an increase in migration-related violence. As a counter example, however, they provided a study assessing the migration of people from northwest Africa to Europe that was in part induced by worsening environmental conditions which found the opposite (194–95). Here, investment by wealthy countries (European presumably) in projects that built resilience to climate change actually led to economic improvement in those areas in the African countries. The real lesson from Brzoska and Fröhlich's study, therefore, is probably not that migration induced by climate change might not cause conflict. Operationally, it is probably better stated as recognising that wealthy countries have the resources to find peaceful solutions to climate-induced migration and avoid conflict – developing countries with fewer resources and governance issues perhaps not. These sorts of factors are of course human-driven, and their consideration requires us to take a much broader approach to this question than simply looking for direct causality.

Buhaug (2016) concludes that although climate change does not cause armed conflict, 'climate variability and change may well influence the dynamics of interaction (including violent contention) between societal actors. However, such an effect will occur in conjunction and sometimes interaction with other prevalent conflict drivers' (333). Therefore, the specific context of a situation will shape the way in which these interactions play out. As an example, whether a drought in a given area has potential to generate conflict will depend on how at-risk the local population is to its effects as a function of their resources and coping mechanisms (such as access to groundwater and alternative livelihoods), the nature of the local land use and the response to the drought by governmental bodies (such as relief aid and subsidies) (333–34).

From an operational perspective, perhaps causality is irrelevant

Importantly, each of these factors identified by Buhaug may be mediated by other overarching human factors, such as ethnic strife. For example, Schleussner and his co-authors (2016) have found a positive correlation between climate-related natural disasters and the outbreak of armed conflict in countries that were ethnically fractionalised. During the period 1980–2010, armed conflict broke out within one month of a climate-related natural disaster 23 percent of the time in the 50 countries that exhibited the highest degree of fractionalisation. Globally, this only occurred 9 percent of the time (9217–19).

Although an earlier study by Bergholt and Lujala (2012) had found that climate-related natural disasters did not increase the chance of armed conflict, they only assessed shock-like natural disasters such as floods and storms without including ethnic fractionalisation (150–61), and Schleussner and his co-authors used what was likely a more accurate set of disaster events through inclusion of other forms of climate-related natural disasters such as heat waves and droughts (9218). This study suggests that the level of ethnic strife in an operational theatre is not just a potential source of conflict itself, but that it also represents a societal risk to climate-change's effects that appears related to the outbreak of armed conflict. From a planning perspective then, ethnic strife can also be seen as a climate change-related factor of which commanders and staffs should be aware in mission areas where the potential for armed conflict exists.

To similar effect, Buhaug (2016) suggests that rather than climate change causing armed conflict, the stronger positive correlation might actually be armed conflict as a factor in accelerating the effects of environmental degradation and thereby the impacts of climate change (335). This suggests that the negative impacts of climate change might register upon at-risk population cohorts such as women and girls more adversely in the wake of armed conflict than in the time preceding the outbreak of war. If this is so, a good case can also be made that planners and operators would need to be mindful of this relationship in stability operations being conducted in theatres where armed conflict has occurred and climate change's effects are beginning to be felt.

From an operational perspective, there is a risk that such factors might be translated into some sort of checklist in conducting a doctrinally consistent operational analysis. Instead, specific and nuanced analysis would need to be conducted to isolate the specific potential causal pathways leading to instability and identify the most pertinent mediating influences upon these pathways in any given operational area. To be useful, this analysis would need to be adjusted along the temporal and geographical scales that would capture the different impacts of the relationships between climate change and armed conflict on different civilian population cohorts. With this in mind, it is useful to look at a particularly complex and severe armed conflict occurring during a time and in an area apparently affected by climate change, and to gauge the depth of factual detail that would be necessary to identify the sources of risk to climate change's impacts and assess the impact of these risks on a gender-differentiated basis.

Syria: a case study in climate change-affected armed conflict?

As noted in the Introduction, the conflict in Syria has been described by senior political officials and in the media as having been caused by climate change, at least in part. Hoerling et al. (2011) have assessed that the Mediterranean region experienced a drying trend during the cold-season months over the period 1902–2010, with increased drought frequency after 1970 or so. They attributed part of this reduction in precipitation to human activity rather than natural variability (2151–52, 2157–58). Focusing specifically on Syria from 1959 until 2008, Skaf and Mathbout (2010) found that rainfall measurements at different collection stations were very variable, but that in general, over time Syria had been experiencing more dry days during the rainy season, and the number of consecutive dry days had increased (110–12).

Thus, there is a reasonable basis to suggest that Syria is experiencing climate change, and that one form of this is less precipitation on fewer days during the time of year the country would ordinarily receive most of its precipitation. Assuming that this is correct, was the severe drought in north-eastern Syria between 2006 and 2008 or 2009 the result of climate change? Even if it was, were the impacts of the drought the proximate cause of the civil war that broke out in 2011, or was the drought only a contributing factor to some degree in the uprising against the regime of Bashar al-Assad? If it was only a contributing factor, perhaps one among many, is it operationally relevant from a gender perspective?

In a succinct survey of the scholarly literature exploring the causes of the Syrian Civil War, Abboud (2016) has identified a number of potential causes, including declining socioeconomic conditions for Syria's young and rapidly expanding population, the political failure of the governing Ba'ath Party, the contagion effect of the Arab Spring revolts in Tunisia and Egypt and regional power politics on the part of Syria's neighbours, as well as climate change and drought (6–7). Because of the depth and severity of the Syrian conflict, and its ripple effects in the international security environment even today, this war stands out as a useful case study to explore the extent to which climate change might have caused it.

Given the conflict's complexity, it is necessary to first understand the recent historical context that preceded the drought before turning to the drought itself and its impact on perhaps the single most important crop grown in Syria – wheat. Especially wheat grown in the north-eastern part of the country, which was hit particularly hard by drought conditions between 2006 and 2008. Next, it is important to consider Syria on the eve of the uprising, and to identify the different human-related factors within which the spark of uprising kindled. Finally, the actual spark itself, the bloody demonstrations in Dar'a in March 2011, will be examined for traces of the drought's impact.

Historical background

McHugo (2015) notes that in the aftermath of a 1963 military coup, the Ba'ath Party effectively came to power in Syria. Promoting Arab unity from a socialist perspective, the Ba'ath Party had begun in Syria before World War II and grew into a transnational Arab political movement and organisation (118–21). Many of the military officers involved in the 1963 coup were party members, and as they consolidated their power the Syrian army came under Ba'athist control (144–45). In 1970, the Ba'athist minister of defence, Hafez al-Assad, launched his own coup, and remained in power as the president of Syria until his death in 2000 (152–53). Of humble origins himself, under Hafez al-Assad the Syrian government paid a great deal of attention to the rural areas, and it made real progress in bringing education, electricity and greater access to water to most of the country (185–86).

Unfortunately, this progress was accompanied by vigorous and often brutal repression of dissent, widespread corruption in the government and a lack of accountability in the security forces (McHugo, 2015: 185; Phillips, 2016: 45, 47; Van Dam, 2017: 45–54). Succeeding his father as president in 2000, Bashar al-Assad at first seemed to be open to political reforms (Abboud, 2016: 50–51), but instead largely maintained the existing power structure (McHugo, 2015: 204–07, 217; Phillips, 2016: 44). Bashar al-Assad did however implement a number of measures intended to continue diversification of the economy, and began to favour urban areas and the services sector of the economy, at the cost of traditional Ba'athist support of the rural areas and the agricultural sector (McHugo, 2015: 204–07, 218; Hokayem, 2017: 42–43).

The wheat crop and the 2006–2008 drought in north-eastern Syria

The Eastern agricultural region of Syria, which includes the governorates, or provinces, of Ar Raqqah, Dayr az Zwarr and Al Hasakah in the north-east, is the largest in the country, and makes extensive use of irrigation in growing wheat (FAO, 2008: 4). This area's dominance in wheat is rooted partially in the privately financed land reclamation projects in the river valley areas in these governorates after independence in 1946, which opened up the plains to greater cultivation of rain-fed cereals (McHugo, 2015: 142). Before the civil war, most of Syria's wheat harvest was consumed domestically, but it did export wheat in certain years. For example, in 2005, Syria consumed 4.1 million metric tonnes of the 4.7 million metric tonnes it produced, or about 87 percent (Index Mundi, n.d.).

Wheat is well suited as a drylands crop, but there is a significant difference in the yield of wheat in Syria between irrigated wheat (3.4–3.5 tonnes/ha) and rain-fed (0.6–0.8 tonnes/ha) (FAO; 2008: 11). Accordingly, irrigated wheat constitutes almost two-thirds of the national crop, whilst rain-fed wheat only accounts for about a third (FAO, 2008: 4), even though about 45 percent of the wheat-growing area is irrigated (Sagardoy and Varela-Ortega, 2010: 130). In Al Hasakah, the 'bread basket' of Syria, about two-thirds of the wheat-growing area is irrigated, and it produced 37 percent of the total Syrian wheat crop in 2013 according to the US Department of Agriculture (USDA) (USDA, 2015: Figure 2). In 2006, Syrian wheat production was about 4.6 million tonnes, but it dropped to about 4.2 million tonnes in 2007. In the 2007–2008 drought years, it crashed to about 2 million tonnes. Production picked back up in 2009 through 2013, reaching 4 million tonnes in 2013, but then crashed again during the 2013–2014 drought (USDA, 2015: Fig. 1).

It is important to note that there is no complete consensus on the scope of the drought in terms of its intensity and duration. Trigo, Gouveia and Barriopedro

(2010) assessed on the basis of measured precipitation that the drought period of 2007–2009 was the driest time for the Fertile Crescent area since 1940, although the drought period of 1998–2000 was close behind (1255). He also noted that the severity of the drought in this area in 2007 amplified the effects of the dryness in the next year, with levels of soil moisture reaching extremely low levels by the spring of 2009, with associated lower levels of vegetative growth as shown by satellite data (1249). Similarly, Eklund and Thompson (2017a) explained that it is not likely sufficient to rely only on a reduction in precipitation to establish whether a drought has occurred. It is also important to assess soil moisture and crop production to get a fuller picture of the socioeconomic impact of the dryness (2). Analysing satellite data to measure these other moisture-related factors, they concluded that croplands in north-eastern Syria show evidence of land degradation over time, and that the drought in 2007–2009 'occurred at a time when land and water resources were already under stress' (7). They further concluded that 2008 was perhaps the only year of deep drought in terms of precipitation (4).

Looking at other information regarding precipitation, the drought would appear to have begun in the autumn of 2006 and continued through the 2008 growing season. Al Hasakah governorate, for example, was hit particularly hard, and received on average less than 50 mm of rain between September 2007 and April 2008 (USDA, 2008). On average, Al Hasakah would have ordinarily received almost 280 mm of rain during this time period (World Bank Group, n.d.). Despite the extensive use of irrigation to water wheat in Al Hasakah, the depth of this drought likely had a profoundly negative impact on the wheat crop – irrigation alone would not likely have provided the full amount of moisture needed to grow an optimal wheat crop in many areas (USDA, 2008). The north-eastern region also experienced drought conditions in the spring of 2010 according to certain reports (Maldanado, 2011: 3).

In some cases, farmers might have even lost access to irrigation water. Syria's groundwater-fed irrigation systems have significantly lowered the water table in many agricultural areas. For example, Sagardoy and Varela-Ortega reported that in 2006, 14,000 fewer hectares were irrigated with groundwater than the year before (2010: 126–27). Al Hasakah governorate in particular has very high groundwater use deficits (130). Starr (2012) reports that the irrigation systems used in the north-eastern governorates consisted generally of open channels made of concrete, which were often in poor condition because of a lack of government investment in repair. This resulted in a significant amount of wasted water (122–23).

Further, wheat requires more moisture at certain critical times in its life cycle (Jin et al., 2016: 4), and therefore variability in the timing of either irrigated water or rain can have significant impacts on plant growth (Maldanado, 2011: 3). Winter crops, such as wheat in Syria, are 'dependent on deeper reserves of soil moisture and their vegetative cycle is the result of combined effect of precipitation (over weeks and months), evaporation and in some regions, temperature' (Trigo, Gouveia and Barriopedro, 2010: 1246). Finally, reports from Syria at the time suggested that there was a shortage of diesel fuel available to farmers to power their water pumps, because the heavily subsidised fuel could be smuggled into neighbouring countries and sold for up to ten times the domestic price (USDA, 2008).

Syria on the uprising's eve

Abboud's (2016) analysis of the Syrian Civil War notes Syria's pronounced religious and ethnic diversity, and the fact that on the eve of the uprising, over half of Syria's people now lived in urban centres (5). Phillips (2016) found that sectarianism was significant in Syria, because political and economic patronage in the Syrian government and economy often reflected the fact that many Syrian governmental, security force and economic leaders belonged to the minority Alawite sect, an offshoot of Shia Islam, whilst the largest Islamic group in the country were the Sunnis. Christian and other minority groups often supported the regime because of its tolerance for religious difference, and fear of persecution in a Sunni-dominated country (51–52). Al-Assad's government maintained its control through the *mukhabarat*, the repressive police and security forces, which quickly quashed expressions of discontent (McHugo, 2015: 185).

Phillips (2016) reported that Syria's population had increased from 3.3 million in 1950 to 21 million in 2011. A combination of higher living standards and government policies encouraging larger families resulted in a youth bulge in the nation's demographics – by 2010, 55 percent of the population was younger than 24. Many of these young people had attended university, but found few jobs in which to employ their talents (45). The unemployment rate was officially stated to be around 10 percent between 2003 and 2011, but other estimates suggested it was closer to 20 percent, and much higher for young people (46).

Abboud (2016) has noted that after achieving independence in 1946, Syria's economy was largely dependent on its agricultural sector and then later also on its oil production (5). Phillips (2016) highlighted the importance of remittances from Syrians working in the oil industry in other Middle Eastern countries to the Syrian economy (12). Beginning in the 1990s, the government sought to diversify the country's economic base, and this process gained speed in the early 2000s as government planners sought to move more economic activity to the services sector. This shift coincided with significant environmental degradation in many agricultural areas and decreased agricultural productivity, and led to a change in 'the social basis around which agricultural activity occurred' (Abboud, 2016: 5).

As investment in rural areas declined, this gap was filled to a degree by charitable efforts by Islamic non-governmental organisations (McHugo, 2015: 218). However, farmers both lost subsidies for fertilisers and were forced to buy fuel at increasingly unsubsidised prices during this time (Phillips, 2016: 46). Further, water resources had been mismanaged for decades, and the government response to the drought was ineffectual. Phillips (2016) observed that the north-eastern region had been particularly hard hit by the drought, and that it was home to approximately 58 percent of Syria's poor (46).

Prior to the drought, the US-led invasion of Iraq in 2003 resulted in perhaps as many 1.2 million Iraqi refugees settling in Syria, straining social services and increasing competition for jobs (McHugo, 2015: 210). There is a lack of consensus on the number of people who were internally displaced by the drought. Starr (2012) cites a figure of 300,000 people leaving drought-stricken agricultural areas in eastern and southern Syria and moving to cities including Damascus, Aleppo and Dar'a (3). Interestingly, despite these influxes, Starr's impression as a journalist was that Syrians were largely content with the direction in which the country was moving, especially in the urban centres (3). In contrast to Starr, Van Dam (2017) states that over a million rural people were displaced by the drought (89). Despite the difficulty in identifying a definitive number of refugees and displaced persons who moved to urban areas in Syria, it is reasonable to assume that their numbers added further strain to an already overburdened system of government services. Mitigating this somewhat, an unknown number of persons displaced by drought likely moved back to their agricultural homes and livelihoods, given the rebound in the Syrian wheat harvest when the 2006–2008 drought ended (USDA, 2015).

The spark in Dar'a

Phillips (2016) recounts that although there were some earlier protests in Damascus and other cities in February and early March 2011, the spark that appeared to ignite the uprising occurred in the southern city of Dar'a in mid-March 2011. Dar'a was relatively poor compared to cities such as Damascus and Aleppo, and is located in a more agrarian governorate that had been negatively impacted by the 2006–2008 drought (49). It had been affected as well by the government policies that had shifted economic support away from agricultural areas to the more populous cities. Further, significant population increases compromised the ability of the state to deliver essential infrastructure and services (Hokayem, 2013: 42–43).

Dar'a was primarily Sunni, and Hokayem (2013) notes that although it had historically been a Ba'athist stronghold (42), its politics were more tribally oriented than nationally oriented (Phillips, 2016: 49). The security forces commander was an Alawite not from the area, and was a cousin to Bashar al-Assad (Phillips, 2016: 49). He and the Damascus-appointed governor were seen as outsiders indifferent to local concerns by the people of Dar'a (Hokayem, 2013: 42). McHugo (2015) describes how in early March 2011, a number of young students, aged 9–15, wrote anti-government graffiti on the walls of their school and were caught by the school principal and government security forces (221). The children were beaten and tortured, and their parents' requests for the children's release were met with insulting refusal. A number of the students were from prominent local families, and when family members took to the streets to protest, they too were arrested (Starr, 2012: 3–4). On 15 March 2011, another protest by thousands of demonstrators was met with gunfire by government security forces (Starr, 2012: 3–4), killing four protestors (McHugo, 2015: 221; Phillips, 2016: 49).

By 23 March 2011, the security forces had launched an even harsher crackdown, surrounding the city and cutting off utilities and mobile phone networks. From that point on, Abboud (2016) notes, protests quickly spread across the country to other peripheral urban areas, including Al Hasakeh and Dayr az Zawr in the northeast (56–57), and these were also often met with lethal force on the part of the Syrian security forces (Starr, 2012: 7–8). Support for the uprising did not break cleanly everywhere along sectarian lines, however, and it was often influenced by historical experiences of repression by the Assad regime and perceptions of favouritism shown to the Alawite minority to which Bashar al-Assad belonged (Hokayem, 2013: 44–45).

The internet and social media figured significantly in both the speed of the uprising's growth and the government's response to it. For those supporting the uprising, social media served as a mobilisation platform and an accelerant in the distribution of news about protests and government crackdowns (Phillips, 2016: 49). Importantly, the themes of the protests that spread across the country were similar – anti-corruption, the oppression of the security forces and a lack of freedom, and the lack of jobs and opportunities (McHugo, 2015: 222). In response, the government used the same resources to spy on the uprising, to identify those who appeared to have anti-government leanings, to hack pro-uprising web pages and to distribute pro-government information (Starr, 2012: 62–63).

Research, analysis and critique

Researchers and analysts are divided on the significance of the drought to the uprising. De Châtel (2014) has noted the complex interplay in Syria between 'social, economic and political factors, including, in this case, growing poverty caused by rapid economic liberalization and cancellation of state subsidies after 2005, a growing rural-urban divide, widespread corruption, [and] rising unemployment' as well as the drought (1). She pointed out that drought, even severe drought, is no stranger to Syria, and that between 1961 and 2009, Syria experienced nearly 25 years of drought conditions. De Châtel acknowledged that farmers in the areas affected by the 2006–2008 drought perceived an increase in the occurrence of droughts, but suggested that this might be a result of a greater population and depleted groundwater resources. She assessed that although climate change likely had some impact on the situation in Syria as the uprising unfolded and Syria fell into civil war, the political, social and economic factors were likely much more important and in numerous instances pre-dated the drought by many years. To de Châtel, focusing on the possible role of climate change in the conflict was not only irrelevant, it was also an unhelpful distraction (2–3, 12).

In a 2015 study, Kelley and his co-authors assessed that although multi-year droughts in the Fertile Crescent area are not uncommon, it was unlikely that the 2007–2010 drought conditions would have been as intense but for the century-long drying process identified in Hoerling's study (3245). They found a statistically significant decrease in rainfall in the region since 1931, and an increase in surface temperature such that each of the years between 1994 and 2009 were above the century-long mean (3242). This had resulted in both less winter precipitation and greater summer evaporation, which impacted winter crops such as wheat in particular because of their reliance on soil moisture reserves in their growth cycles (3243). The drought, combined with the Syrian government's promotion of unsustainable agricultural practices and its failure to adequately respond to the needs of a displaced population that had been placed at greater risk to the drought's impacts, suggested to the study's authors 'a connected path running from human interference with climate to severe drought to agricultural collapse and mass human migration' (3245).

Other researchers have found a more attenuated set of linkages between the drought and the uprising. Eklund and Thompson's (2017b) evaluation of the connections between the drought, migration and the subsequent civil war did not find strong relationships between these events. They found that the data regarding out-migration from the drought-affected areas to Syrian cities was based on limited reports and established only that the two phenomena occurred at the same time. Because people migrate for different reasons and the act of migration itself can require a certain amount of capital, Eklund and Thompson pointed out that the links between the two remain unproven scientifically. They concluded that connections between the risks faced by the farming populations to the drought, including land degradation through over-irrigation and loss of economic subsidies from the government, did lead to dissatisfaction with the government response to the drought. Unhappiness with the government had been brewing for years in other parts of the country as well, however, and although the 'economic struggles stemming from drought vulnerability . . . did contribute to widespread dissatisfaction with the government', it was unclear what role migration related to the drought actually played in promoting the uprising (Eklund and Thompson, 2017b).

Some researchers have gone further in finding no positive correlation between the drought and the Syrian Civil War. Selby et al. (2017) have argued that even though it is impossible to say that the drought did not have some impact on the situation in Syria, 'there is . . . no good evidence to conclude that global climate change-related drought in Syria was a contributory casual factor in the country's civil war' (241). Their study assessed the Kelley study's findings, and they argued that on the basis of available data, the scale of the out-migration from the affected agricultural areas had not been reliably quantified, and that although the drought might have contributed to it, the migration was more likely caused by the Assad regime's economic liberalisation than the drought itself. This study raises important points regarding the potentially counterproductive oversimplification of climate change as a 'threat multiplier' on the part of certain parts of the international security community and the media (241). However, an examination of certain of the data upon which the analysis in Selby's study are based suggests that a degree of caution might be prudent in applying its conclusions to any operational gender analysis of a mission theatre in seeking to identify relationships between climate change and conflict.

For example, Selby and his co-authors doubted that existing data and research established an apparent drying trend in the Eastern Mediterranean area (237). They further expressed concern about the inconsistencies and variability in the rainfall data used as the basis for Kelley's and other studies suggesting a link between the drought and the uprising. Their study questioned the extent of the drought, citing for example official Syrian government statistics that show Dar'a received close to average rainfall between 2006 and 2009 (234). According to other official statistics, Dar'a did receive 210 mm of rainfall in 2006, and 209.8 mm in 2007 (Central Bureau of Statistics, 2008), which the US Department of Agriculture (USDA, 2008) assessed as approximately 100 percent of the normal expected rainfall for the period September 2006 through April 2007. However, between September 2007 and April 2008, Dar'a received less than 50 percent of its expected normal rainfall (USDA, 2008). To characterise the time period in Dar'a as one of almost average rainfall is to dilute the granularity of the data and its possible significance given Dar'a's role in the uprising – a critique levelled by Selby and his co-authors of the Kelley study in general.

Another example is the Selby study's conclusion that the economic nature of agriculture in the north-east shows that the wheat crop during the drought years should not have been highly susceptible to drought. Because wheat yields are 'relatively stable' when irrigated, and that 'the vast majority of Al Hasakah's production was irrigated' (239), Selby and his co-authors concluded that the more important factor in reduced wheat production during the drought years was the Syrian government's sudden withdrawal of price supports and subsidies for fuel and fertiliser in 2008 and 2009 (238). This conclusion appears premised on the assumption that an efficient irrigation system with sufficient supplies of water provided all of the irrigated wheat crop's water needs, and that rainfall was not important to ensuring adequate moisture availability at key times in its growth cycle. In its minimisation of the drought's impact, the Selby study did not address the difference between what the wheat yield was during the drought and what it could have been had there been normal rainfall. Although the study's general points about the need to be precise in assessing the relationship between climate change and conflict remain valid, its approach in arriving at some of its conclusions with regard to Syria as a case study tend to undercut the strength of some of the study's arguments.

Summary

Climate change and armed conflict in general

One general premise of arguments supporting a positive correlation between climate change and conflict is that in times of resource scarcity, competition for those resources will lead to conflict. There are of course numerous historical examples of resource scarcity, whether real or perceived, resulting in armed conflict (Kemp, 1978: 396–98). Research indicates this proposed relationship is not always valid for the effects of climate change, however. For example, studies of pastoral people in northern Kenya, such as the Turkana and Pokot peoples first mentioned in Chapter 1, 'Armed conflict and gender', suggest that whether conflict occurs depends upon the resource and its condition. In times of drought when water is scarce, these people traditionally tend to cooperate to get through the hard times. When water is plentiful, however, it is worthwhile to engage in cattle-raiding against each other (Butler and Gates, 2012: 30–31), perhaps because the cattle are plumper and worth stealing, and it would be easier to water and feed them as they were herded away to their new home.

In a later study focused on pastoralists in East Africa, Ayana and his co-authors (2016) confirm that conflict events, vegetative cover and abundant precipitation tend to occur simultaneously in the same areas. They point out, however, that it is important to consider the importance of how the herd animals are used (dairy versus meat) and even the type of herd animals (camels being more drought-resistant than cattle) in examining conflict and climate change relationships (609–10). The level of analysis in this example is quite detailed, and it suggests that there might be no useful general operational assumption that could be made regarding climate change and armed conflict. Instead, what is likely required to generate actionable intelligence regarding climate change's effects on an operational theatre is a thorough understanding of how human-related factors in the theatre influence the risks related to climate change's negative impacts. Mindful that these risks can be sex and gender-differentiated, this provides a starting point for developing a gender analysis that could assess whether there is the potential for any compounding effect of climate change upon the negative impacts of armed conflict to at-risk civilians in the operational area.

Climate change and armed conflict in Syria

It is important to note that although Dar'a had suffered negative impacts from the nationwide drought of 2006–2008, the actual spark of the conflict was local and apparently political, social and cultural in nature, and sectarianism appears to have played a role. The anti-government protests only grew after the security forces rebuffed requests for accountability on the part of the government officials and for an honourable solution, and then responded with lethal force against the demonstrators (Hokayem, 2013: 42; McHugo, 2015: 218). The themes of the protests appear to be only indirectly related to the drought – accountability for the security forces, greater toleration of dissent, and increased jobs and opportunities. Of these main themes, perhaps only the demand for increased jobs and opportunities sets a clear linkage between drought and population displacement.

Starr (2012) reported that thousands of farmers moved to Dar'a looking for work in the import/export business in that city near the Jordanian border as a result of the 2006–2008 drought (122). Fröhlich, one of the Selby study's co-authors (2017), conducted interviews with a total of 32 Syrian nationals who were refugees in Jordan in late 2014 and early 2015. The interviewees had all worked full-time or part-time in the agricultural area (242). Four interviewees from the Dar'a area stated that the displaced persons from the north-east in general had little to do with local politics and largely left Dar'a when the troubles began (240). Ironically, however, Starr (2012) also reports that many of the men displaced by drought went to work for the government security forces in putting down demonstrations (123). With little to lose and not being from the urban areas to which they had migrated, perhaps they were more likely to use violence against local demonstrators, which might have escalated the spiral of repression and rebellion. The grievances of some persons displaced by the drought who remained behind in cities might in fact have helped fuel the uprising, but not in the way one might ordinarily have expected.

Further, it is important to not discount the impact communications technology had upon the spread of the uprising. Past repression, such as the draconian crackdown of Hafez al-Assad against the Muslim Brotherhood in Hama in 1982, which might have resulted in as many as 20,000 casualties, had been successful and did not result in any real national-level uprising (McHugo, 2015: 193–94, 247). In 2011, however, YouTube videos of demonstrations and security forces' repression allowed the uprising to avoid the state-controlled media and inform the rest of the country of what was happening very quickly (Starr, 2013: 55). These postings appeared to encourage local demonstrations in other cities – but it is difficult to draw meaningful, proximate connections from the drought to the skilful use of Western-developed social media platforms.

Finally, it is possible that the Syrian uprising and the resulting civil war would have eventually concluded in some fashion years ago but for the involvement of neighbours and regional and international actors in supporting the government or the rebels. Turkey and influential Gulf states such as Saudi Arabia and Qatar opposed al-Assad and provided political and financial support as well as light arms to the outgunned rebels. From Lebanon, Hezbollah provided fighters to the government, and Iran and later Russia provided arms and fighters as the conflict dragged on (Hokayem, 2013: 112–28, 172–77; Phillips, 2016: 213–23). It is not easy to trace a line between the 2006–2008 drought, or even the possible general drying trend in this part of the Middle East, and the outsized role that foreign actors played even in the earlier parts of the Syrian Civil War.

In sum, the severity and the extent of the dryness and the resulting impacts on the livelihoods of large numbers of people at-risk to the drought's effects make it illogical to assume that it had no impact on the situation which resulted in the uprising. By the same token, no evidence supports the idea that the drought was the proximate and direct cause of the Syrian Civil War. Even with the global attention that the conflict in Syria has received since it began, there is a lack of data at this time from which definitive conclusions could be drawn to establish any causal pathways between the drought and the civil war. Conversely, Syria's political, economic and sectarian complexities, coupled with the role of foreign influences, present a large array of independent variables with clear impacts from which causal threads can be more convincingly drawn.

This does not mean that Syria is without value as a case study on the relationship between climate change and conflict – quite the contrary. From an operational perspective, it serves as a warning of how incomplete any analysis of the human environment in a civilian-centric mission theatre would be if it neglects to include climate change as a potential factor. It forces us to consider the problem from the vantage point of analytical economy – where and in what fashion are the destabilising effects of climate change likely to express themselves in a measurable way (if even by proxy), such that military organisations can recognise and meaningfully address it in their activities and operations in situations of armed conflict? The roles played by women and girls in many developing countries are essential in maintaining economic and social stability, and the potentially greater risks they face of the compounding impacts of armed conflict and climate change present a possible answer to this question.

Climate change, armed conflict and women

As noted in the beginning of this chapter, there is strong support for the notion that armed conflict can result in significant environmental degradation. Weakened environments are at greater risk to climate change's effects, as are the human populations that dwell within them, and depend upon them for their livelihoods. One study (Zhang et al., 2018) suggests that as plant species in an environment adapt to the effects of climate change, such as more frequent drought conditions, the composition of a forest for example might begin to favour species that invest greater energy in their root systems rather than leaves and trunks that can act as larger carbon sinks (99–102). If only incrementally, a climate change response on the part of these tree species is then the net decrease in a forest's ability to mitigate the force causing its distress. As to the reverse relationship, recent analyses of climate change and armed conflict studies suggest that although the causal evidence is not as strong, the effects of climate change can only be indirectly correlated with armed conflict, as they are strongly mediated by existing cultural, social and economic conditions.

Assuming that these descriptions of the relationships between climate change and armed conflict are correct, this means that military organisations need to be aware of trends in operational environments such that climate change might be a signal of increasing likelihood of instability and even conflict at some point. They also need to be aware that it could have a compounding effect on the risks already faced by women and girls in armed conflict situations. Further, military organisations should recognise that fostering climate change adaptation and mitigation measures in post-conflict operational environments could be useful in reducing the likelihood of recurring armed conflict in the long run. Finding the appropriate analytical framework to understand this and to then act upon this understanding is important, because otherwise the risk of 'paralysis by analysis' is too great in this complex area of intersecting forces and effects.

Noting critiques of quantitative research on the relationships between climate variability and armed conflict, Seter (2016) argues that the solution to this gap does not necessarily consist of adjusting the technical aspects of analytical methodologies, or even generating new datasets to analyse. Instead, she sets out an analytical framework focusing on what she identifies as the three most likely proximate drivers of climate change-related conflict (economic hardship, favourable economic conditions and migration). She then proposes evaluating these mechanisms from the perspectives of the likely actors, the nature of the variability in climate to which they might be reacting, the likely form of conflict that might result, and the time scales and geographic areas over which these conflicts might occur (2–9). For Seter, the 'interesting contexts when it comes to the climate-conflict relationship are those that make the population vulnerable to climate variability' (2–3).

Although Seter intended this approach to help scientists conduct more focused research, it could also serve as a template for military headquarters to begin to analyse the possible impacts of climate change in civilian-centric operational environments. Particular risks faced by population cohorts because of their inferior socioeconomic status have immediate operational significance – whilst climate change that would have a general destabilising effect on a civilian-centric theatre outside the near-term future will unlikely be a priority for a deployed headquarters. Importantly, Seter's analysis format lends itself to incorporating a useful gender aspect, given our understanding of the potential differentiated impacts of armed conflict and climate change upon women and girls as a functional expression of the increased risks they face because of their potentially inferior social, political and economic conditions.

Particularly in developing countries, and even in countries that are fairly well along the path of development but still have significant portions of their populations living in poverty (such as Syria was prior to the civil war), underlying gender-differentiated risks could be amplified by armed conflict, and then re-amplified by climate change, all in a non-linear fashion and irrespective of any direct causal relationship between climate change and armed conflict. As Eklund and Thompson (2017a) point out in the case of Syria, the real question is not whether there was drought, but rather what the population's risks to the drought conditions were (6–7). These potential cascading negative effects on women and girls could have a significant impact on developing stability in a civilian-centric operation because females in these populations likely still bear the burden of keeping homes operating and caring for other family members, among their many other chores. Addressing their needs in times of climate change and armed conflict, and providing them the tools to build resilience in their home and community economies and ecologies, could pay handsome dividends in long-term stability far greater than the initial investments.

For this to occur, however, military organisations (especially those of developed states that deploy forces frequently) must first be cognizant of the potential gender-differentiated impacts of armed conflict and climate change upon female civilians in different parts of the host nation, and the potential range of relationships between armed conflict, women and girls, and climate change. In any given operational theatre, these militaries might be engaged in missions in which they are sharing operational space with national governmental, international and non-governmental organisations, all ostensibly working towards similar goals. Even if the civilian agencies are working in ways that account for the gender-differentiated impacts of armed conflict and climate change, synchronised actions between them and military organisations are not likely to prove productive in this regard if the military does not recognise the full operational relevance of gender.

Finally, even though non-governmental organisations might not be willing to work directly with military forces in theatre, any information exchange between them that might enable complementary actions will likely be sub-optimal unless the military understands how gender issues ripple through civilian-centric theatres of operation. Otherwise, there simply might not be similar-enough categories of data and analysis collected and developed by both to allow for the expeditious transfer of information. With this in mind, it is time to examine the strategy, policies and military doctrine of NATO, the UK, the US and Australia to better understand how they incorporate their understandings of gender's operational relevance into the guidance that they issue to their military forces.

References

Abboud, SN (2016) *Syria*. Malden, MA: Polity Press.
Adams, C; Ide, T; Barnett, J and Detges, A (2018) 'Sampling Bias in Climate-conflict Research'. *Nature Climate Change* 8: pp. 200–03. DOI: 10.1038/s41558-018-0068-2.
Anwari, R and Siddique, A (2017) 'Afghan Lawmakers Investigate Illegal Logging'. *Gandhara* (Radio Free Europe/Radio Liberty) (9 January). Available at: http://gandhara.rferl.org/a/pakistan-kunar-illegal-logging/28222069.html (Accessed 2 March 2018).
Austin, JE and Bruch, CE (2000) *The Environmental Consequences of War – Legal, Economic and Scientific Perspectives*. Cambridge: Cambridge University Press.
Ayana, EK; Ceccato, P; Fisher, JRB and DeFries, R (2016) 'Examining the Relationship Between Environmental Factors and Conflict in Pastoralist Areas of East Africa'. *Science of the Total Environment* 557–58: pp. 601–11. DOI: 10.1016/j.scitotenv.2016.03.102.
Bergholt, D and Lujala, P (2012) 'Climate-Related Natural Disasters, Economic Growth, and Armed Civil Conflict'. *Journal of Peace Research* 49(1): pp. 147–62. DOI: 10.1177/0022343311426167.
Bonds, E (2015) 'Challenging Global Warming's New 'Security Threat' Status'. *Peace Review* 27(2): pp. 209–16. DOI: 10.1080/10402659.2015.1037630.
Borger, J (2007) 'Darfur Conflict Heralds Era of Wars Triggered by Climate Change, UN Report Warns'. *The Guardian* (23 June). Available at: www.theguardian.com/environment/2007/jun/23/sudan.climatechange (Accessed 2 November 2018).
Brzoska, M and Fröhlich, C (2016) 'Climate Change, Migration and Violent Conflict: Vulnerabilities, Pathways and Adaptation Strategies'. *Migration and Development* 5(2): pp. 190–210. DOI: 10.1080/21632324.2015.1022973.
Buhaug, H (2016) 'Climate Change and Conflict: Taking Stock'. *Peace Economics, Peace Science and Public Policy* 22(4): pp. 331–38. DOI: 10.1515/peps-2016-0034.
Butler, CK and Gates, S (2012) 'African Range Wars: Climate, Conflict, and Property Rights'. *Journal of Peace Research* 49(1): pp. 23–34. DOI: 10.1177/0022343311426166.
Central Bureau of Statistics (2008), 'Annual Precipitation in MM, 2003–2007'. *2008 Yearbook* [online]. Available at: www.cbssyr.sy/yearbook/2008/Data-Chapter1/TAB-14-1-2008.htm (Accessed 4 July 2018).
Davidson, K (2004) 'Pentagon-Sponsored Climate Change Report Sparks Hullabaloo in Europe/But New Ice Age Unlikely, Bay Area Authors of Study Say'. *San Francisco Chronicle* (25 February). Available at: www.sfgate.com/green/article/Pentagon-sponsored-climate-report-sparks-2791555.php.
De Châtel, F (2014) 'The Role of Drought and Climate Change in the Syrian Uprising: Untangling the Triggers of the Revolution'. *Middle Eastern Studies* [online]: pp. 1–15. DOI: 10.1080/00263206.2013.850076.
Denyer, S (2014) 'Kerry Calls Climate Change a Weapon of Mass Destruction, Derides Skeptics'. *The Washington Post* (16 February). Available at: www.washingtonpost.com/

world/asia_pacific/kerry-calls-climate-change-a-weapon-of-mass-destruction-derides-skeptics/2014/02/16/1283b168-971a-11e3-ae45-458927ccedb6_story.html?utm_term=.2fa1dfc47dbd (Accessed 21 March 2018).
Department of Defense (US) (2010) *Quadrennial Defense Review* (1 February). Washington, DC: Department of Defense. Available at: www.defense.gov/Portals/1/features/defenseReviews/QDR/QDR_as_of_29JAN10_1600.pdf (Accessed 25 June 2018).
Department of Defense (US) (2014) *Quadrennial Defense Review* (4 March). Washington, DC: Department of Defense. Available at: http://archive.defense.gov/pubs/2014_Quadrennial_Defense_Review.pdf (Accessed 25 June 2018).
Directorate of Intelligence, Office of Political Research (1974) 'Potential Implications of Trends in World Population, Food Production, and Climate'. Central Intelligence Agency (Original on file with the University of Illinois at Urbana-Champaign). Available at: https://babel.hathitrust.org/cgi/pt?id=uiug.30112018865672;view=1up;seq=5 (Accessed 16 June 2017).
Eklund, L and Thompson, D (2017a) 'Differences in Resource Management Affects Drought Vulnerability across the Borders between Iraq, Syria, and Turkey'. *Ecology and Society* 22(4): pp. 1–11. DOI: 10.5751/ES-09179-220409.
Eklund, L and Thompson, D (2017b) 'Is Syria Really a "climate war"? We Examined the Links between Drought, Migration and Conflict'. *The Conversation*. Available at: http://theconversation.com/is-syria-really-a-climate-war-we-examined-the-links-between-drought-migration-and-conflict-80110 (Accessed 27 January 2018).
Fawkes, C (2017) 'Is Climate Change Making Hurricanes Worse?' *BBC* (30 December). Available at: www.bbc.com/news/world-us-canada-42251921 (Accessed 16 June 2018).
Food and Agriculture Organisation of the UN (FAO) (2008) 'Aquastat: Syrian Arab Republic'. FAO.org. Available at: www.fao.org/nr/water/aquastat/countries_regions/SYR/SYR-CP_eng.pdf (Accessed 25 June 2018).
Formoli, TA (1995) 'Impacts of the Afghan-Soviet War on Afghanistan's Environment'. *Environmental Conservation* 22(1): pp. 66–69. DOI: 10.1017/S037689290003409.
Fountain, H (2015) 'Researchers Link Syrian Conflict to a Drought Made Worse by Climate Change'. *New York Times* (2 March). Available at: www.nytimes.com/2015/03/03/science/earth/study-links-syria-conflict-to-drought-caused-by-climate-change.html?_r=0 (Accessed 13 January 2018).
Groninger, J (2012) 'Reforestation Strategies Amid Social Instability: Lessons from Afghanistan'. *Environmental Management* 49: pp. 833–45. DOI: 10.1007/s00267-012-9817-6.
Her Majesty's Government (2010) 'A Strong Britain in an Age of Uncertainty: The National Security Strategy'. GOV.UK (October). Available at: https://assets.publishing.service.gov.uk/government/uploads/system/uploads/attachment_data/file/61936/national-security-strategy.pdf (Accessed 30 June 2018).
Her Majesty's Government (2015) 'National Security Strategy and Strategic Defence Review and Security Review 2015 – A Secure and Prosperous United Kingdom'. GOV.UK (November). Available at: www.gov.uk/government/uploads/system/uploads/attachment_data/file/478936/52309_Cm_9161_NSS_SD_Review_PRINT_only.pdf (Accessed 13 November 2018)
Hoerling, M; Eischeid, J; Perlwitz, J; Quan, X; Zhang, T and Pegion, P (2011) 'On the Increased Frequency of Mediterranean Drought'. *Journal of Climate* 25: pp. 2146–61. Available at: https://journals.ametsoc.org/doi/pdf/10.1175/JCLI-D-11-00296.1 (Accessed 24 June 2018).
Hokayem, E (2013) *Syria's Uprising and the Fracturing of the Levant*. New York, NY: Routledge.
Ide, T; Schilling, J; Link, JSA; Scheffran, J; Ngaruiya, G and Weinzierl, T (2014) 'On Exposure, Vulnerability and Violence: Spatial Distribution of Risk Factors for Climate Change

and Violent Conflict across Kenya and Uganda'. *Political Geography* 43: pp. 68–81. DOI: 10.1016/j.polgeo.2014.10.007.

Index Mundi (n.d.) 'Syria – Wheat – Production, Consumption, Imports & Exports'. *Index Mundi.com*. Available at: www.indexmundi.com/syria/agriculture/wheat.html (Accessed 23 December 2017).

IPCC (Intergovernmental Panel on Climate Change) (2007a) *Climate Change 2007 – Impacts, Adaptation, and Vulnerability*. UN IPCC. Available at: www.ipcc.ch/pdf/assessment-report/ar4/wg2/ar4_wg2_full_report.pdf (Accessed 18 December 2017).

IPCC (Intergovernmental Panel on Climate Change) (2007b) *Climate Change 2007 – Synthesis Report*. UN IPCC. Available at: www.ipcc.ch/pdf/assessment-report/ar4/syr/ar4_syr_full_report.pdf (Accessed 16 June 2018).

IPCC (Intergovernmental Panel on Climate Change) (2014) Chapter 12, Human Security. In: *Climate Change 2014: Impacts, Adaptation, and Vulnerability*. UN IPCC. Available at: www.ipcc.ch/pdf/assessment-report/ar5/wg2/WGIIAR5-Chap12_FINAL.pdf (Accessed 18 December 2017).

Jin, N; Tao, B; Ren, W; Feng, M; Sun, R. and Liang, Z (2016) 'Mapped Irrigated and Rainfed Wheat Areas Using Multi-Temporal Satellite Data'. *Journal of Remote Sensing* (3 March): pp. 1–19. Available at: file:///C:/Users/jmprescott/Downloads/remotesensing-08-00207.pdf (Accessed 21 December 2017).

Kelley, CP; Mohtadi, S; Cane, MA; Seager, R. and Kushnir, Y (2015) 'Climate Change in the Fertile Crescent and Implications of the Recent Syrian Drought'. *Proceedings of the National Academy of Sciences* 112(11): pp. 3241–46. DOI: 10.1073/pnas.1421533112.

Kemp, G (1978) 'Scarcity and Strategy'. *Foreign Affairs* 56(2): pp. 396–414.

Lunn, J and Scarnell, E (2015) 'The 2015 UK National Security Strategy, Briefing Paper 7431'. *House of Commons Library* (14 December). Available at: http://researchbriefings.parliament.uk/ResearchBriefing/Summary/CBP-7431#fullreport (Accessed 21 December 2018).

Maldanado, J (2011) 'Syria: Grain and Feed Annual Report. Department of Agriculture Foreign Agricultural Service'. USDA. Available at: https://gain.fas.usda.gov/Recent%20GAIN%20Publications/Grain%20and%20Feed%20Annual_Damascus_Syria_2-17-2011.pdf (Accessed 29 December 2017).

McHugo, J (2015) *Syria*. New York: The New Press.

Ministry of Defence (UK) (2014a) *Global Strategic Trends – Out to 2045*, 5th Ed. Shrivenham, UK: Development, Concepts and Doctrine Centre (29 August). Available at: www.gov.uk/government/publications/global-strategic-trends-out-to-2045 (Accessed 16 June 2018).

Ministry of Defence (UK) (2014b) *Future Operating Environment 2035*. Shrivenham, UK: Development, Concepts and Doctrine Centre (30 November). Available at: https://assets.publishing.service.gov.uk/government/uploads/system/uploads/attachment_data/file/646821/20151203-FOE_35_final_v29_web.pdf (Accessed 16 June 2018).

Ministry of Defence (UK) (2015) 'Sustainable MOD Strategy – Act & Evolve 2015–2025'. GOV.UK (September). Available at: www.gov.uk/government/uploads/system/uploads/attachment_data/file/498482/Sustainable_MOD_Strategy_2015-2025.pdf (Accessed 17 June 2016).

Ministry of Defence (UK) (2017) 'Defence Infrastructure Organisation Estate and Sustainable Development'. GOV.UK (January). Available at: https://www.gov.uk/guidance/defence-infrastructure-organisation-estate-and-sustainable-development (Accessed 22 September 2018).

Nagel, J (2015) 'Gender, Conflict, and the Militarization of Climate Change Policy'. *Peace Review* 27(2): pp. 202–08. DOI: 10.1080/10402659.2015.1037629.

National Intelligence Council (2008) 'National Intelligence Assessment on the National Security Implications of Global Climate Change to 2030, Statement for the Record of

Dr. Thomas Fingar' (25 June). Office of the Director of National Intelligence. Available at: https://fas.org/irp/congress/2008_hr/062508fingar.pdf (Accessed 24 June 2018).

National Intelligence Council (2016) 'Implications for US National Security of Anticipated Climate Change'. Office of the Director of National Intelligence (21 September). Available at: www.dni.gov/files/documents/Newsroom/Reports%20and%20Pubs/Implications_for_US_National_Security_of_Anticipated_Climate_Change.pdf (Accessed 24 June 2018).

Obama, BH (2015) 'Remarks by the President at the United States Coast Guard Academy Commencement (20 May)'. The White House. Available at: https://obamawhitehouse. archives.gov/the-press-office/2015/05/20/remarks-president-united-states-coast-guard-academy-commencement (Accessed 13 April 2018).

Phillips, C (2016) *The Battle For Syria – International Rivalry In The New Middle East*. New Haven: Yale University Press.

Prescott, JM (2014) 'Climate Change, Gender and Rethinking Military Operations'. *Vermont Journal of Environmental Law* 15(4): pp. 766–802. Available at: http://vjel.vermontlaw.edu/files/2014/06/Prescott_ForPrint.pdf (Accessed 24 June 2018).

Sagardoy, JA and Varela-Ortega, C (2010) 'Present and Future Roles of Water and Food Trade in Achieving Food Security, Reducing Poverty and Water Use'. In: Luis Martínez-Cortina, L; Alberto, G and López-Gunn, E (eds.) *Re-thinking Water and Food Security: Fourth Botín Foundation Water Workshop*. London: CRC Press, pp. 109–43.

Salehyan, I (2014) 'Climate Change and Conflict: Making Sense of Disparate Findings'. *Political Geography* 43: pp. 1–5. DOI: 10.1016/j.polgeo.2014.10.004.

Schleussner, C-F; Donges, JF; Donner, RV and Schellnhuber, HJ (2016) 'Armed-Conflict Risks Enhanced by Climate-Related Disasters in Ethnically Fractionalized Countries'. *Proceedings of the National Academy of Sciences* 113(33): pp. 9216–21. DOI: 10.1073. pnas.1601611113.

Schwartz, P and Randall, D (2003) 'An Abrupt Climate Change Scenario and Its Implications for United States National Security'. Columbia University. Available at: http://eesc. columbia.edu/courses/v1003/readings/Pentagon.pdf (Accessed 16 June 2018).

Secretary of Defense (US) (2018) *Summary of the 2018 National Defense Strategy of the United States of America: Sharpening the American Military's Competitive Edge*. Department of Defense. Available at: www.defense.gov/Portals/1/Documents/pubs/2018-National-Defense-Strategy-Summary.pdf (Accessed 24 June 2018).

Selby, J; Dahi, O; Fröhlich, C and Hulme, M (2017) 'Climate Change and Syrian civil war revisited'. *Political Geography* 60: pp. 232–44. DOI: 10.1016/j.polgeo.2017.05.007.

Seter, H (2016) 'Connecting Climate Variability and Conflict: Implications For Empirical Testing'. *Political Geography* 53: pp. 1–9. DOI: 10.1016/j.polgeo.2016.01.002.

Shabecoff, P (1990) 'Senator Urges Military Resources Be Turned to Environmental Battle'. *New York Times* (29 June). Available at: www.nytimes.com/1990/06/29/us/senator-urges-military-resources-be-turned-to-environmental-battle.html (Accessed 18 November 2017).

Skaf, M and Mathbout, S (2010) 'Drought changes over last five decades in Syria'. In: López-Francos, A (ed.) *Economics of Drought and Drought Preparedness in a Climate Change Context*. Zaragoza, Spain: CIHEAM/FAO/ICARDA/GDAR/CEIGRAM/MARM, pp. 107–12.

Starr, S (2012) *Revolt in Syria: Eyewitness to the Uprising*. New York: Columbia University Press.

Townsend, M and Harris, P (2004) 'Now the Pentagon tells Bush: Climate Change Will Destroy Us'. *The Guardian* (21 February). Available at: www.theguardian.com/environment/2004/feb/22/usnews.theobserver (Accessed 21 November 2018).

Trigo, RM; Gouveia, CM and Barriopedro, D (2010) 'The Intense 2007–2009 Drought in the Fertile Crescent: Impacts and Associated Atmospheric Circulation'. *Agricultural and Forest Meteorology* 150: pp. 1245–57. DOI: 10.1016/j.agformet.2010.05.006.

Trofimov, Y (2010) 'Taliban Capitalize on Afghan Logging Ban'. *Wall Street Journal* (10 April). Available at: www.wsj.com/articles/SB10001424052702303960604575157683859247368 (Accessed 3 August 2015).

UN Environment Programme (UNEP) (2003) *Afghanistan – Post-Conflict Environmental Assessment*. Switzerland. UNEP. Available at: https://postconflict.unep.ch/publications/afghanistanpcajanuary2003.pdf (Accessed 16 June 2018).

USDA (US Department of Agriculture), Foreign Agricultural Service (2008) 'Commodity Intelligence Report, Syria: Wheat Production in 2008/09 Declines Owning to Season-Long Drought' *USDA* (9 May). Available at: https://ipad.fas.usda.gov/highlights/2008/05/Syria_may2008.htm (Accessed 21 December 2017).

USDA (US Department of Agriculture), Foreign Agricultural Service (2015) 'Commodity Intelligence Report, Syria: 2015–2016 Wheat Production Up from Last Year due to Favorable Precipitation'. *US Department of Agriculture* (23 July). Available at: https://ipad.fas.usda.gov/highlights/2015/07/Syria/Index.htm (Accessed 21 December 2017).

Van Dam, Nikolaos (2017) *Destroying A Nation: The Civil War In Syria*. London: I.B. Tauris & Co, Ltd.

Vidal, J (2011) 'Sudan – Battling the Twin Forces of Civil War and Climate Change'. *The Guardian* (21 November). Available at: www.theguardian.com/environment/2011/nov/21/sudan-civil-war-climate-change (Accessed 24 June 2018).

Von Uexkull, N (2014) 'Sustained drought, vulnerability and civil-conflict in Sub-Saharan Africa'. *Political Geography* 43: pp. 16–26. DOI: 10.1016/j.polgeo.2014.10.003.

The White House (2015) 'National Security Strategy'. *The White House* (February). Available at: http://nssarchive.us/wp-content/uploads/2015/02/2015.pdf (Accessed 30 June 2018).

The White House (2017) 'National Security Strategy of the United States of America'. The White House (December). Available at: www.whitehouse.gov/wp-content/uploads/2017/12/NSS-Final-12-18-2017-0905.pdf (Accessed 23 March 2017).

Wischnath, G and Buhaug, H (2014) 'Rice or Riots: On Food Production and Conflict Severity across India'. *Political Geography* 43: pp. 6–15. DOI: 10.1016/j.plgeo.2014.07.004.

World Bank Group (n.d.) 'Climate Change Portal: Syria'. World Bank.org. Available at: http://sdwebx.worldbank.org/climateportal/index.cfm?page=country_historical_climate&ThisCCode=SYR (Accessed 30 June 2018).

Zhang, T; Niinemets, Ü; Sheffield, J and Lichstein, J (2018) 'Shifts in Tree Functional Composition Amplify the Response of Biomass to Climate'. *Nature* 556: pp. 99–102. DOI: 10.1038/nature26152.

4
NATO STRATEGY, DOCTRINE AND GENDER

Strategy and doctrine have no universally agreed upon definitions, and each have different levels with their own respective areas of application. From the perspective of the US Joint Chiefs of Staff, 'Strategy is about how nations use the power available to them to exercise control over people, places, things, and events to achieve objectives in accordance with their national interests and policies' (Joint Chiefs of Staff, 2018: v). Doctrine is an organising tool that militaries use to implement strategies in their activities and operations in a consistent and effective fashion at the strategic, operational and tactical levels.

This chapter will begin a comparative case study approach of the ways in which NATO, the UK (Chapter 5), the US (Chapter 6) and Australia (Chapter 7) have worked gender considerations into their respective strategies and doctrine. Although Australia is not a NATO member, it works closely with NATO, the UK and the US in defence matters. Importantly, these three countries, along with Canada and New Zealand, are also part of the so-called 'Five Eyes Community', originally developed during World War II as an intelligence sharing arrangement amongst the English-speaking countries allied against the Axis Powers, and it has continued into the present as a broader collaborative defence platform (Tossini, 2017).

As an example of this, these countries coordinate with each other in their development of doctrine, and every year the Quinquepartite Combined Joint Warfare Conference (QCJWC) meets 'to identify commonalities and differences between the joint doctrine, lessons learned, and concept development program of work of the member nations' (QCJWC, 2015: 1). Further, the UK essentially adopts NATO doctrine as its own, making changes only when necessary to accommodate its national approach on certain issues (Ministry of Defence, 2014: iii). The effort to ensure coherence between UK and NATO doctrine appears to have taken on greater importance as the UK prepares to leave the EU (QCJWC, 2017: E4–3 n. 3). For its part, the US adopts NATO doctrine when engaged in NATO operations (Headquarters, Department of the Army, 2014: 2-4).

Because this high degree of doctrinal commonality is designed to increase interoperability between the forces of these nations, a comparative case study of their different treatments of UNSCR 1325's goals and principles is particularly useful from an operational perspective. The national case studies will examine their respective strategies with regard to gender, how their military doctrines link to NATO's, and whether those strategies are reflected in joint-level doctrine. Finally, each national case study will examine how strategy and joint doctrine related to UNSCR 1325 is captured in the lower-level land force doctrine that would most likely be used to implement the national strategies and joint-level doctrine in civilian-centric environments. Before exploring NATO's treatment of gender in its strategy and doctrine, however, a brief discussion of doctrine's substance and function is worthwhile.

The nature of doctrine and its purpose

Representing a global perspective on the meaning of the term, the UN Department of Peacekeeping Operations and the Department of Field Support (UN DKPO and DFS) (UN DPKO and DFS, 2008) define doctrine as the 'evolving body of institutional guidance that provides support and direction to personnel preparing for, planning and implementing . . . operations' (93). From the perspective of the New Zealand Defence Force (NZDF), a Five Eyes Community military that is meaningfully engaged in international security operations despite being relatively small (Young, 2017), doctrine is guidance 'formulated and based on our own national experiences of making strategy and conducting military operations' (Chief of Defence Force, 2012: iii). Whilst acknowledging doctrine as authoritative, the NZDF also notes that its doctrine 'conveys broad principles that require judgment in their application, according to situational imperatives. Doctrine is not mandatory dogma to be applied in all circumstances; that is simply not the NZDF's way of doing business' (Chief of Defence Force, 2012: iii).

The process of doctrine creation and even its fundamental premise (better doctrine and compliance with it will lead to more effective military operations) are not immune to criticism. Paparone (2017) expresses concern that doctrine writing moves too slowly to truly capture the value of lessons learned in operational theatres, and that the decisions as to what to validate and include in doctrine are top-driven rather than debated, refereed and based on critical analysis. Similarly, doctrinal changes that do occur often happen in an incremental fashion, and rarely challenge underlying assumptions in a revolutionary way (522–23).

Paparone (2017) also notes, however, that irrespective of its actual operational efficacy, doctrine takes on very important roles in the explanation of military actions to civilian political leaders. It becomes immensely important as a planning base for determining what human and materiel resources are needed to accomplish the tasks assigned to the military within allocated budgets (524–25). Further, it serves as a content base for military education curricula, training exercises and planning for contingencies (Belde, 2004: 30–31; Beth, 2004: 13–15; Izquierdo-Navarette,

2004: 22–27; Martin, 2004: 28–29). In this way, Posen (2016) notes that 'In modern democracies, doctrine also reassures society that the military is focused only on those tasks that civilians have specified, that it is not a foreign policy or domestic menace' (160).

Military doctrine generally nests within international and national security strategies approved by a nation's or a defence organisation's political leadership. Similarly, military doctrine itself is like a Russian *matryoshka* doll, containing within in it layers of subordinate doctrine that become more focused, more specific and often more prescriptive the deeper one goes. These subordinate layers generally reflect the different tasks applicable to different levels of operations, from the strategic level, through the operational level, and down to the tactical level where soldiers are working in small units in the field. In structure and in content, doctrine works to mitigate risk by reducing military uncertainty (Posen, 2016: 159–60, 171–72). In terms of practical application, doctrine works best when considered in the context of the strategies it is being used to implement.

NATO security strategy regarding gender

NATO-Euro Atlantic Partnership Council

NATO-EAPC joint policy

At the strategic level, NATO has been active in seeking to incorporate gender perspectives in its external diplomacy and in its operations and activities since UNSCR 1325's promulgation. In 2007, for example, in partnership with the NATO-Euro Atlantic Partnership Council (NATO-EAPC) (a group of 27 non-NATO countries from Asia, Europe, the Middle East and Oceania), NATO issued its first joint high-level strategy document on the implementation of UNSCR 1325, which it updated in 2011 and most recently again in 2014 (NATO-EAPC, 2014). Although the 2010 NATO security strategy included no mention of women or gender, the declaration issued at the end of the 2010 Lisbon Summit noted NATO's joint policy with the EAPC and endorsed a NATO action plan 'to mainstream the provisions of UNSCR 1325 into ... current and future crisis management and operational planning, into Alliance training and doctrine, and into all relevant aspects of the Alliance's tasks' (NATO, 2010: para. 7).

The joint policy recognises that for the necessary changes in mind-sets to occur among military personnel regarding the operational significance of gender, there must first be concrete steps at the national level by the partner states. The policy reaffirms that the NATO and partner nations have the primary responsibility for implementing the goals and principles of UNSCR 1325, and that the two products that national force contributors are expected to provide for missions are troops trained on gender issues and experts on gender issues (NATO-EAPC, 2014: para. 17). Nations are advised to build this capacity through including a gender perspective in their education and training programmes, and by 'developing institutions accessible and responsive to the needs of both men and women and . . . including the promotion of women's equal participation in national armed forces' (NATO-EAPC, 2014: para. 19). Finally, nations are encouraged to develop their own national action plans consistent with UNSCR 1325 to promote the capacity-building initiatives (NATO-EAPC, 2014: para. 20).

Given these requirements in these three specific focus areas, it is no surprise that NATO and the EAPC focused on human resource policies; education, training and exercises; and public diplomacy as the primary drivers of the changes they are seeking to bring about. For example, the policy requires that NATO and its EAPC partners 'give specific consideration to the recruitment and support of female leaders in both civilian and military structures in the defence and security area' (NATO/EAPC, 2014: para. 23). It takes concerted and systematic institutional effort to build leaders. It means having the right human resource policies in place to deliver the proper students to the educational and training institutions that will help grow them into successful leaders. If the desired product is a seasoned, senior woman leader, consistent public diplomacy supporting the implementation of UNSCR 1325 will encourage national decisions to create the conditions that will allow her to emerge at different operational levels at ever-increasing rank with its increased attendant responsibilities. An example of this sort of approach is the recent exchange programme bringing Jordanian (EAPC member) female military personnel to the US (NATO member) to train with the Colorado State National Guard as part of the larger continuing partnership between the two that began back in 2004 (O'Bryan, 2017).

NATO-EAPC joint action plan

In 2016, NATO and the EAPC released an updated joint action plan to support their revised policy for implementing UNSCR 1325. The 2016 Action Plan establishes two strategic outcomes to be achieved. The first is the reduction of barriers to 'the active and meaningful participation of women in NATO's, Allies' and partners' defence and security institutions, and within NATO-led operations, missions and crisis management' (NATO-EAPC, 2016: 1). The second is the integration of NATO-identified priorities under UNSCR 1325 into 'policies, activities and efforts under-taken by NATO, Allies and partners to prevent and resolve conflicts' (1). To accomplish these tasks, the action plan sets out 19 separate supporting outcomes nested within the areas of Institutional Policies and Structures; Human Resource Policies; Education, Training and Exercises; Public Diplomacy; Monitoring and Reporting; Cooperative Security; Crisis Management and NATO-led Operations and Missions; and Defence Planning and Policy (2–14).

Although there were no specific deadlines set out other than the overall two-year period for the action plan, it did establish very specific actions and indicators tied to these actions to help determine whether the efforts will have been successful. For example, under the area of Crisis Management and NATO-led Operations and Missions, Outcome 15 is stated as achieving 'Increased operational effectiveness by including Gender perspectives in policies, exercises, conflict analysis, planning, execution, assessment and evaluation' in these sorts of activities (NATO-EAPC, 2016: 11). Supporting this outcome are eight specific measures, including Action 15.2, which is defined as deploying 'trained, full-time Gender Advisors to operations' at all levels, and ensuring 'that Gender Advisors are positioned in Command Groups with clearly defined roles and responsibilities set in job description' (11). The metrics established for these actions are very clear and easily measured – the numbers of Gender Advisors (GENADs) that report directly to military leadership and the number of GENADs positioned in command groups whose responsibilities are set out in their job descriptions (11). Formulating the action plan in this manner will likely have a positive effect on holding headquarters and units accountable for their implementation efforts, and it could provide much of the substantive content for the NATO Secretary General's annual report on the action plan (NATO-EAPC, 2011: para. 5).

Bi-strategic command directives

In 2009, the year before the Lisbon Summit, NATO's two strategic-level headquarters, Allied Command Operations (ACO) located in Mons, Belgium and Allied Command Transformation (ACT) located in Norfolk, Virginia, had already issued its first military strategy addressing UNSCR 1325 – *Bi-Strategic Command Directive 40–1, Integrating UNSCR 1325 And Gender Perspectives In The NATO Command Structure Including Measures For Protection During Armed Conflict* (*Bi-SCD 40–1*) (ACO and ACT, 2009). The directive was very important, because the two commands issuing it are the four-star general-officer level commands that oversee NATO's operational components and its educational, training and research efforts, respectively. This document was significantly updated in 2012 *by Bi-SCD 40–1 Rev. 1, Integrating UNSCR 1325 and Gender Perspective Into The NATO Command Structure (Bi-SCD 40–1 Rev. 1)* (ACO and ACT, 2012), and the latest version, *Bi-SCD 040–001, Integrating UNSCR 1325 and Gender Perspective into the NATO Command Structure (Bi-SCD 040–001)* was issued in October 2017 (ACO and ACT, 2017). In many respects, the newest version is a marked improvement over the 2012 directive, and it demonstrates a significant level of organisational maturity in dealing with the operationalisation of gender in NATO. Some areas could still use further development, however, and it is difficult to tell at this point whether some of the changes, such as those in the Standards of Behaviour which will be discussed later in this chapter, will actually result in improved compliance with the directive by military personnel.

The aim of the 2017 directive is to ensure the effective implementation of UNSCR 1325 and related resolutions, the NATO-EAPC Policy and Action Plan on Women, Peace and Security, and the *Military Guidelines on the Prevention of, and Response to, Conflict-Related Sexual and Gender-Based Violence* (*Military Guidelines*) throughout the Alliance's core tasks of collective defence, cooperative security and crisis management at all levels of operations (ACO and ACT, 2017: 3). The directive's requirements are premised on the recognition that although 'men, women, boys and girls are components of a gendered system' influenced by armed conflict, 'women and girls are disproportionately affected and thus, have a unique perspective to share and solutions to offer' (3). Interestingly, the directive notes that the failure to address gender issues negatively impacts 'conflict prevention, conflict-resolution, post-conflict reconstruction and peace-building' (3), but it omits any impact on actual conflict itself. This is likely a troubling continuation of the earlier directives' focus on the human rights aspects of UNSCR 1325 to the exclusion of looking at gender and the use of force in kinetic operations through the lens of IHL (Prescott, 2017b: 208, 216).

The 2017 directive emphasises the need for the Alliance members to increase women's participation in the units made available to the Alliance for operations and exercises, as well as gender-related requirements such as GENADs and the composition of engagement teams and medical units (3–4). NATO members are also encouraged to incorporate gender perspectives into their operational 'planning at all levels to better understand the societal and structural processes, context and expected effects which influence how military operations and missions are conducted' (4). The directive recognises the counterproductive effect of 'any form of exploitation and harassment by NATO and NATO-led forces' on Alliance efforts and credibility, and establishes a complementary linkage between NATO forces complying with the NATO Standards of Behaviour and Code of Conduct, and responding to instances of 'Conflict-Related Sexual and Gender-Based Violence' (4) in accordance with the NATO *Military Guidelines* published in 2015 (NATO Secretary General, 2015).

Implementing a gender perspective

'Gender mainstreaming', defined as 'a strategy used to achieve gender equality by assessing the implications for women and men of any planned action, in all areas and at all levels, . . . to assure that the concerns and experiences of both sexes are taken into account' (ACO and ACT, 2017: 5), is the vehicle for integration of gender perspectives into Alliance activities and operations. Very usefully from an operational perspective, the directive also defines 'gender analysis' in a straight-forward fashion as 'the systematic gathering and examination of information on gender differences and on social relations between men and women . . . to identify and understand inequities based on gender' (5). The directive then provides an explanation of the definition, describing in practical terms what such analysis might look like and how it might be used (5–6). Similarly, 'exploitation' is helpfully defined by examples that could realistically occur in an operational environment, such as prostitution, rather than an abstract description.

Implementing gender perspectives into operations planning is accomplished through use of Allied Command Operations' 'Gender Functional Planning Guide' that was published in 2015 (ACO and ACT, 2017: 7), which will be discussed later in Chapter 9, 'Gender in military activities and operations', as an example of a well-developed planning measure that implements many of the goals of UNSCR

1325. Unfortunately, as with earlier versions, the 2017 directive comes close to recognising the significance of considering gender in the kinetic parts of kinetic operations, but ultimately fails to grapple with this important topic directly. For example, it states that 'Women's participation in conflict resolution has proven essential and thus, the differing gendered experiences need to be considered during operations planning and execution' (7). The footnote to this statement begins in promising fashion, noting that 'There is a commonly accepted misperception that gender perspective is deemed less important during kinetic operations' (7 n. 14). The next sentence is a bit of a disappointment though – 'kinetic operations often cause significant damage to the social and cultural fabric of a given society and thus, the integration of gender perspectives become ever more important in the stabilisation and reconstruction processes that follow' (7 n. 14). This explanation avoids the application of gender considerations to the kinetic actions themselves.

The section of the 2017 directive that discusses the importance of developing gender-related 'early warning indicators' to use in intelligence assessments of potential conflict areas of interest to the Alliance (ACO and ACT: 7) is positive, however. By making this a requirement, NATO begins to promote the building of an intelligence analyst base accustomed to searching for this sort of gender-disaggregated data and incorporating it into analysis used by planners. Over time, this could lead to an expectation on the part of commanders and staffs that this sort of information is both relevant and available. This in turn could drive the resourcing and use of data collection teams in-theatre specifically seeking this sort of information and specially constructed to obtain it, such as Female Engagement Teams (Prescott, 2013: 59–60), that will be discussed in greater detail in Chapter 9.

Importantly, this section also illustrates the areas of research that could contribute to developing this sort of understanding. These areas include collecting gender-disaggregated data in the dimensions of politics, military matters, socioeconomics, education, issues involving refugees and internally displaced persons, human rights violations and conflict-related sexual and gender-based violence. This data is then used to analyse the security situation in a sex, age and gender-disaggregated way, which serves as a bridge to engage with non-NATO organisations working on gender issues in coordination with the J-9 staff section (Civil-military cooperation, or 'CIMIC') (ACO and ACT, 2017: 8).

The 2017 directive also notes an area of research that is in keeping with the definition of gender mainstreaming as being concerned with all genders, but which is not well-developed in the directive itself – 'masculinities'. Specifically, this section describes the importance of understanding 'roles played by women in different parts of social groups in society; with specific focus on masculinities' (8). 'Masculinities' are not defined in the directive, and the term is only mentioned four times. Although Dowd (2010) shows that masculinity in general is not a new topic in the field of gender studies, and Chisholm and Tidy (2017) describe in some detail how the concept of 'military masculinities' in particular has developed as a result of scholarly work in this area (99), masculinities are not currently a common area for planning analysis in most militaries, nor are they recognised by many militaries as part of military culture in general. For example, the US Army promotes loyalty, duty, respect, selfless service, honour, integrity and personal courage as its official values. Introspection as to the qualities of the male-normative nature of the US Army or of its male soldiers as men is not set out as an element of the definitions of these terms (US Army, 2017).

Operational impact of gender

Building on these illustrations, however, the 2017 directive then describes functional impacts of conducting an operational gender analysis and its components, based on Allied Command Operations' 'Gender Field Planning Guidance'. This section notes how integrating a gender perspective into the analysis of the operational theatre can help identify information gaps regarding the prevention of and the response to conflict-related sexual and gender-based violence, and help refine the intelligence collection plan (ACO and ACT, 2017: 9). Its description of some of the components of a gender analysis is very useful. It notes that these could include the significance of conflict-related sexual and gender-based violence to human trafficking, the movements of refugees, smuggling, changes in gender relations and roles caused by conflict, steps NATO could take to promote 'gender equality and women's empowerment' during all phases of a conflict (9), and considering gender in internal planning, such as through force composition, logistics and 'Standards of Behaviour or Codes of Conduct' (10).

Importantly, the directive carries the incorporation of a gender perspective forward by requiring continuing monitoring, assessment and reporting on gender issues throughout all levels and phases of military operations (ACO and ACT, 2017: 10–12). The directive minimises the chance that gender-related data will become analysis orphans by encouraging that the reporting of this information become embedded in the reporting system, and not ordinarily treated as a stand-alone item. Further, it specifically notes that gathering this sort of information will require information exchange with non-NATO organisations (10), which is an implicit recognition that NATO intelligence sources alone will not likely be sufficient to provide the necessary information or analysis for an effective gender analysis.

This section of the directive is very practical. For example, it directs that sex and age-disaggregated data be used in all reporting requirements. Although mere establishment of a reporting content requirement will not by itself drive perfect collection of this data, the very fact that there is an information field established in a reporting format means that it is more likely to be completed with the appropriate information. The directive reinforces this by suggesting different ways to collect this information, such as through the use of Female Engagement Teams, supported by sufficient numbers of female interpreters and female military assessment personnel (12).

Standards of conduct

In terms of military discipline, the 2017 directive sets out what the Alliance expects as minimum standards of behaviour on the part of the personnel provided by the

NATO members, whilst acknowledging that the investigation and prosecution of disciplinary matters remain the province of the members themselves. The directive notes that these standards 'are not intended to replace or restrict national policies or NATO personnel regulations, but are provided to depict the standards of professionalism and the high expectations' the Alliance and its partners have as to the performance of NATO forces (ACO and ACT, 2017: 13). In the event NATO personnel are suspected of having breached the NATO Standards of Behaviour or Code of Conduct, these allegations will be reported to the J-1 staff section (Personnel), the LEGAD, the installation Provost Marshal, 'and when required, GENAD' (14).

If the relevant NATO commander deems an initial fact-finding inquiry appropriate, the inquiry will commence within 72 hours of the original notification of the allegations. Once complete, the results of any inquiry will be reported to Supreme Headquarters Allied Powers Europe (SHAPE) (the command element of Allied Command Operations). National investigations are given primacy, and if a NATO inquiry is initiated, the LEGAD is expected to advise the NATO commander on the coordination between the two. Coordination issues between investigations are handled by the LEGAD and the J-1, who are supported by the Provost Marshal. If required, the GENAD will also support the LEGAD and J-1. The NATO commander and national authorities are expected to address 'remedies and reparations available' to victims, and the final resolution of the cases will 'be reported in parallel, through the national and the NATO chains of command' (14).

This arrangement is similar to the incident handling procedure set out in the 2012 version of the directive, and it reflects a very realistic and practical recognition of the sensitivities of investigating and prosecuting indiscipline in the multinational setting. In the 2012 version, however, the GENAD was expected to fulfil a larger role regarding inquiries, supporting the commander, J-1 and LEGAD 'with any inquiry or investigation initiated by the Commander concerning a breach of NATO Standards of Behaviour, or an allegation of violence, rape, or other forms of sexual abuse' (ACO and ACT, 2012: A-2). This was criticised at the time as unrealistic given the GENAD's role and likely level of expertise (Prescott, 2013: 58). Further, under the status of forces agreements between NATO and the host nations that were receiving NATO forces, such as the agreement between the Interim Administration of Afghanistan (IAA) and ISAF (IAA and ISAF, 2002: Annex A), local authorities were unlikely to have criminal jurisdiction over the deploying soldiers. Thus, GENADs would have been venturing a bit far afield in an area ripe for political controversy were they to have sought to play any meaningful role in this area.

The Standards of Behaviour and Code of Conduct set out in the directive are consistent with the NATO Code of Conduct and the *Military Guidelines* (NATO Secretary General, 2015). The NATO Code of Conduct covers much more than just issues of military discipline. Rather than list a series of prohibited activities, it instead sets out five core values (integrity, impartiality, loyalty, accountability and professionalism) and then describes how characteristics of daily activities promote those values. For example, to foster loyalty, NATO personnel are expected to 'Contribute to the development and maintenance of a positive team spirit' (NATO,

2013: 2). The *Military Guidelines* define conflict-related sexual and gender-based violence as 'Any sexual and/or gender-based violence against an individual or group of individuals, used or commissioned in relation to a crisis or an armed conflict' (NATO Secretary General, 2015: 3), and direct military commanders to implement measures to prevent and respond to this sort of violence (3–5).

Consistent with the NATO Code of Conduct, the 2017 directive does not set out a specific list of prohibited activities. It instead notes the negative operational impacts of disrespect and unprofessionalism, and in the area of sexual relations, states that 'Sexual relationships when based on inherently unequal power dynamics are strongly discouraged and may not only undermine the credibility and integrity of the work of the Alliance' (ACO and ACT, 2017: 14). Also, commanders are strongly encouraged to 'foster an environment that prevents sexual exploitation of men, women, girls and boys – no matter their perceived age, consent or maturity' (15), but these are really the only specific references to prohibited sexual activities. The 2012 version of the directive was more direct in this regard. It explicitly stated that NATO personnel should not abuse detainees by 'any illegal act of unnecessary violence or threat', 'commit any act that could result in physical, sexual or psychological harm or suffering, especially related to women and children', and not 'abuse alcohol, use or traffic drugs' (ACO and ACT, 2012: B-2). Language in the 2009 version was even more explicit – acceptance or participation 'in activities that support human trafficking, including prostitution' was specifically prohibited (ACO and ACT, 2009: 2-2).

The language describing prohibited activities in the 2017 version is perhaps more inclusive in terms of potential victims because of its abstract nature, but it uses vocabulary that is often euphemistic or jargon-like, and is therefore not as likely to register fully with military personnel as clear guidance would be. On the other hand, perhaps this approach is actually more likely to be successful in pre-deployment training. Although NATO forces are not immune to the horrifying familiar depredations of forces involved in UN and regional organisation-sponsored missions against women and children, this has not been as significant an issue in NATO operations to date. There is a possibility that the more explicit approach might cause soldiers receiving instruction on preventing sexual and gender-based violence in pre-deployment training to ignore the instruction because they see themselves as more professional and better-disciplined in general than those other troops, and therefore might tend to discount the experiences of these forces as being inapplicable to them (Prescott, 2013: 58).

Education and training

The 2017 directive also addresses the maturing NATO educational and training system for gender matters within the Alliance. Importantly, it notes that 'To promote interoperability, this education and training is consistent' with UN Department of Peacekeeping Operations and EU standards and action plans (ACO and ACT, 2017: 16). Although Sweden is not a NATO member, its Nordic Centre for

Gender in Military Operations is designated as the NATO department head for gender education and training. Gender training itself is recognised as a NATO member responsibility, and it is to be included in pre-deployment, in-theatre and individual training for their personnel (16). The directive sets out clear minimum requirements for the implementation of gender training programmes, including 'Coaching and mentoring programmes to support competence development of all personnel', inclusion of gender perspective training in Alliance and national education centres' curricula, the development of a standard NATO training materials package, and the inclusion of gender perspective issues in NATO exercises and collective training (17). Further, those personnel appointed as GENADS and Gender Field Advisors are required to undergo training on the NATO implementation of UNSCR 1325 and related resolutions (18). These are all positive and concrete steps towards mainstreaming gender perspectives in NATO's activities and operations.

Gender advisors

The role of the GENAD is much better defined and more realistic in the 2017 directive than it was in the 2012 version, although there are still certain troubling deficiencies. For example, in the earlier directive, the standard operating procedure for a GENAD or Gender Field Advisor at any headquarters was expected to include responsibility for supporting and enabling 'local law, directives and commitments related to UNSCR 1325 and related resolutions, women and gender perspective' (ACO and ACT, 2012: A-2). Such authority might have been tenable if the NATO mission was essentially an occupation under IHL or operating under a UN mandate that provided similar authority, but it could prove awkward in a mission into an allied sovereign country, such as ISAF in Afghanistan. Further, as noted earlier, GENADs and Gender Field Advisors were tasked with supporting the 'Commander, J1 and LEGAD with any inquiry or investigation initiated by the Commander concerning a breach of NATO Standards of Behaviour, or an allegation of violence, rape, or other forms of sexual abuse' (A-2). It is difficult to conceive what such support might have looked like given the primacy of NATO members in handling allegations of troop misconduct. Perhaps the GENAD or Gender Field Advisor could serve as a conduit to relay allegations received to a headquarters provost marshal, but they would not likely be able to ethically or practically function both as a victim advocate and a command advisor.

Conversely, certain of the assigned gender staff responsibilities were too narrow in the 2012 version. For example, GENADs and Gender Field Advisors were also tasked with providing LEGADs with 'gender dimensions in the judicial system' and 'relevant information where women, girls and boys legal rights are neglected and/or violated' (ACO and ACT, 2017: A-3). This information could have been useful if the LEGAD was working on rule of law issues perhaps (Prescott, 2017a: 13), but it would have been of no value to the advice that the LEGAD provided to the commanders on the factors to be considered in their proportionality analyses before deciding to engage with armed force. Because armed conflict has a differentiated and disparate effect on women and girls as compared to men and boys, it is in precisely this application of IHL that a gender perspective is needed (Prescott, 2017b: 208, 216). In the end, regardless of their scope, the large number of significant staff responsibilities given to the GENAD to effect implementation of a gender perspective in the 2012 directive was unrealistic (Prescott, 2013: 58). This impracticality was further compounded by the fact that the GENAD was not required to be a direct report to the commander, but instead was supposed to have 'direct access' (ACO and ACT, 2012: A-1) – meaning that with great responsibility came little actual authority, either direct or derived.

A number of these issues have been resolved in the 2017 version. Now, the GENAD reports to the commander, 'and the GENAD office is organisationally placed' (ACO and ACT, 2017: 19) within the group of the commander's staff advisors, such as the Political Advisor and the LEGAD. The GENAD's tasks are clearer and more flexible, and include the provision of technical guidance and advice to the commander, the command group and staffs and to affiliated headquarters as necessary. The GENAD is required to maintain a working relationship with gender advisors in higher and subordinate headquarters, and to do the same with higher-level cross-functional groups in a headquarters, such as those involved in strategic planning or joint operations. The GENAD is also responsible for creating a gender analysis of the operational theatre to support planning efforts, and to collect data on crisis areas and conflicts. Finally, the GENAD is responsible for providing support for the education and training of headquarters personnel, including those who are designated as 'Gender Focal Points' (19).

Gender Focal Points had been identified in the 2012 version as 'dual-hatted' positions that supported the commander in gender perspective implementation, but appear to have been located down at the tactical level, focusing on things such as patrolling, providing humanitarian aid and assisting in developing procedures for searching local nationals (presumably women) (ACO and ACT, 2012: A-3). A very significant improvement in the 2017 version is the creation of Gender Focal Points within the different staff sections of the headquarters, which should both take some of the transformational load off of the GENAD's shoulders and co-locate responsibility for implementation of a gender perspective in the staff sections as well. Importantly, the GENAD still retains a degree of technical oversight on their activities, because the Gender Focal Points are required to maintain 'functional dialogue with the GENAD including providing support to the gender analysis, monitoring and reporting functions; reports within the chain of command' (ACO and ACT, 2017: 20).

For the most part, the illustrations of Gender Focal Point duties for the different staff sections make sense. For example, for the J-9 (CIMIC), the Gender Focal Point on that section's staff is expected to support efforts to engage with non-NATO organisations to promote implementation of UNSCR 1325 and related resolutions, to exchange information with these organisations, and 'engage with local women in conflict resolution and peace building processes' (ACO and ACT, 2017: B-2). However, certain of the illustrations show that NATO still has work to do to fit a gender perspective into kinetic operations.

For example, in the J-3 (Operations) staff section, that group's Gender Focal Point is appropriately expected to 'Support assessments regarding women's security situations and gender analysis, supporting the planning and execution of operations' (B-1). Unfortunately, the examples given of this are 'Info Ops, PSYOPS, patrols, and search operations' (B-1) – things with which a J-3 shop will often not be involved, particularly in a large deployment. In addition, at an operational level, a deployed NATO headquarters might find itself involved in dynamic targeting, but unless the Gender Focal Points are personnel from Five Eyes Community countries, they might not be allowed to participate in the actual dynamic targeting process because of security classifications regarding the handling of nationally provided intelligence and technology (Prescott, 2016a: 146).

Handling sex and gender-based violence

The 2009 version of the directive required operations to be analysed to determine which measures were available to protect against gender-based violence, particularly 'rape and other forms of sexual abuse and violence in situations of armed conflict' (ACO and ACT, 2009: 1–3), and then to implement these measures. For example, rules of engagement (ROE) might need to be developed to provide troops with the ability to not only use force for mission accomplishment, but also to use force to shield civilians, in particular women and children, from sexual violence and other serious crimes (Kjelgaard, 2011: 53). Under the NATO ROE, women and children could fall within the rule that allows the use of force to protect persons with designated special status (PDSS) (NATO Military Committee, 2003: A-1-1).

Conceivably, however, certain NATO members and partners might be unwilling to exercise the force authorised by such protective ROE, whether for political or capability reasons (House of Commons, 2006: 13), and this could lead to confusion among local civilians, international organisations and non-governmental organisations as to who would be protected by whom and when. Proper understandings of ROE are incredibly important in an operational setting, and are surprisingly complex in a multinational operation (Prescott, 2016b: 251–74). This GENAD responsibility was not carried forward in the 2017 version, which suggests that at the strategic level, NATO is continuing to take a more human-rights oriented approach to operationalising gender, and perhaps by omission, not engaging on kinetic force issues. However, as will be discussed in detail in Chapter 9, 'Gender in military activities and operations', lower-level NATO headquarters' standard operating procedures apparently still track the possible use of ROE for gender protective purposes.

The 2017 directive also provides a practical framework for assessing the significance of sexual and gender-based violence, and developing measures to counter it. It identifies this form of violence as a military threat, because of the negative impact it has on communities and groups involved, which can lead to a conflict lasting longer. It suggests analysing whether the violence is being used as a strategy by a side in the conflict, or is a result of a breakdown of public order, or perhaps 'A result of gender inequalities and/or masculinities' (ACO and ACT, 2017: A-1-1). The directive then provides a list of possible solutions, such as using armed patrols, expedited economic development projects or even 'Gender-Sensitive Camp Design and Management' to reduce the incidence of sexual and gender-based violence (A-1-1–A-1-2).

Although consideration of these measures could certainly be useful in the overall scheme of implementing UNSCR 1325's requirements, this focus might overshadow other aspects of armed conflict's differentiated impact on women and girls over which NATO forces would likely have more control. First, the exploration of 'masculinities' in light of the under-developed nature of the term in the directive is not likely to be fruitful in an operational sense, and it would therefore be just a distraction. Second, the Human Rights Unit of the UN Assistance Mission in Afghanistan (UNAMA HRU) reports that perhaps hundreds of Afghan women and girls have died or been injured from the use of air dropped munitions over the course of the lengthy conflict there (2017: 6). Those weapons dropped on the basis of some form of dynamic targeting will have used a targeting methodology that does not explicitly take account of the attacks' contribution to the disparate impact of armed conflict upon women and girls as part of any proportionality analysis (Prescott, 2016a, 144–45).

Further, based on recent statistics reported by the US military (Department of Defense, 2017: 8–9), what is worrisome from a violence perspective is not the externally directed threat posed by NATO actors using sexual and gender-based violence as a weapon of war. Rather, it is the unacceptable number of sexual assaults against female and male soldiers largely by their male comrades among US forces. As noted earlier, dealing with this internal threat is the responsibility of the nations contributing troops to the mission, but struggling to effectively police one's own forces likely undermines the legitimacy and credibility of efforts to promote change in other countries' units and troops.

NATO doctrine

Capstone doctrine

Because of the scope of its potential application in military operations involving many of the world's most technologically advanced and deployable military forces, it is worthwhile to analyse NATO doctrine and its treatment of women and gender in detail. NATO doctrine was originally published by the NATO Standardization Agency (NSA), which was reorganised as the NATO Standardization Office (NSO) in 2014 (NATO, 2017). NATO defines 'doctrine' as the 'fundamental principles by which military forces guide their actions in support of objectives. It is authoritative but requires judgement in application' (NSO, 2017c). *Allied Joint Publication (AJP) -01*, *Allied Joint Doctrine*, is the capstone doctrinal document for NATO joint operations, and is intended to provide commanders and their staffs 'with a common framework for understanding the approach to all Alliance operations' (NSO, 2017a:

IX). Because interoperability is the key to the Alliance forces working together effectively, doctrine serves as a 'common language' for NATO operations (NSO, 2017a: 1-2).

Below *AJP-01* are several important 'keystone' doctrines, including *AJP-3(B), Allied Joint Doctrine for the Conduct of Operations* (NSA, 2011); *AJP-5, Allied Joint Doctrine for Operational Level Planning* (NSA, 2013e); and *AJP-6, Allied Joint Doctrine for Communication and Information Systems* (NSO, 2017b). Nested within each of these doctrines are subordinate doctrines that deal with more specific aspects of each function area. For example, *AJP-3.22, Allied Joint Doctrine for Stability Policing* (NSO, 2016d), falls under *AJP-3(B), Allied Joint Doctrine for the Conduct of Operations*, as does *AJP-3.4.5, Allied Joint Doctrine for the Military Contribution to Stabilization and Reconstruction* (NSO, 2015b).

The capstone doctrine is a recent document, published in 2017. In the section that deals with NATO and international law, it includes a paragraph that notes that 'Gender inequalities are often exacerbated during periods of crisis and conflict and if not addressed, may continue after the end of the conflict' (NSO, 2017a: 1-7). Therefore, 'Gender perspectives must be considered during all stages of a NATO operation; men and women must participate equally to achieve a comprehensive and enduring resolution' (1-7). On this basis, NATO is committed to fully implementing UNSCR 1325 'across all three of its core tasks' (1-7) (collective defence, cooperative security and crisis management). Importantly, referencing the work of General Sir Rupert Smith, it also notes that in 'so-called war amongst the people', positive identification of adversaries is difficult, particularly in urban areas, and that this 'will heighten the risk of collateral damage, especially to vulnerable groups' such as women, children and refugees (2-15 n. 46).

Although women and gender are mentioned in *AJP-01* infrequently, the significance of these terms' inclusion should not be downplayed. First, this is NATO's capstone doctrine. If writers of subordinate doctrine wanted to pick up the gender thread in operations and amplify it, these statements would provide sufficient doctrinal authority for doing so. Second, the scope of these statements as to gender's broad applicability in operations and gender's linkages to the core Alliance tasks means that writers of subordinate doctrine should not perceive that gender's relevance is confined to only civilian-centric operations or the civilian-centric parts of kinetic operations. Third, the direct linkage of gender to collateral damage directly implicates the principle of proportionality in the use of armed force by commanders. If this thread were fully taken up by writers of operational and targeting doctrine, this concept could be the start of an important adjustment in the application of IHL.

Subordinate doctrine

Perhaps as a result of NATO's doctrine review and revision cycle, however, these gender-perspective threads cannot yet be traced through a number of the subordinate keystone doctrines. For example, *AJP-3 (B), Allied Doctrine for the Conduct of Operations*, was last revised in 2011. It contains no mention of women, gender or UNSCR 1325. Unfortunately, in a number of the doctrinal documents nested within *AJP-3 (B)* that have been revised fairly recently, several still contain little or no information with regard to the operational significance of women or gender. For instance, *AJP-3.2, Allied Joint Doctrine for Land Operations* (NSO, 2016a), makes no mention of women or gender. Similarly, in a sister doctrine, *AJP-3.22, Allied Joint Doctrine for Stability Policing* (NSO, 2016d), there is no mention of either. In contrast, however, *AJP-5, Allied Joint Doctrine for Operational Level Planning* (NSA, 2013e), was published three years earlier and it at least mentions the need to incorporate women's perspectives and gender mainstreaming into mission and problem analysis and the development of concepts of operations (1-7).

Kinetic operations

In terms of specific kinetic operations, *AJP-3.4.4, Allied Doctrine for Counter-Insurgency (COIN)* references UNSCR 1325 in noting that 'Employment of gender identical personnel might help address specific requirements of that specific populace segment' in assisting counter-insurgents understand the needs of the local population (NSO, 2016e: 3-6–3-7). In the section titled 'Civil-military cooperation' it states that in terms of helping establish good governance, 'Gender Advisory Teams could support local women to participate in the peace process, protect them from sexual violence, promoting women's inclusion in electoral systems and engaging their voices in legal and judicial procedures' (4-18). It also notes that NATO forces must coordinate with local law enforcement personnel in their training, including upon the gender aspects of their duties (4-23), but this is the extent of its treatment of women and gender.

AJP-3.9, Allied Joint Doctrine for Joint Targeting, makes no mention of women, girls, gender or UNSCR 1325 (NSO, 2016b). NATO's failure to address the operational significance of gender in targeting is no surprise really. The doctrinal supplement to the high-level *Standardization Agreement 2449, Training in the Law of Armed Conflict* (NSA, 2013c) basically repeats in great detail existing IHL in a lengthy series of slides (NSA, 2013b) that are likely to be of only modest training value. It neither mentions UNSCR 1325 nor does it reference the applicable bi-strategic command directive. Its explanation of the application of IHL makes no note of the differentiated impact of armed conflict on women and girls and how that might factor into a commander's proportionality analysis. In this regard, this document is not in keeping with NATO's commitment to promote gender mainstreaming through implementing the Comprehensive Approach in all of its activities and operations (Prescott, 2013: 58). As will be discussed in Chapter 8, 'International humanitarian law and gender', however, NATO is not alone in this regard.

Finally, at the low end of kinetic intensity, *AJP-3.16, Allied Joint Doctrine for Security Force Assistance*, does reference a number of points in which gender is to be incorporated into efforts to assist friendly nations develop their security forces. In noting the importance of the legitimacy of the mission to its overall success, it recognises that it 'comprises legal, cultural, historical, religious, social, gender, moral and political aspects' (NSO, 2016c: 1-5). Further, it also recognises that security force assistance activities 'should promote the rule of law, including applicable human rights laws and the gender perspective' (1-6). Also, that to be effective, *AJP-3.16* recognises that NATO forces need to understand the nature of the host nation forces being trained, the 'gender balance and the organizational structures and culture as well as women's and men's roles in society' (1-13–1-14, 3-13), the local culture from a gender perspective and what motivates local actors (1-14, 2-10). Although the treatment of women and gender is quite modest, it is at least at the same depth as that found in the counterinsurgency doctrine discussed earlier.

Civilian-centric operations

The archetypical civilian-centric doctrine is that covering the field of Civil-Military Cooperation, or 'CIMIC'. CIMIC is defined as 'The coordination and cooperation, in support of the mission, between the NATO Commander and civil actors, including national population and local authorities, as well as international, national and non-governmental organizations and agencies' (NSA, 2013a, 2-1). Unfortunately, this NATO doctrine still hews to a male-normative view of the populations with whom NATO soldiers will interact. Although it identifies women as a particularly at-risk part of the population in conflict or disaster situations (6-2), it only recommends that cultural and gender competence is a proficiency CIMIC personnel should have, and that humanitarian assistance must be provided without discrimination as to gender (5-2, 6-2, 6-3).

Similarly, doctrine focused on civilian-centric missions in the areas of humanitarian assistance and post-conflict reconstruction also shows signs of uneven progress. *AJP-3.4.3, Allied Joint Doctrine for the Military Contribution to Humanitarian Assistance*, includes *Bi-SC Directive 40–1* as a reference (NSO, 2015a: REF-1), but does not specifically mention the resolution in the text. It does note that women are a particularly at-risk portion of the population in conflict and disaster situations, and that in these situations there may be increased inequality, such as the marginalisation of women (1-1, 1-7). Further, in terms of medical planning, it only states that NATO-led forces should consider the composition of the affected population, 'such as fragile and weak infants, pregnant women' and the elderly (3-4).

AJP-3.4.2, Allied Joint Doctrine for Non-Combatant Evacuation Operations, notes that a medical consideration in the planning for these operations is the unique composition of the evacuee population, 'such as infants, pregnant women, and the elderly' (NSA, 2013d: 5-7, A-8, E-9). It also reminds planners that in terms of a search protocol for evacuees, they need to consider who is designated to search women (C-1, D-3). Lastly, in setting up a comfort station for evacuees prior to their boarding of evacuation transportation, it states the need to 'Provide sufficient protection for women and children, especially for those families where only mothers are present during the evacuation operation' (E-11). As a final example of largely ignoring women or gender in the non-kinetic area, *AJP-3.10.1, Allied Joint Doctrine for Psychological Operations*, only states that in terms of creating themes and messages to be used to address host nation audiences, they 'must be culturally and socially attuned', and 'account for gender concerns' (NSO, 2014a: 5-5). The sparseness of gender-related references in these fairly recent documents is troubling.

On the other hand, great progress in working with a gender perspective is shown in peace support and in stabilisation and reconstruction doctrine. *AJP-3.4.1, Allied Joint Doctrine for the Military Contribution to Peace Support*, includes an extensive treatment of gender, and refers the reader to *AJP-3.4.5, Allied Joint Doctrine for Military Support to Stabilization and Reconstruction* (NSO, 2014b: 3-12). For example, *AJP-3.4.1* specifically notes that the bi-strategic command directive 'is applicable to all international military headquarters or any organizations operating with NATO chains of command' (3-12, n. 38). It explains how women and children are particularly at risk in conflict situations, the significance of UNSCR 1325 and that although gender refers to both men and women, that the doctrine focuses primarily on women and children (3-12). It also describes in detail the impacts of sexual and gender-based violence against women and girls on them as individual victims and upon their families and societies, and how rape is often used as a weapon in armed conflict to destabilise communities as an aim of war (3-13). Further, *AJP-3.4.1* also notes that burden placed on women as they assume roles in conflict-afflicted societies that absent men had previously filled, and points out that their roles are often neglected in the peace negotiation process, which causes their experiences and needs to be overlooked (3-13).

The fullest treatment of women, peace and security in the operational family of doctrine is found in *AJP-3.4.5, Allied Joint Doctrine for the Military Contribution to Stabilization and Reconstruction*. It addresses the need to develop a comprehensive analysis of the operational environment including gender issues (NSO, 2015b: 3-2, 3-4). It contains a detailed annex that describes the role of women in peace and security with great thoroughness. Annex E introduces UNSCR 1325, and distils the resolution into the three core areas of action – increasing the participation in efforts to prevent and resolve conflict, protecting vulnerable population cohorts especially women and girls, and preventing sexual and gender-based violence.

In doing so, *AJP-3.4.5* provides important analytical context to help understand the negative impacts to societies of gender inequality and sexual and gender-based violence (E-1). It also points out that stability and reconstruction efforts may provide important opportunities to amend host nation governance documents and processes to provide for greater equality between men and women, and to find practicable solutions to women's security issues (E-2 – E-3). Finally, NATO forces are directed to adopt a 'more interventionist approach' in establishing measures to prevent the occurrence of sexual or gender-based violence, such as promoting investments in infrastructure such as forensic laboratories and the development of human capital to assist victims of this type of violence (E-3).

Summary

The very significant strategic treatment of operational gender issues shown by NATO through its joint policy and action plan with the EAPC and the different iterations of the bi-strategic command directive is reflected in the institutionalisation of gender in NATO's headquarters structure. This includes the establishment of the position of the special representative on gender matters, the positions of full-time GENADS throughout its headquarters structures and the work done by the gender office (Kennedy, 2016: 1059–65). NATO's strategic-level documents have been increasingly better developed with regard to gender over time, and they will likely prove to be very influential in driving NATO member states and partners to operationalise gender in their own militaries to enhance interoperability, and in shaping the mechanisms through which it will occur, such as through the use of GENADs. As to doctrine for non-kinetic operations, work in this area shows significant but uneven progress, as shown by the practically genderless and therefore male-normative 2013 CIMIC doctrine for example. Finally, NATO seems to have avoided incorporating gender considerations into kinetic doctrine in any meaningful way.

This might be due in part to the significant influence of the Nordic approach to incorporating gender in operations, which tends to view it from more of a peacekeeping perspective in important respects. In 2013, at NATO's behest, a review was conducted by the well-respected Swedish Defence Research Agency on the operationalisation of gender in NATO operations in Afghanistan and Kosovo. Significant improvements have occurred since the time of the survey, but at that time the analysts identified significant deficits in implementing even *Bi-SCD 40–1 Rev. 1*'s less challenging parts, such as developing awareness of the directive's requirements among headquarters staff members (Lackenbauer and Langlais, 2013: 67–68). Although the study is well written and well documented, it did not enquire into any operationalisation of gender from the kinetic perspective, and therefore made no findings in this regard. Regardless, on 23 October 2013, 'the Defence Ministerial Meeting at NATO Headquarters in Brussels, endorsed the review's implementation plan, which focused on training, gender advisors, planning and assessment tools, and reporting' (Department of Defence, n.d.). It is not realistic to expect an organisation to fix things that have not been identified as being broken – but if it does not look for them, it will not find them.

As noted earlier, however, one should not be too hasty in drawing conclusions about the way a military organisation is handling an issue merely on the basis of its doctrine, and Chapter 9 will describe some of the important initiatives that NATO has undertaken to incorporate UNSCR 1325 into its activities and operations. However, in light of NATO's reorientation towards the threat posed by a resurgent Russia on its eastern flank (Prescott, 2017a: 6–7, 18–19), kinetic operations doctrine and that non-kinetic operations doctrine which more directly supports it is more likely to get attention in NATO planning and exercises. The non-kinetic humanitarian assistance and stabilisation and reconstruction doctrine might therefore become marginalised to some degree, and this doctrine's important treatment of women and gender might thereby become less mainstreamed than expected.

On the other hand, the doctrinal thread set out in *AJP-01* identifying injuries to at-risk population cohorts such as women and girls as a form of collateral damage might find its way into revised operations and targeting doctrine over time. If this were to happen, it would represent a significant step forward in more fully realising the goals and principles of UNSCR 1325 in incorporating a gender perspective in all operations, including those that are kinetic. For this to happen, however, it would require a reassessment by the Alliance as to the relationship between gender and the portions of IHL related to the use of armed force, which will be set out in detail in Chapter 8, 'International humanitarian law and gender'. Before we get there though, we must first work our way through the national case studies on strategy, military doctrine and gender, starting with the UK.

References

Allied Command Operations and Allied Command Transformation (ACO and ACT) (2009) *Bi-Strategic Command Directive 40–1, Integrating UNSCR 1325 and Gender Perspectives in the NATO Command Structure Including Measures For Protection During Armed Conflict* (2 September). Mons and Norfolk, VA: NATO. Available at: www.nato.int/nato_static_fl2014/assets/pdf/pdf_2009_09/20090924_Bi-SC_DIRECTIVE_40-1.pdf (Accessed 24 June 2018).

Allied Command Operations and Allied Command Transformation (ACO and ACT) (2012) *Bi-Strategic Command Directive 40–1 Rev. 1, Integrating UNSCR 1325 And Gender Perspective Into the NATO Command Structure* (8 September). Mons and Norfolk, VA: NATO. Available at: www.nato.int/issues/women_nato/2012/20120808_NU_Bi-SCD_40-11.pdf (Accessed 24 June 2018).

Allied Command Operations and Allied Command Transformation (ACO and ACT) (2017) *Bi-Strategic Command Directive 040-001, Integrating UNSCR 1325 and Gender Perspective into the NATO Command Structure* (16 May). Mons and Norfolk, VA: NATO. Available at: www.nato.int/issues/women_nato/2017/Bi-SCD_40-1_2Rev.pdf (Accessed 24 June 2018).

Belde, H (2004) 'The Doctrine, Education and Training Synergy in Germany'. *Doctrine, General Military Review* 2 (March). Paris: C.D.E.S (French Army Doctrine and Higher Military Education Command), pp. 30–31 (Copy on file with author).

Beth, (2004) 'Role of Simulation in the Doctrine, Education and Training Synergy'. *Doctrine, General Military Review* 2 (March). Paris: C.D.E.S (French Army Doctrine and Higher Military Education Command), pp. 13–15 (Copy on file with author).

Chief of Defence Force (New Zealand) (2012) *New Zealand Defence Force Doctrine Publication – Doctrine*, 3rd Ed. (June). Wellington: New Zealand Defence Force. Available at: www.nzdf.mil.nz/downloads/pdf/public-docs/2012/nzddp_d_3rd_ed.pdf (Accessed 24 June 2018).

Chisholm, A and Tidy, J (2017) 'Beyond the Hegemonic in the Study of Militaries, Masculinities, and War'. *Critical Military Studies* 3(2): pp. 99–102. DOI: 10.1080/23337486.2017.1328182.

Department of Defence (Australia) (n.d.) 'Women, Peace and Security in Defence'. Department of Defence. Available at: www.defence.gov.au/Women/NAP/GenderPerspective.asp (Accessed 21 June 2018).

Department of Defense (US) (2017) *Annual Report on Sexual Assault in the Military, Fiscal Year 2016* (1 May). Washington, DC: Department of Defense. Available at: http://sapr.mil/public/docs/reports/FY16_Annual/FY16_SAPRO_Annual_Report.pdf (Accessed 21 June 2018).

Dowd, NE (2010) 'Asking the Man Question: Masculinities Analysis and Feminist Theory'. *Harvard Journal of Law & Gender* 33: pp. 415–30. Available at: www.law.harvard.edu/students/orgs/jlg/vol332/415-430.pdf (Accessed 25 June 2018).

Headquarters, Department of the Army (US) (2014) *Army Doctrine Publication 1–01, Doctrine Primer* (2 September). Washington, DC: Department of the Army. Available at: https://armypubs.army.mil/epubs/DR_pubs/DR_a/pdf/web/adp1_01.pdf (Accessed 24 June 2018).

Interim Administration of Afghanistan and International Security Assistance Force (ISAF) (2002) *Military Technical Agreement Between the International Security Assistance Force and the Interim Administration of Afghanistan* (January). Kabul, Afghanistan: ISAF. Available at: www.bits.de/public/documents/US_Terrorist_Attacks/MTA-AFGHFinal.pdf (Accessed 24 June 2018).

Izquierdo-Navarette, J (2004) 'Synergy Between Military Doctrine and Training in Spain'. *Doctrine, General Military Review* (March). Paris: CDES (French Army Doctrine and Higher Military Education Command: pp. 22–27 (Copy on file with author).

Joint Chiefs of Staff (US) (2018) *Joint Doctrine Note 1–18, Strategy* (25 April). Washington, DC: Department of Defense. Available at: www.jcs.mil/Portals/36/Documents/Doctrine/jdn_jg/jdn1_18.pdf?ver=2018-04-25-150439-540 (Accessed 21 June 2018).

Kennedy, K (2016) 'Gender Advisors In NATO: Should The U.S. Military Follow Suit?' *Military Law Review* 224: pp. 1052–72. Available at: www.jagcnet.army.mil/DOCLIBS/MILITARYLAWREVIEW.NSF/20a66345129fe3d885256e5b00571830/d8d774904fe66a42852580d400676503/$FILE/By%20LTC%20Keirsten%20H.%20Kennedy.pdf (Accessed 25 June 2018).

Kjelgaard, L (2011) 'Gender Dimension'. *The Three Swords* 20: pp. 50–54. Available at: www.jwc.nato.int/images/stories/threeswords/THREE_SWORDS_20.pdf (Accessed 24 June 2018).

Lackenbauer, H and Langlais, R (2013) *Review of the Practical Implications of UNSCR 1325 for the Conduct of NATO-led Operations and Missions*. Stockholm: Swedish Defence Research Agency (FOI). Available at: www.nato.int/nato_static/assets/pdf/pdf_2013_10/20131021_131023-UNSCR1325-review-final.pdf (Accessed 24 June 2018).

Martin, V (2004) 'Doctrine, Training and Military Education in the U.S. Army'. *Doctrine, General Military Review* 2 (March). Paris: C.D.E.S (French Army Doctrine and Higher Military Education Command), pp. 28–29 (Copy on file with author).

Ministry of Defence (UK) (2014) *Joint Doctrinal Publication (JDP) 0–01, UK Defence Doctrine* (November). Shrivenham, UK: Development, Concepts and Doctrine Centre. Available at: https://assets.publishing.service.gov.uk/government/uploads/system/uploads/attachment_data/file/389755/20141208-JDP_0_01_Ed_5_UK_Defence_Doctrine.pdf (Accessed 21 June 2018).

NATO (2010) *Active Engagement, Modern Defence: Strategic Concept for the Defence and Security of the Members of the North Atlantic Treaty Organization* (19 November) Brussels: NATO. Available at: www.nato.int/cps/ua/natohq/official_texts_68580.htm (Accessed 21 June 2018).

NATO (2010) 'Lisbon Summit Declaration'. NATO (20 November). Available at: www.nato.int/cps/en/natohq/official_texts_68828.htm (Accessed 27 December 2017).

NATO (2013) 'No. 1235–13, NATO Code of Conduct'. NATO (2 December). Available at: www.nato.int/structur/recruit/info-doc/code-of-conduct.pdf (Accessed 27 December 2017).

NATO (2017) 'NATO Standardization Office'. NATO (9 June). Available at: www.nato.int/cps/ua/natohq/topics_124879.htm (Accessed 21 June 2018).

NATO and Euro-Atlantic Partnership Council (NATO-EAPC) (2014) 'NATO/EAPC Policy for the Implementation of UNSCR 1325 on Women, Peace and Security and Related Resolutions'. NATO (1 April). Available at: www.nato.int/cps/en/natolive/official_texts_109830.htm (Accessed 27 December 2017).

NATO and Euro-Atlantic Partnership Council (2016) 'NATO/EAPC Action Plan for the Implementation of the NATO/EAPC Policy on Women Peace and Security'. NATO.

Available at: www.nato.int/nato_static_fl2014/assets/pdf/pdft_2016_07/160718-wps-action-plan.pdf (Accessed 27 December 2018).

NATO Military Committee (2003) *MC 362/1, NATO Rules of Engagement* (30 June). Brussels: NATO. Available at: www.act.nato.int/images/stories/budfin/rfp016046.pdf (pp. 327–429 of RFP-ACT-SACT-16-46) (Accessed 24 June 2018).

NATO Secretary General (2015) 'MCM-0009-2015, Military Guidelines on the Prevention of, and Response to, Conflict-Related Sexual and Gender-Based Violence' NATO (1 June). Available at: www.nato.int/issues/women_nato/2015/MCM-0009-2015_ENG_PDP.pdf (Accessed 3 August 2017).

NATO Standardization Agency (NSA) (2011) *AJP-3(B), Allied Joint Doctrine for the Conduct of Operations*, Ed. B, Ver. 1 (March). Brussels: NATO. Available at: https://assets.publishing.service.gov.uk/government/uploads/system/uploads/attachment_data/file/623172/doctrine_nato_conduct_op_ajp_3.pdf (Accessed 24 June 2018).

NATO Standardization Agency (2013a) *AJP-3.4.9, Allied Joint Doctrine for Civil-Military Cooperation*, Ed. A, Ver. 1 (February). Brussels: NATO. Available at: https://assets.publishing.service.gov.uk/government/uploads/system/uploads/attachment_data/file/625816/doctrine_nato_cimic_ajp_3_4_9.pdf (Accessed 24 June 2018).

NATO Standardization Agency (2013b) *NATO Standard, ATrainP-2, Training in the Law of Armed Conflict*, Ed. A, Ver. 1 (March). Brussels: NATO. Available at: file:///C:/Users/jmprescott/Downloads/ATrainP-2%20EDA%20V1%20E%20(3).pdf (Accessed 24 June 2018).

NATO Standardization Agency (2013c), *Standardization Agreement 2449, Training in the Law of Armed Conflict*, Ed. 2 (20 March). Brussels: NATO. Available at: file:///C:/Users/jmprescott/Downloads/2449efed02%20(1).pdf (Accessed 24 June 2018).

NATO Standardization Agency (2013d) *AJP-3.4.2, Allied Joint Doctrine for Non-Combatant Evacuation Operations*, Ed. A, Ver. 1 (May). Brussels: NATO. Available at: https://assets.publishing.service.gov.uk/government/uploads/system/uploads/attachment_data/file/625781/doctrine_nato_noncombatant_evacuation_ajp_3_4_2.pdf (Accessed 24 June 2018).

NATO Standardization Agency (2013e) *AJP-5, Allied Joint Doctrine for Operational-Level Planning* (June). Brussels: NATO. Available at: https://assets.publishing.service.gov.uk/government/uploads/system/uploads/attachment_data/file/393699/20141208-AJP_5_Operational_level_planning_with_UK_elements.pdf (Accessed 24 June 2018).

NATO Standardization Office (NSO) (2014a) *AJP-3.10.1, Allied Joint Doctrine for Psychological Operations*, Ed. A, Ver. 1 (September). Brussels: NATO. Available at: https://assets.publishing.service.gov.uk/government/uploads/system/uploads/attachment_data/file/450521/20150223-AJP_3_10_1_PSYOPS_with_UK_Green_pages.pdf (Accessed 24 June 2018).

NATO Standardization Office (2014b) *AJP-3.4.1, Allied Joint Doctrine for the Military Contribution to Peace Support*, Ed. A, Ver. 1 (December). Brussels: NATO. Available at: https://assets.publishing.service.gov.uk/government/uploads/system/uploads/attachment_data/file/624153/doctrine_nato_peace_support_ajp_3_4_1.pdf (Accessed 24 June 2018).

NATO Standardization Office (2015a) *AJP-3.4.3, Allied Joint Doctrine for the Military Contribution to Humanitarian Assistance*, Ed. A, Ver. 1 (October). Brussels: NATO. Available at: https://assets.publishing.service.gov.uk/government/uploads/system/uploads/attachment_data/file/625788/doctrine_nato_humanitarian_assistance_ajp_3_4_3.pdf (Accessed 24 June 2018).

NATO Standardization Office (2015b) *AJP-3.4.5, Allied Joint Doctrine for the Military Contribution to Stabilization and Reconstruction*, Ed. A, Ver. 1 (December). Brussels: NATO. Available at: https://assets.publishing.service.gov.uk/government/uploads/system/uploads/

attachment_data/file/625763/doctrine_nato_stabilization_reconstruction_ajp_3_4_5. pdf (Accessed 24 June 2018).

NATO Standardization Office (2016a) *AJP-3.2, Allied Joint Doctrine for Land Operations*, Ed. A, Ver. 1 (March). Brussels: NATO. Available at: https://assets.publishing.service.gov.uk/ government/uploads/system/uploads/attachment_data/file/624149/doctrine_nato_ land_ops_ajp_3_2.pdf (Accessed 24 June 2018).

NATO Standardization Office (2016b), *AJP-3.9, Allied Joint Doctrine for Joint Targeting*, Ed. A, Ver. 1 (April). Brussels: NATO. Available at: https://assets.publishing.service.gov.uk/gov ernment/uploads/system/uploads/attachment_data/file/628215/20160505-nato_tar geting_ajp_3_9.pdf (Accessed 24 June 2018).

NATO Standardization Office (2016c) *AJP-3.16, Allied Joint Doctrine for Security Force Assistance*, Ed. A, Ver. 1 (May). Brussels: NATO. Available at: https://assets.publishing.service. gov.uk/government/uploads/system/uploads/attachment_data/file/637583/doctrine_ nato_sfa_ajp_3_16.pdf (Accessed 24 June 2018).

NATO Standardization Office (2016d) *AJP-3.22, Allied Joint Doctrine for Stability Policing*, Ed. A, Ver. 1 (July). Brussels: NATO. Available at: https://assets.publishing.service.gov.uk/ government/uploads/system/uploads/attachment_data/file/628228/20160801-nato_ stab_pol_ajp_3_22_a_secured.pdf (Accessed 24 June 2018).

NATO Standardization Office (2016e) *AJP-3.4.4, Allied Joint Doctrine for Counter-Insurgency (COIN)*, Ed. A, Ver. 1 (July). Brussels: NATO. Available at: https://assets.publishing. service.gov.uk/government/uploads/system/uploads/attachment_data/file/625810/ doctrine_nato_coin_ajp_3_4_4.pdf (Accessed 24 June 2018).

NATO Standardization Office (2017a) *AJP-01, Allied Joint Doctrine*, Ed. E, Ver. 1 (February). Brussels: NATO. Available at: https://assets.publishing.service.gov.uk/government/ uploads/system/uploads/attachment_data/file/602225/doctrine_nato_allied_joint_ doctrine_ajp_01.pdf (Accessed 24 June 2018).

NATO Standardization Office (2017b) *AJP-6, Allied Joint Doctrine for Communication and Information Systems*, Ed. A, Ver. 1 (February). Brussels: NATO. Available at: https://assets. publishing.service.gov.uk/government/uploads/system/uploads/attachment_data/ file/602827/doctrine_nato_cis_ajp_6.pdf (Accessed 24 June 2018).

NATO Standardization Office (2017c) 'NATO Term, the Official NATO Terminology Database'. NATO Standardization Office (29 September). Available at: https://nso.nato. int/natoterm/Web.mvc (Accessed 3 December 2017).

O'Bryan, E (2017) 'Partners Colorado and Jordan Explore Military Women's Evolving Leadership Roles'. *US Army* (2 February). Available at: www.army.mil/article/181756/ partners_colorado_and_jordan_explore_military_womens_evolving_leadership_roles (Accessed 3 December 2017).

Paparone, C (2017) 'How We Fight: A Critical Exploration of US Military Doctrine'. *Organization* 24(4): pp. 516–33. DOI: 10.1177/1350508417693853.

Parliament. House of Commons, Defence Committee (2006) *The UK Deployment to Afghanistan: Fifth Report of Session 2005–06*. UK Parliament. Available at: https://publications. parliament.uk/pa/cm200506/cmselect/cmdfence/558/558.pdf (Accessed 21 June 2018).

Posen, BR (2016) 'Foreword: Military Doctrine and the Management of Uncertainty'. *Journal of Strategic Studies* 39(2): pp. 159–73. DOI: 10.1080/01402390.2015.1115042.

Prescott, JM (2013) 'NATO Gender Mainstreaming: A New Approach to War Amongst the People?' *The Royal United Service Institute Journal* 158: pp. 56–62. DOI: 10.1080/03071847.2013.847722.

Prescott, JM (2016a) 'NATO Gender Mainstreaming, LOAC, and Kinetic Operations'. *NATO Legal Gazette* 36: pp. 140–47. Available at: www.act.nato.int/images/stories/ media/doclibrary/legal_gazette_36special.pdf (Accessed 24 June 2018).

Prescott, JM (2016b) 'Tactical Implementation of Rules of Engagement in a Multinational Force Reality'. In Corn, G; VanLandingham, R and Reeves, S (eds.) *U.S. Military Operations: Law, Policy, and Practice*. Oxford: Oxford University Press, pp. 249–74.

Prescott, JM (2017a) 'Lawfare: Softwiring Resilience into the Network'. *The Three Swords* 32: pp. 6–20. Available at: www.jwc.nato.int/images/stories/_news_items_/2017/LAW-FARE_JWCThreeSwordsJuly17_ColPrescott.pdf (Accessed 24 June 2018).

Prescott, JM (2017b) 'The Law of Armed Conflict and the Operational Relevance of Gender: The Australian Defence Force's Implementation of the Australian National Action Plan'. In: Babie, P and Stephens, D (eds.) *Imagining Law: Essays in Conversation With Judith Gardam*. Adelaide: University of Adelaide Press, pp. 195–216.

QCJWC (Quinquepartite Combined Joint Warfare Conference) (2015) *Annual Report 2014–2015* (14 April). Joint Education and Doctrine Directorate. Available at: https://wss.apan.org/2411/QCJWC%20Document%20Library/QCJWC%202015%20(UK)/Annual%20Report%202014-15.pdf (Accessed 4 July 2018).

QCJWC (Quinquepartite Combined Joint Warfare Conference) (2016) *Annual Report 2015–2016* (14 March). Development, Concepts and Doctrine Centre. Available at: https://wss.apan.org/2411/QCJWC%20Document%20Library/QCJWC%202016%20(AS)/QCJWC_Annual_Report_Final_2015-16.pdf (Accessed 21 June 2018).

QCJWC (Quinquepartite Combined Joint Warfare Conference) (2017) *Annual Report 2016–2017* (27 April). Joint Doctrine Directorate. Available at: https://wss.apan.org/2411/QCJWC%20Document%20Library/QCJWC%202016-17%20Annual%20Report%20Final%2027%20Apr%2017.pdf (Accessed 21 June 2018).

Tossini, V (2017) 'The Five Eyes – The Intelligence Alliance of the Anglosphere'. *UK Defence Journal* [online] (14 November). Available at: https://ukdefencejournal.org.uk/the-five-eyes-the-intelligence-alliance-of-the-anglosphere/ (Accessed 20 July 2018).

UNAMA HRU (UN Assistance Mission in Afghanistan Human Rights Unit) (2017) *Afghanistan: Protection of Civilians in Armed Conflict, Midyear Report 2017* (July). UNAMA. Available at: https://unama.unmissions.org/sites/default/files/protection_of_civilians_in_armed_conflict_midyear_report_2017_july_2017.pdf (Accessed 21 June 2018).

UN DPKO and DFS (UN Department of Peacekeeping Operations and Department of Field Support) (2008) *United Nations Peacekeeping Operations – Principles and Guidelines*. New York: UN. Available at: www.un.org/en/peacekeeping/documents/capstone_eng.pdf (Accessed 24 June 2018).

US Army (2017) 'The Army Values'. US Army.mil. Available at: www.army.mil/values/ (Accessed 7 January 2018).

Young, A (2017) 'NZ Has Avoided United Nations Peacekeeping Missions Because of Safety Concerns: McCully'. *New Zealand Herald* (5 April). Available at: www.nzherald.co.nz/nz/news/article.cfm?c_id=1&objectid=11832306 (Accessed 3 December 2017).

5

UK STRATEGY, DOCTRINE AND GENDER

The UK presents a very useful starting point for the three national cases studies on the interplay between strategy, doctrine and gender. First, as discussed in Chapter 3, 'Climate change and armed conflict', the Ministry of Defence has developed forecasting documents that are neither strategy nor doctrine, but which sketch out perhaps as far as is reasonably possible the potential impacts of climate change on the international security environment at discrete time horizons, 2035 and 2045. This provides a basis for making planning assumptions that can be checked for accuracy as certain events or time milestones occur. Second, as will be described in detail in this chapter, there is a high degree of commonality between NATO and UK doctrine, and this provides a vetted platform of military thought and guidance accessible both to other countries and the public as a whole. Finally, the UK acted quickly after UNSCR 1325 was promulgated and the Foreign and Commonwealth Office, the Department for International Development and the Ministry of Defence (FCO, DFID and MOD) produced the first UK national action plan in 2006 (FCO, DFID and MOD, 2006). This has given the UK experiential depth in its continuing work to implement its successive national action plans, and this suggests that the UK's approach and implementation measures are well worth studying by nations seeking to create or improve their own national action plans.

National security strategy

The first national-level UK security strategy document that appears to have addressed the situation of women and girls in conflict situations, and UNSCR 1325's goals and principles implicitly, was the *Building Stability Overseas Strategy* published in 2011 (FCO, DFID and MOD, 2011). First, it noted that 'Conflict and violence have a particularly negative impact on women, children and young people' (7). It further stated that women needed to be included in political systems to reduce exclusion from decision-making and provide 'legitimacy to mediate conflict peacefully' (11, 12), that violence against women and children were seen as indicators of failed states, and that this violence could 'fuel grievances of the population' against governments of these states (14). The *Building Stability Overseas Strategy* recognised that 'Women have a central role in building stability', and described the UK efforts under the first UK national action plan 'to address violence against women and support women's role in building peace' (26). It noted the UK's use of Female Engagement Teams in Afghanistan and the UK's support of 'Afghan women's civil society organisations to strengthen the influence of Afghan women in public life and to enhance their protection from violence through support to legal reforms', and the UK's efforts in the UN, EU and NATO 'to drive international action in support of UN Security Council Resolutions relating to women and conflict' (26).

The UK's most recent national-level security strategy dates from 2015. The *National Security Strategy and Strategic Defence and Security Review* (*National Security Strategy*) mentions gender only once, in the sense that religious extremists often discriminate on the basis of gender in conducting their activities (Her Majesty's Government, 2015: 37). In terms of women and girls, however, it first notes that it is consistent with core British values such as democracy, the rule of law and equality of opportunity to seek the 'empowerment of women and girls' as one of 'the building blocks of successful societies' (10). The *National Security Strategy* also sets a goal to recruit at least 15 percent women into the UK armed forces by 2020, who would be able to serve in all military roles including combat (33). In terms of fostering international development, it notes that the UK development programme focuses assistance on 'improving peace, security and governance; equality of opportunity for girls and women; access to basic services for the poorest; and building resilience to crises and responding to disasters when they occur' (48).

In terms of strengthening the rules-based international order and its institutions, the *National Security Strategy* specifically addresses the goals set out for greater protection of women and girls under UNSCR 1325. It makes the 'full attainment of political, social and economic rights for women' a UK priority, and identifies it as 'central to greater peace and stability overseas' (Her Majesty's Government, 2015: 63). It specifically states that this priority will be taken into account in new military doctrine. Finally, the *National Security Strategy* states that the UK will reinforce its pioneering efforts to prevent sexual violence in conflict situations by tackling 'impunity for sexual violence crimes', securing 'widespread implementation of the International Protocol on the Documentation and Investigation of Sexual Violence in Conflict, and encouraging greater international support for survivors' (63).

Strategy and strategy-related documents issued since the *National Security Strategy*'s publication also address the topics of women, peace and security favourably, but without much detail. For example, the government's *First Annual Report* on the strategy noted that 'all UK troops deploying on an overseas mission receive training on WPS and Preventing Sexual Violence in Conflict', and that the UK had developed a pre-deployment training module for non-UK troops as well (Her Majesty's Government, 2016: 20). The latest UK strategy document, the *International Defence*

Engagement Strategy published by the Foreign and Commonwealth Office and the Ministry of Defence (FCO and MOD) (2017), identifies the UK's efforts on women, peace and security and on the Prevention of Sexual Violence Initiative as key to its international security engagement efforts (3). It notes the UK decision to expand its support to UN operations, 'notably Peacekeeping and the Women, Peace and Security agenda' (6), and that the UK is 'leading on the development of action plans such as those relating to Women, Peace and Security and preventing sexual violence in conflict' (15), but this is the extent of its treatment of gender issues.

UK national action plans

2014 UK national action plan

Although the treatment of UNSCR 1325's goals and principles in high-level UK strategy is not extensive, there has been very significant investment in the creation and assessment of the UK's different national action plans. The latest and fourth version of the UK national action plan was published in 2018. Because it retains many of the general features of the 2014 national action plan, it is worthwhile to first examine that earlier plan in some detail, as well as the continuing assessments that were being made of it prior to the delivery of the 2018 version.

The 2014 *National Action Plan on Women, Peace & Security 2014–2017* (*National Action Plan 2014–2017*) was based on three general principles. First, women's participation was required to make and build peace and to prevent conflict from breaking out. Second, it recognised that conflicts and emergencies result in women and girls suffering specific forms of violence. Third, as a result of this particularised suffering, it recognised 'women and girls have specific needs which need to be met' (FCO, DFID and MOD, 2014: 3). Importantly, because the 2014 plan focuses on work in conflict-affected states, the responsibility for executing its requirements belonged jointly with its authors, with the Stabilisation Unit providing cross-departmental support (8). The Stabilisation Unit is a cross-government, civil-military-police unit that supports integrated coordination of UK government activities in fragile and conflict-affected states (Stabilisation Unit, n.d.).

To meet these three general principles, the 2014 plan set out five key outcomes that were to be achieved: increased participation of women in peace processes and decision-making, the prevention of conflict and violence against women and girls, the greater protection of women's and girls' human rights, addressing women's and girls' needs in relief and recovery and building national capacity in focus states. Each of the five key outcomes was supplemented by a rationale for its use, the specific outputs expected as a result of government activities being conducted in support of the outcome and draft indicators that would provide some initial metrics to assess progress (FCO, DFID and MOD, 2014: 10–19).

Although it recognised the UK's continuing involvement with international organisations and partners on gender issues in general, the 2014 plan focused its efforts on six countries where the UK was already supporting the goals of

UNSCR 1325 – Afghanistan, Myanmar, Democratic Republic of the Congo, Libya, Somalia and Syria (FCO, DFID and MOD, 2014: 20). The 2014 national action plan was supplemented by a detailed Country-Level Implementation Plan (Implementation Plan), also jointly created by the Foreign and Commonwealth Office, the Department for International Development and the Ministry of Defence that same year. The Implementation Plan was organised by UK national action plan key outcome or 'pillar', and for each pillar it described an 'outcome' that was more detailed than established in the national action plan. The outputs were to be measured by 'high-level indicators', and to be driven by specific activities, similar to the scheme set forth in the national action plan. Each activity was then analysed using baseline (October 2014) and target (September 2017) 'indicators'. Each of the focus countries was then analysed within each of the pillars (FCO, DFID and MOD, 2014: 6–24).

UK personnel working with the Implementation Plan in their respective focus countries might have found the framework for its delivery somewhat complex. There were significant changes in nomenclature between the 2014 national action plan and the Implementation Plan, such as indicators becoming 'high-level indicators' and 'activities' turning into 'outputs', which does not seem to have been necessary and was not likely helpful (FCO, DFID and MOD, 2014: 3, 8). Further, the inconsistencies in the measurements of the baseline indicators and the target indicators, as shown with Somalia, where courses being taught to African Union personnel were linked to individual student outcomes that would have been difficult if not impossible to measure with rigour, were not likely helpful in quantifying and qualifying assessments (18).

Assessment of the 2014 national action plan

Prior to the publication of the 2018 national action plan, the UK undertook a very comprehensive assessment of its progress in meeting the goals of *National Action Plan 2014–2017* and potential paths forward to enhance its efforts in implementing the plan. The UK commissioned three reports by an external evaluator, the non-governmental organisation Social Development Direct (SDDirect), which specialises in expert social development assistance and research services in the areas of gender inclusion (SDDirect, n.d.). The first report, published in 2015, assessed the baseline of government activity related to women, peace and security that existed at the time the 2014 national action plan was promulgated (SDDirect, 2015: 5). The second, or 'midline', report was published in 2016, and it focused more on lessons learned since the baseline evaluation, and discussed global best practices to consider for the next version of the UK national action plan (SDDirect, 2016: 6). The third, or 'endline', report was published in 2017 just prior to the conclusion of the final work on the 2018 UK national action plan, and it evaluated progress on the 2014 plan and made concrete recommendations for the 2018 version (SDDirect, 2017: 5).

The first report found that although the different UK government agencies were actively engaged in efforts to promote the goals of UNSCR 1325, the national action plan itself was not actually leading the way. With its work plan being largely driven by the efforts already under way in the six focus countries, the 2014 national action plan tended to showcase these initiatives instead. The first report also found that efforts to categorise the different initiatives as being under one pillar or another tended to create 'artificial distinctions between programmes which typically have an impact across a range of WPS priorities' (SDDirect, 2015: 5). In sum, however, the first report found that 'The UK's efforts on WPS in the six focus countries studied are for the most part strategic, diverse, thoughtfully implemented and often innovative' (6).

SDDirect's second report on the 2014 plan noted the importance of gender issues having in the interim been included in the 2015 *National Security Strategy* (SDDirect, 2016: 6). It noted that there was a concern among the gender experts SDDirect had interviewed that 'an explicit focus on women (rather than gender) in the WPS agenda has led to a lack of nuanced gender analysis' and 'insufficient consideration of the role of men and boys' (21). The report recommended ten possible improvements to be made in the 2018 national action plan, such as adopting a thematic framework that was more broadly aligned to the four pillars of the 2014 plan, and extending the life of the national action plan from three to five years (7).

Another recommendation made by SDDirect in the midline report was to make a greater commitment to gender and conflict analysis in threat assessments and planning, and it specifically noted the UK's inclusion of gender in the use of the Joint Analysis for Conflict and Stability (JACS) tool in country assessments (SDDirect, 2016: 35). JACS was introduced by the earlier *Building Security Overseas Strategy*, and it is 'a tool to strengthen cross-government approaches to tackling overseas conflict and instability' (Stabilisation Unit, 2017: 1). JACS is generic and could conceivably be commissioned by any government agency including the Ministry of Defence, and it contemplates the use of military intelligence assets and reports in its methodology as well as gender inequality and social exclusion analyses conducted by the Department for International Development (Stabilisation Unit, 2017: 20). It is not clear that the JACS methodology actually expects gender analysis from defence sources, however.

SDDirect's final, or 'endline', report was published in the summer of 2017. In contrast to the midline report, it recommended retaining the basic structure and content of *National Action Plan 2014–2017*, finding that it served 'a useful purpose in providing shared cross-departmental vision on the UK's WPS agenda that is in keeping with the UK's commitments under UNSCR 1325' (SDDirect, 2017: 7). The endline report favourably noted the implementation of the Ministry of Defence's action plan, and the support and interest it had garnered among senior defence leaders (8). For example, the Ministry of Defence had previously designated the Vice Chief of Defence Staff as the department's 'gender champion' (FCO, DFID and MOD, 2016: 5), a fact that highlights the seriousness with which the Ministry of Defence takes the requirement to include gender considerations in its activities.

Similarly, the endline report noted significant changes in the Ministry of Defence's Single Departmental Plan (its business plan), that reflected tangible progress towards the 2014 national action plan's goals. For example, the departmental plan had moved from having no references to women, the UK national action plans or the goals of UNSCR 1325 in the 2012 version, to including 'ambitious targets' such as promoting the participation of women in peacekeeping roles in the 2015 plan (SDDirect, 2017: 25). The endline report also highlighted the concrete steps the Ministry of Defence was taking to mainstream gender in the UK armed forces, to increase the percentage of women in the armed forces, and in reviewing the training provided by Ministry of Defence assets to international partners that was related to women, peace and security (28).

2018 UK national action plan

The final internal report on the 2014 national action plan written by the Foreign and Commonwealth Office, the Department of International Development and the Ministry of Defence (FCO, DFID and MOD, 2017) noted SDDirect's largely favourable evaluation of the work that had been conducted to reach the plan's goals (16), and it assessed that 'Overall the UK has made good progress in delivering its 2014–2017 NAP commitments . . . except for those we have had to postpone due to political, security or operational constraints' (2). The report also noted that SDDirect's recommendations had informed the development of the 2018 national action plan, and that a number of its recommendations had been taken on board. As examples, the time period of the new plan was to be extended from three to five years, and the static implementation plans that the 2014 national action plan had featured were to be removed (15).

The report specifically highlighted contributions by the Ministry of Defence in meeting the 2014 plan's goals. It noted that the UK military had integrated a module on Preventing Sexual Violence in Conflict into the gender awareness training that was now part of any large-scale deployment's pre-deployment training, and that it had increased its GENAD capacity with more trained personnel and with as many as 50 Gender Focal Points designated throughout the armed forces. It also noted that the UK had begun planning to develop its 'own specialist military gender courses' (FCO, DFID and MOD, 2017: 14). Further, as shown by the example of the UK deployment to the UN mission in South Sudan, each troop rotation contained at least three Gender Focal Points 'to maintain our gender capability in that country' (11). The report also noted that 'Military doctrine continues to be updated to be made more gender sensitive on a rolling basis where applicable and appropriate' (11).

The *National Action Plan on Women, Peace & Security 2018–2022* (*National Action Plan 2018–2022*) sets out seven strategic outcomes that provide the vision for UK efforts in implementing the goals and principles of UNSCR 1325 in conflict situations. These outcomes are based on their relevance to the focus countries identified by the UK as its primary targets for its national action plan implementation, and pragmatically, on the UK's assessment of its ability to lead or contribute significantly to these efforts. Importantly, the 2018 plan increases the number of focus countries from the previous plan by three, and it now includes Afghanistan, Myanmar, Democratic Republic of Congo, Iraq, Libya, Nigeria, Somalia, South Sudan and Syria (FCO, DFID and MOD, 2018: 5).

The seven strategic outcomes are: greater participation by women in decision-making, increased efforts in peacekeeping, preventing gender-based violence, providing a humanitarian response, enhancing the delivery of security and justice, preventing and countering violent extremism, and the increased development of UK capabilities to deliver on its commitments regarding women, peace and security (FCO, DFID and MOD, 2018: 6). Of these outcomes, peacekeeping, preventing gender-based violence, humanitarian response, security and justice, and preventing and countering violent extremism appear most relevant for purposes of this chapter. Significantly, the UK is not requiring all of its specific country-level efforts to address each of the seven strategic outcomes, which addresses a criticism that had been raised about the implementation of the 2014 national action plan. For example, as to the UK's efforts to empower women in the opposition-held areas of Syria to address sexual violence, the 2018 plan assesses that these efforts will address decision-making (raising awareness of women's rights at the community level), security and justice (raising awareness of sexual and gender-based violence and connecting survivors to support services), and gender-based violence itself (through the awareness training) (9), rather than attempting to address each outcome.

Regarding peacekeeping under *National Action Plan 2018–2022*, the UK commits itself to supporting approaches which are survivor-centred, promote the taking of police, judicial and disciplinary actions against peacekeepers who engage in sexual exploitation and abuse, provide better training for deploying personnel (both civilian and military), and result in the deployment of greater numbers of women on peacekeeping missions (FCO, 2018: 10). For the strategic outcome of gender-based violence, its practical measures overlap strongly with those set out under peacekeeping for taking care of survivors (12). In conducting its humanitarian responses, the UK commits to identifying and meeting the needs of women and girls from the beginning of operations, in large part by including women in the design and delivery of these missions, and meeting the pertinent international standards for integrating gender-based violence interventions into the mission (14). As to security and justice, the UK's efforts are to be particularly geared to assisting female victims in both formal and informal processes, increasing recruiting, retention and promotion of women in security forces, and developing institutions that have the capability to identify gaps in gender equality in the provision of security (16).

The most ambitious strategic outcome is the prevention and countering of violent extremism. The 2018 national action plan identifies the different ways in which women and girls are impacted by violent extremism, including roles as participants, and it assesses that 'The roles of women and girls in relation to violent extremism have been far less visible than those of men and boys, and are often overlooked' (FCO, DFID and MOD, 2018: 18). It acknowledges that 'The evidence base in this area is emerging', and that 'More evidence is needed on how efforts to prevent or counter violent extremism can effectively integrate gender perspectives' (18).

Not surprisingly, it provides little specificity as to how such programmes might actually be implemented, other than noting that they 'should incorporate an increasing sensitivity to the rights and priorities of women and girls and engage with women on the design and delivery of strategies to prevent and counter violent extremism' (18). This is a bit vague, but given the state of current research and experience in this area, it is at least realistic.

To better assess its progress under the 2018 national action plan, the UK will both use external scrutiny in the form of an independent assessment and collaborative oversight with civil society actors, and it has adopted what is described as 'a more strategic approach to monitoring, evaluation and learning that maintains accountability to Parliament and streamlines reporting for implementing teams' (FCO, DFID and MOD, 2018: 24–25). This approach largely uses indicators of progress that have been employed by international organisations in assessing gender issues, and whilst they are certainly quantifiable, they are also modest in important respects. For example, there are three indicators for Strategic Outcome 2, 'Peacekeeping'. The first is the number of peacekeeping mandates that include language on women, peace and security; the second is the percentage of sexual exploitation and abuse cases reported against uniformed or civilian peacekeepers that are acted upon; and third, the percentage of UN peacekeeping troops who are female (25).

These are measures of performance, not measures of effectiveness, and the UK should be able to register as being successful in all of these indicator areas without significant effort on its part. Many UN Security Council resolutions have been passed since UNSCR 1325 was promulgated that further drive the international community's efforts to achieve greater protection of women and girls in situations involving armed conflict. Given the ordinarily civilian-centric nature of UN peacekeeping operations, it seems highly likely that all mandates will deal in some fashion with gender and operations. Because the UK is a permanent member of the UN Security Council and therefore has veto power, the only way a mandate would not deal with women, peace and security is if the UK allowed it to go through that way. This is obviously not the best metric for charting real progress.

By the same token, assessing progress in judicial and disciplinary accountability by the percentage of sexual exploitation and abuse cases 'acted upon' may not tell us very much. Realistically, one form of action in the military justice system of a rule-of-law abiding troop contributing nation will be that the allegations are not found to be substantiated (McAllister, 1998) or that a court will find that due process protections for the accused were not complied with in the prosecution of an alleged violator (*United States v. Riesbeck*, 2018). Such findings may be seen as positive indicators that the rights of the accused have been properly observed. Finally, the percentage of female personnel on peacekeeping missions is not by itself a meaningful metric. Unless these women are in leadership positions and occupational specialties that are outside the traditional medical and logistical roles military women have often played, or are made part of a gendered engagement team, it is difficult to assess that gender considerations in operations are actually being fully reflected in any given mission on this basis.

Joint doctrine in general

As noted in the previous chapter on NATO, the UK has a sophisticated doctrinal regime that relies significantly on NATO doctrine, which simplifies review of its treatment of gender. *Joint Doctrine Publication (JDP) 0–01, UK Defence Doctrine*, notes that UK armed forces will use 'NATO's operations, planning and execution doctrine for both national and Alliance operations, augmenting it with national text where necessary' (MOD, 2014: iii). In the UK doctrine hierarchy, *JDP 0–01* is the strategic-level capstone document (MOD, 2011: 3-2). Immediately underneath *JDP 0–01* is the operational-level Joint Keystone document *JDP 01, Campaigning*, and below it are Joint Supporting Publications that are categorised as 'environmental', such as *Army Doctrine Publication (ADP) Operations*; 'thematic', such as the former *JDP 3–40, Security and Stabilisation – the Military Contribution*; or 'functional', such as doctrine for the Joint Staff Sections J-1 (Personnel) through J-9 (CIMIC). Below these at the tactical level, for land operations for example, are British Army field manuals, such as *Army Field Manual (AFM) Volume 1, Part 10, Countering Insurgency*, and handbooks (MOD, 2011: 3-3).

Certain relevant and recent joint UK doctrine contains no mention of women or gender at all, such as *JDP 0–20, UK Land Power* (MOD, 2017a). Other doctrine occasionally mentions women, but not the goals of UNSCR 1325 regarding women, peace and security. For example, *JDP 3–51, Non-combatant Evacuation Operations*, notes certain evacuees such as pregnant women may have specific requirements. It also notes that military units might need to establish requirements for searching women, which might require male and female searchers (MOD, 2013b: 3-15, 4A-5). The now-withdrawn *JDP 3–40, Security and Stabilisation: The Military Contribution* (QCJWC, 2017: E4–2) appeared to have located responsibility for gender matters largely with the Department for International Development, noting for example that it was this department that was ordinarily responsible for Gender and Social Exclusion Analyses (Ministry of Defence, 2009: 100). This document was scheduled to be merged with the newer *JDP-05, Shaping a Stable World: the Military Contribution*, in 2018 (MOD, 2016a: iii), which will be discussed later.

In part because of its reliance on NATO doctrine, there are some noticeable gaps in other joint civilian-centric doctrine. For example, *JDP 3–90, Civil-Military Cooperation (CIMIC)* (Ministry of Defence, 2017b), is essentially the NATO *Allied Joint Publication (AJP) 3.4.9, Allied Doctrine for Civil-Military Cooperation* published by the NATO Standardization Agency (NSA) with a UK cover page. This is unfortunate, because as noted earlier, even though the NATO CIMIC doctrine explains the importance of working with international and host nation actors as part of NATO's Comprehensive Approach, it merely identifies women as a particularly at-risk part of the population in conflict or disaster situations, notes only that a recommended proficiency for CIMIC personnel is to have cultural and gender competence, and states that humanitarian assistance must be provided without discrimination as to gender (NSA, 2013: 1-5, 5-2, 6-2, 6-3). On the other hand, as noted in the last chapter, there has been significant progress in dealing with gender issues in other more recent NATO doctrine. For example, the NATO Standardization Office's (NSO's) *AJP-3.4.1, Allied Joint Doctrine for the Military Contribution to Peace Support* includes a very decent treatment of the relevant women, peace and security issues (NSO, 2014b). The UK adopted *AJP-3.4.1* as its peace operations doctrine in June 2017 (MOD, 2017c).

Joint disaster relief and stability doctrine

Other recently revised UK joint doctrine also shows significant progress in incorporating operational gender considerations, and has moved in a very tangible way forward from the general NATO doctrinal baseline in incorporating the goals of UNSCR 1325. This is shown particularly by *JDP 3–52, Disaster Relief Operations Overseas: The Military Contribution* (MOD, 2016b) and *JDP -05, Shaping a Stable World: the Military Contribution* (MOD, 2016a). The progress shown by these publications might be directly related to a document published by the Development, Concepts and Doctrine Centre in 2013, *Joint Doctrine Note (JDN) 4/13, Culture and Human Terrain*, and it is therefore useful to examine this document in some detail first. In the British doctrine hierarchy, joint doctrine notes occupy a unique position. They do not go through the full endorsement process that all other doctrinal documents do prior to being released, and they are intended to 'contain national doctrine that meets an urgent need, or ... promote debate on an emerging idea' (MOD, 2013a: 2-1).

JDN 4/13 (MOD, 2013a) explains the differences between sex and gender, and how different gender roles for men and women in different societies play out in different and important ways (1-9). In sum, 'Gender relations affect all aspects of society, including the ways in which men and women work, the right of inheritance or the way that land and resources are used' (1-10). In terms of human security, it recognises that men and women will experience armed conflict and reconstruction differently, and that 'Although not formally recognised as a security problem, sexual violence will affect the security environment' (1-10). It notes in particular that 'Rape and other acts of sexual-based violence (subjected to females and males) can occur where vulnerable individuals are working in isolation, for example, in camps for internally displaced persons and refugees' (1-10).

The doctrine note also states that 'Implementing UNSCRs is a command responsibility and most headquarters have gender advisors. It is, however, incumbent upon everyone to consider gender perspectives during the planning process' (MOD, 2013a: 1-10 n. 13). It helpfully describes a Stabilisation Unit, the interagency unit described earlier in this chapter which is composed of Department for International Development, Foreign and Commonwealth Office, and Ministry of Defence personnel, which can provide analysis on 'community-based approaches to conflict, monitoring and evaluation, and cultural context, religion and gender' (3-3). Finally, *JDN 4/13* also describes what a human terrain analysis looks like, and explains the subcategory of analysis of social exclusion and gender analysis (3B-2).

JDP 3–52, Disaster Relief Operations Overseas: The Military Contribution, and *JDP -05, Shaping a Stable World: the Military Contribution*, reflect this baseline of gender considerations in operational environments, and build on them in very specific ways. *JDP 3–52* notes that humanitarian aid must be distributed impartially, and that there should be no discrimination based on gender (MOD, 2016b: 5). In terms of gender-based violence, it notes that 'Historically, women and girls make up a higher proportion of the death tolls of natural disasters, and are often more vulnerable than men and boys in the aftermath' and that in these situations, 'gender-based violence, including sexual assault, domestic violence and transactional sex, often increases' (17). On this basis, *JDP 3–52* states that the UK's role in preventing sexual exploitation is prioritised in all types of operations, and that this includes ensuring that all personnel receive appropriate training (17).

This doctrine also reminds UK forces of the way gender roles and the experiences of the populations being served in disaster relief can impact the delivery of aid. 'When disaster relief is being provided by military actors', *JDP 3–52* notes,

> there can sometimes be additional difficulties for women and girls in interacting with the relief efforts. Cultural constraints and reputations of local and international soldiers can prevent women from feeling able to raise issues, seek help and participate fully in disaster relief programmes.
>
> *(MOD, 2016b: 17)*

Conversely, where

> there has been forced conscription, men and boys may also be more reluctant. As with many scenarios, it is important to understand the gender dynamics of the particular situation to provide the necessary assistance to all members of an affected population.
>
> *(17)*

Pragmatically, *JDP 3–52* encourages UK forces to not become paralysed through analysis – although needs assessments are useful in distinguishing between emergency and chronic needs, units are advised that these products are not necessary before acting to prevent gender-based violence and provide reproductive health support (52).

Published in 2016 like *JDP 3–52*, *JDP-05, Shaping a Stable World: the Military Contribution*, states that in terms of understanding stability and instability from a gender perspective, 'Conflict affects women, men, girls and boys differently. It is important to analyse the implications of conflict and stabilisation activities in each specific context' (MOD, 2016a: 42). In terms of contextual analysis, this document points out that in violent conflicts, 'Underlying causes will generally relate to issues such as, "social, political or economic exclusion based on ethnicity, religion or gender or unequal power relations between the centre and periphery" or cross-border influences' (48). For the most part, however, *JDP-05*'s gender analysis is focused on women.

It notes that 'females may be particularly vulnerable to abuse, gender specific corruption, exploitation and exclusion' (MOD, 2016a: 42), particularly in fragile and conflict-affected states, 'where: key infrastructure has been destroyed; existing (often limited) protective legislation and social networks are disrupted; and insecurity or societal tradition limits female participation in decision-making forums' (43). Building stability requires deployed units to 'understand the particular viewpoints, needs and challenges of women and girls' (43) and to take them into consideration wherever possible. Importantly, from the perspective of staff planners figuring out how gender perspectives factor into operations, *JDP-05* explains the UK's political commitment to the goals of UNSCR 1325 in the 2014 UK national action plan, and sets out the plans four pillars (participation, prevention, protection and addressing women's and girls' needs in relief and recovery). Finally, the document provides planners with the UK's political priorities on gender, noting that 'Our Government has placed particular emphasis on the *Preventing Sexual Violence Initiative* which seeks to eradicate gender-based violence within conflicts. The positive significance of women filling key roles throughout society should also be taken into consideration' (43).

Army doctrine

The gender perspective threads woven within *JDP 3–52* and *JDP-05* do not yet appear to continue down into British Army tactical doctrine, unfortunately. *Army Doctrine Publication Land Operations*, the capstone document for UK land forces, mentions women only in the sense of female Army personnel being part of the force (Land Warfare Development Centre, 2017: 3-12). The lower tactical level *Army Field Manual* consists of two volumes, and neither appears to be in the public domain in full. The one civilian-centric doctrine in the *Army Field Manual* that can be found online deals with counterinsurgency, and if it is indicative of the general treatment of gender issues at the tactical level in civilian-centric UK land force doctrine, then there is likely little about gender considerations included in doctrine at this level.

Army Field Manual Volume 1, Part 10, Countering Insurgency (AFM, Part 10), notes only that when conducting meetings with key local national leaders, thought should be given as to the gender of the person who was designated to conduct these meetings, because 'In some cultures and in certain situations, the selection of a female may produce sub-optimal results or even prove to be counter-productive' (Land Warfare Centre, 2009: 8-A-2). Regarding women, it notes that in analysing social structures, 'the role of women in a particular society' is a significant area to consider and understand (3-6), but it does not provide a sketch of how this might be done. In terms of threats, *AFM, Part 10* notes insurgents will 'employ women and children in intelligence gathering roles' (5-13), but it provides no context to help commanders and soldiers understand that such people have very different roles and needs in their respective societies than those of male insurgents.

Summary

At the national security strategy and national military strategy levels, the UK has both marked out a space for meeting the goals of UNSCR 1325 regarding women, peace and security and provided a significant level of political commitment to realising its aims in UK activities and operations abroad. The UK national action plan assessment and review process is deliberate and measured, and the inclusion of an external civil society organisation evaluator promotes transparency. At the joint doctrine level this strand is picked up particularly in the more recent publications dealing with civilian-centric operations, such as the stability and disaster relief documents. The Development, Concepts and Doctrine Centre has registered some concern that the pace of joint doctrine review and revision is not fast enough, and that in part because of the high operational tempo UK forces have found themselves in for a number of years, the lessons that should be learned from these operations are not being fed back into doctrine quickly enough (QCJWC, 2017: E4-1, E4-7–E4-8). If the Development, Concepts and Doctrine Centre is able to streamline the review and revision process, then gender perspectives might more quickly become embedded in joint doctrine as a whole. Realistically, 'streamlining' translates to assigning additional doctrine writers and reviewers to the task.

The absence of gender content in the older British Army doctrine is concerning, because making these non-traditional concepts familiar and useful to military personnel really needs to begin early in their careers when they are more focused on the tactical level. From the perspective of fully realising the goals of UNSCR 1325 regarding IHL, the current arc along which both UK joint-level doctrine and British Army doctrine are proceeding is marginalising the consideration of the women, peace and security concepts in the kinetic parts of kinetic operations. The disaster relief and stability operations documents focus more on the human rights aspects of gender considerations, and those areas where international human rights and IHL intersect, such as in dealing with sexual and gender-based violence. As the UK positions itself to deal with possible kinetic threats arising on NATO's eastern flank (BBC, 2017), there might be a tendency to refocus on kinetic-oriented doctrine without assessing how gender might play a role in more effectively working through operations that feature the use of lethal armed force.

The UK intended to co-lead a 2017–2018 project with the US that sought to develop how to identify, deter and respond to hybrid warfare threats with a view towards developing responsive doctrine in this area (QCJWC, 2017: E4-6). Although the results of this work might not be known for some time, this effort is both timely and important. Non-kinetic NATO and UK doctrine that would appear to be particularly pertinent to this sort of conflict, such as *AJP 3–10.1, Allied Joint Doctrine for Psychological Operations*, notes only that psychological operations must account for gender concerns in developing socially and culturally well-attuned messages and themes (NSO, 2014a: 5-5). There is nothing different in the UK version (MOD, 2015: 5-5). Hybrid conflicts, fought amongst the people and leveraged by cyber capabilities, would appear to be a promising field in which the UK and the US could develop doctrine that included gender considerations from the beginning, rather than fitting it in on an ad hoc basis or adding it later in some drawn-out review and revision process. To understand how this might occur, and the challenges and opportunities that such efforts might encounter, it is time to turn to the US treatment of gender considerations in its strategy and doctrine.

References

BBC Staff (2017) 'UK Troops in Estonia to Deter "Russian Aggression"'. *BBC News* (18 March). Available at: www.bbc.com/news/uk-39311670 (Accessed 27 December 2018).

Foreign and Commonwealth Office, Department for International Development and Ministry of Defence (FCO, DFID and MOD) (2006) 'UK national action plan to implement UNSCR 1325'. Foreign and Commonwealth Office. Available at: www.un.org/womenwatch/feature/wps/UK_action_plan_public_version_Sept_06.pdf (Accessed 3 July 2018).

Foreign and Commonwealth Office, Department for International Development and Ministry of Defence (2011) *Building Stability Overseas Strategy* (July). London: GOV.UK. Available at: www.gov.uk/government/uploads/system/uploads/attachment_data/file/27370/bsos_july2011.pdf (Accessed 21 June 2018).

Foreign and Commonwealth Office, Department for International Development and Ministry of Defence (2014) *National Action Plan on Women, Peace & Security 2014–2017*. London: GOV.UK. Available at: https://assets.publishing.service.gov.uk/government/uploads/system/uploads/attachment_data/file/319870/FCO643_NAP_Printing_final3.pdf (Accessed 21 June 2018).

Foreign and Commonwealth Office, Department for International Development and Ministry of Defence (2014) 'Country-Level Implementation Plan'. GOV.UK. Available at: http://actionplans.inclusivesecurity.org/wp-content/uploads/2015/01/UK-National-Action-Plan-on-Women-Peace-and-Security-2014-2017-Country-Level-Implementation-Plan-1.pdf (Accessed 27 December 2017).

Foreign and Commonwealth Office, Department for International Development and Ministry of Defence (2017) *UK National Action Plan on Women, Peace and Security 2014–2017, Report to Parliament, December 2017*. GOV.UK. Available at: https://assets.publishing.service.gov.uk/government/uploads/system/uploads/attachment_data/file/577409/Uk_National_Action_Plan_on_Women__Peace_and_Security_-_2016_report.pdf (Accessed 21 June 2018).

Foreign and Commonwealth Office and Ministry of Defence (FCO and MOD) (2017) *UK's International Defence Engagement Strategy*, London: GOV.UK. Available at: https://assets.publishing.service.gov.uk/government/uploads/system/uploads/attachment_data/file/596968/06032017_Def_Engag_Strat_2017DaSCREEN.pdf (Accessed 21 June 2018).

Foreign and Commonwealth Office, Department for International Development and Ministry of Defence (2018) *National Action Plan on Women, Peace & Security 2018–2022*. London: GOV.UK. Available at: www.gov.uk/government/publications/uk-national-action-plan-on-women-peace-and-security-2018-to-2022 (Accessed 21 June 2018).

Her Majesty's Government (2015) *National Security Strategy and Strategic Defence and Security Review 2015, A Secure and Prosperous United Kingdom*. London: GOV.UK (23 November). Available at: www.gov.uk/government/uploads/system/uploads/attachment_data/file/478933/52309_Cm_9161_NSS_SD_Review_web_only.pdf (Accessed 21 June 2018).

Her Majesty's Government (2016) *National Security Strategy and Strategic Defence and Security Review 2015, First Annual Report 2016*. GOV.UK (December). Available at: https://assets.publishing.service.gov.uk/government/uploads/system/uploads/attachment_data/

file/478933/52309_Cm_9161_NSS_SD_Review_web_only.pdf (Accessed 21 June 2018).
Land Warfare Centre (2009) *Army Field Manual Volume I, Part 10, Countering Insurgency* (October). Warminster, UK: British Army. Available at: http://news.bbc.co.uk/2/shared/bsp/hi/pdfs/16_11_09_army_manual.pdf (Accessed 21 June 2018).
Land Warfare Development Centre (2017) *Army Doctrine Publication AC 71940, Land Operations* (31 March). Warminster, UK: British Army. Available at: https://assets.publishing.service.gov.uk/government/uploads/system/uploads/attachment_data/file/605298/Army_Field_Manual__AFM__A5_Master_ADP_Interactive_Gov_Web.pdf (Accessed 25 June 2018).
McAllister, B (1998) 'McKinney Trial Shows Difference Between Military, Civilian Courts'. *The Washington Post* (15 March). Available at: www.washingtonpost.com/archive/politics/1998/03/15/mckinney-trial-shows-difference-between-military-civilian-courts/8b8ea4a2-5112-44c3-b37c-36b7c9a1a923/?noredirect=on&utm_term=.c38b1845b255 (Accessed 17 January 2018).
Ministry of Defence (2009) *Joint Doctrine Publication (JDP) 3–40, Security and Stabilisation: The Military Contribution* (November). Shrivenham, UK: Development, Concepts and Doctrine Centre. Available at: www.gov.uk/government/publications/security-and-stabilisation-the-military-contribution – 2 (Accessed 21 June 2018).
Ministry of Defence (2011) *Army Doctrine Publication, Army Doctrine Primer* (May). Shrivenham, UK: Concepts and Doctrine Centre (for Force Development and Training). Available at: https://assets.publishing.service.gov.uk/government/uploads/system/uploads/attachment_data/file/33693/20110519ADP_Army_Doctrine_Primerpdf.pdf (Accessed 21 June 2018).
Ministry of Defence (2013a) *Joint Doctrine Note 4/13, Culture and Human Terrain* (September). Shrivenham, UK: Development, Concepts and Doctrine Centre. Available at: https://assets.publishing.service.gov.uk/government/uploads/system/uploads/attachment_data/file/256043/20131008-_JDN_4_13_Culture-U.pdf (Accessed 21 June 2018).
Ministry of Defence (2013b) *JDP 3–51, Noncombatant Evacuation Operations*, 2nd ed (February). Shrivenham, UK: Development, Concepts and Doctrine Centre. Available at: https://assets.publishing.service.gov.uk/government/uploads/system/uploads/attachment_data/file/142584/20130301-jdp3_51_ed3_neo.pdf (Accessed 21 June 2018).
Ministry of Defence (2014) *JDP 0–01, UK Defence Doctrine*, 5th Ed (November). Shrivenham, UK: Development, Concepts and Doctrine Centre. Available at: https://assets.publishing.service.gov.uk/government/uploads/system/uploads/attachment_data/file/389755/20141208-JDP_0_01_Ed_5_UK_Defence_Doctrine.pdf (Accessed 21 June 2018).
Ministry of Defence (2015) *AJP-3.10.1, Allied Joint Doctrine for Psychological Operations*, Ed. B, Ver. 1 (With UK National Elements) (July). GOV.UK. Available at: https://assets.publishing.service.gov.uk/government/uploads/system/uploads/attachment_data/file/450521/20150223-AJP_3_10_1_PSYOPS_with_UK_Green_pages.pdf (Accessed 21 June 2018).
Ministry of Defence (2016a) *JDP-05, Shaping a Stable World: The Military Contribution* (March). Shrivenham, UK: Development, Concepts and Doctrine Centre. Available at: https://assets.publishing.service.gov.uk/government/uploads/system/uploads/attachment_data/file/516849/20160302-Stable_world_JDP_05.pdf (Accessed 21 June 2018).
Ministry of Defence (2016b) *JDP 3–52, Disaster Relief Operations Overseas: The Military Contribution*, Ed. 3 (November). Shrivenham, UK: Development, Concepts and Doctrine Centre. Available at: https://assets.publishing.service.gov.uk/government/uploads/system/

uploads/attachment_data/file/574033/doctrine_uk_dro_jdp_3_52.pdf (Accessed 21 June 2018).

Ministry of Defence (2017a) *JDP 0–20, UK Land Power* (June). Shrivenham, UK: Development, Concepts and Doctrine Centre. Available at: https://assets.publishing.service.gov.uk/government/uploads/system/uploads/attachment_data/file/619991/doctrine_uk_land_power_jdp_0_20.pdf (Accessed 21 June 2018).

Ministry of Defence (2017b) *JDP 3–90, Civil-Military Cooperation (CIMIC) (AJP-3.4.9, Allied Joint Doctrine For Civil-Military Cooperation)* (June). Shrivenham, UK: Development, Concepts and Doctrine Centre. Available at: https://assets.publishing.service.gov.uk/government/uploads/system/uploads/attachment_data/file/625816/doctrine_nato_cimic_ajp_3_4_9.pdf (Accessed 21 June 2018).

Ministry of Defence (2017c) *AJP-3.4.1, Allied Joint Doctrine for the Military Contribution to Peace Support* (June). Shrivenham, UK: Development, Concepts and Doctrine Centre. Available at: www.gov.uk/government/publications/allied-joint-doctrine-for-the-military-contribution-to-peace-support-ajp-341 (Accessed 21 June 2018).

NATO Standardization Agency (NSA) (2013) *AJP-3.4.9, Allied Joint Doctrine for Civil-Military Cooperation*, Ed. A, Ver. 1 (February). Brussels: NATO. Available at: https://assets.publishing.service.gov.uk/government/uploads/system/uploads/attachment_data/file/625816/doctrine_nato_cimic_ajp_3_4_9.pdf (Accessed 24 June 2018).

NATO Standardization Office (NSO) (2014a) *AJP-3.10.1, Allied Joint Doctrine for Psychological Operations*, Ed. A, Ver. 1 (September). Brussels: NATO. Available at: https://assets.publishing.service.gov.uk/government/uploads/system/uploads/attachment_data/file/450521/20150223-AJP_3_10_1_PSYOPS_with_UK_Green_pages.pdf (Accessed 24 June 2018).

NATO Standardization Office (2014b), *AJP-3.4.1, Allied Joint Doctrine for the Military Contribution to Peace Support*, Ed. A, Ver. 1 (December). Brussels: NATO. Available at: https://assets.publishing.service.gov.uk/government/uploads/system/uploads/attachment_data/file/624153/doctrine_nato_peace_support_ajp_3_4_1.pdf (Accessed 24 June 2018).

QCJWC (Quinquepartite Combined Joint Warfare Conference) (2017) *Annual Report 2016–2017*. Joint Doctrine Directorate. Available at: https://wss.apan.org/2411/QCJWC%20Document%20Library/QCJWC%202016-17%20Annual%20Report%20Final%2027%20Apr%2017.pdf (Accessed 21 June 2018).

SDDirect (Social Development Direct) (n.d.) 'About Us. Social Development Direct'. *Social* Development Direct.org. Available at: www.sddirect.org.uk/about/ (Accessed 17 December 2017).

SDDirect (Social Development Direct) (2015) 'Evaluation of the National Action Plan on Women, Peace and Security: Baseline Study'. Social Development Direct.org (August). Available at: www.gov.uk/government/uploads/system/uploads/attachment_data/file/488461/NAP_Baseline_FINAL_DRAFT_-_December_2015.pdf (Accessed 17 December 2017).

SDDirect (Social Development Direct) (2016) 'Midline Report: UK National Action Plan on Women, Peace and Security'. Social Development Direct.org (April). Available at: www.gov.uk/government/uploads/system/uploads/attachment_data/file/533592/External_evaluation_of_the_National_Action_Plan_on_women__peace_and_security_-_midline_evaluation_2016.pdf (Accessed 17 December 2017).

SDDirect (Social Development Direct) (2017) 'Endline Evaluation, The UK National Action Plan on Women, Peace and Security 2014–2017'. Social Development Direct.org (June).

Available at: www.gov.uk/government/uploads/system/uploads/attachment_data/file/631120/NAP_ENDLINE.pdf (Accessed 17 December 2017).

Stabilisation Unit (n.d.) 'About Us'. *GOV.UK*. Available at: www.gov.uk/government/organisations/stabilisation-unit/about (Accessed 21 June 2018).

Stabilisation Unit (2017) 'Joint Analysis of Conflict and Stability, Guidance Note'. GOV.UK (June). Available at: https://sclr.stabilisationunit.gov.uk/publications/programming-guidance/1232-jacsguidance/file (Accessed 21 June 2018).

United States v. Riesbeck (2018) *Military Justice Reporter* 77. US Court of Appeals for the Armed Forces: pp. 154–77.

6
US STRATEGY, DOCTRINE AND GENDER

A close ally of the UK and a fellow NATO member, the US and its approach to UNSCR 1325's goals regarding women, peace and security in its national-level security strategy, its national action plan and its military doctrine merit extensive examination, if for no other reason than the outsized role the US military plays in the international security environment. Unlike the UK, the US has stepped back from making the goals and principles of UNSCR 1325 an explicit part of its national security strategy under the Trump Administration. However, it is not obvious that even under the prior Obama Administration that the US had really embraced the full scope of UNSCR 1325 in its highest level security policy to the extent that it would have been a powerful driver of change in the way the US defence establishment approached the operationalisation of gender.

On the other hand, President Trump signed the *Women, Peace, and Security Act of 2017* (*WPS Act, 2017*), which requires US government agencies, including the US Department of Defense (DoD), to undertake efforts to embed these goals and the mechanisms into their activities and operations. As will be discussed in this chapter, US security strategy and military doctrine is generally lacking in this regard, although there has been some progress at the doctrinal level. Whether the *Women, Peace and Security Act of 2017* will serve as an external driver of continued and significant change in US military strategy and doctrine with regard to gender remains an open question at this point in time. It would be disappointing if compliance with the law results in the US military merely viewing its requirements as blocks to be checked on some annual report form.

National security strategy

The 2015 *National Security Strategy* published by the Obama Administration stated that the US would prioritise women and youth in 'eliminating extreme poverty and promoting sustainable development' (The White House, 2015: Foreword, 9–11), and that the US would continue efforts to promote governance in Afghanistan to increase opportunities for Afghan women and girls (10, 18, 20). The White House also noted that the US would continue to work 'to ensure women serve as mediators of conflict and in peacebuilding efforts, and they are protected from gender-based violence' (11). From a perspective of values, the 2015 *National Security Strategy* noted that the US 'will support women, youth, civil society, journalists, and entrepreneurs as drivers of change' (19). Finally, it stated a US commitment 'to reducing the scourge of violence against women around the globe by providing support for affected populations and enhancing efforts to improve judicial systems so perpetrators are held accountable' (20).

The Trump Administration took a different approach in the 2017 *National Security Strategy of the United States of America (2017 National Security Strategy)*. As previously noted in Chapter 3, 'Climate change and armed conflict', the only references to 'climate' it contains relate to national and international business climates, and recognition that 'Climate policies will continue to shape the global energy system' (The White House, 2017: 20–22). Women fare better in the new strategy than does the global climate, but only in a more general sense. The *2017 National Security Strategy* speaks of empowering women and youth, and that the US 'will support efforts to advance women's equality, [and] protect the rights of women and girls' (38, 42). It also states that 'governments that fail to treat women equally do not allow their societies to reach their potential' (41). Gone, however, are the specific markers of national political intent to reduce violence against women and to include them in conflict resolution processes.

Located just below the overarching national security strategy in the US security policy hierarchy, it is important to note that the Department of State published the updated US strategy to deal with gender-based violence across the world in 2016, as required by US law (Department of State, 2016: iv). Although there are references to the Department of Defense in the document, these tend to relate to interdepartmental work rather than actual initiatives for which the department has responsibilities in this regard (12, 14, 15, 20, 30). The 2015 *National Military Strategy of the United States of America, 2015 – The United States Military's Contribution to National Security* issued by the Chairman of the Joint Chiefs of Staff (Chairman, JCS) mentioned women only in the sense of US female military personnel having greater opportunities in the force because of full integration (Chairman, JCS, 2015: 14), and considerations of gender were not mentioned at all. Similarly, although the 2014 *Quadrennial Defense Review* did mention gender and women, it did so only in the sense of women serving in US forces rather than any incorporation of women, peace and security issues (DoD, 2014: XII, 43–45, 48, 61, 63). The unclassified summary of the 2018 *National Defense Strategy of the United States of America (National Defense Strategy)* makes no mention of women, gender, UNSCR 1325 or climate change (DoD, 2018).

From one perspective, this lack of discussion of the goals and principles of UNSCR 1325 with regard to women, peace and security is arguably not determinative of how the US will operationalise gender. These documents are at such a high policy level they are likely divorced from the types of guidance documents that would be consulted by military planners and operators at the operational level and below. However, these documents likely have great weight in making resource allocation decisions within the Department of Defense. Within the Department of Defense, this silence on gender also suggests that there would be little political will for including these sorts of factors into doctrine if they cannot be linked back to strategic requirements, and therefore the service components of the department, such as the US Army, might not be inclined to assign gender factors the priority they deserve.

US national action plan: Department of Defense tasks and efforts

The first US national action plan was issued in 2011 (The White House, 2011), and an updated version was issued in 2016 (The White House, 2016). Similar to the UK national action plans, the US documents assigned different tasks to different government agencies in implementing their requirements, including the military. Before examining the 2016 version of the US plan, it is useful to first assess what progress had been made in the five years prior to its publication under the first US national action plan.

Assessment of Department of Defense efforts under the 2011 NAP

Conveniently, the 2016 *United States National Action Plan on Women, Peace, and Security* summarises the results of the Department of Defense's implementation of the 2011 national action plan. Foremost, these results include 'the integration of WPS objectives into key policy, strategy, and planning documents to include the National Security Strategy, the Guidance for the Employment of the Force, and the Joint Service Capabilities Plan' (The White House, 2016: 5). It also highlights the work done by the different US Combatant Commands, such as US Africa Command, based in Stuttgart, Germany, in incorporating gender considerations in theatre campaign plans, updated staff instructions and command orientation briefings (5).

In addition, the 2016 plan identifies the important work being done by the Department of Defense's five regional education and education centres, such as the Daniel K. Inouye Asia-Pacific Center for Security Studies (DKI APCSS) in Hawaii. It also notes that the 'US Marine Corps' Female Engagement Teams and the Army's Cultural Support Teams are providing new avenues for female Marines and soldiers to support ongoing operations and engage women in local populations' (The White House, 2016: 5). This short list of accomplishments does not reflect all the important work being done within the Department of Defense that supports the integration of UNSCR 1325's goals into military operations, but since these were set forth as examples, it is important to look at each of these items and see what was actually accomplished.

First, regarding the 2015 *National Security Strategy*, as previously noted it did include some modest (if somewhat vague) guideposts on incorporating considerations of women, peace and security into US government activities, but the *2017 National Security Strategy* now makes no meaningful mention of them at all. Below the national security strategy, the first documents in the US strategy hierarchy that are military similarly make no mention of these activities. Given this awkward disconnect, the next question then is whether these high-level lines of thinking are translated into the other two documents mentioned in the 2016 plan's assessment, the *Guidance for the Employment of the Force* and the *Joint Service Capability Plan*.

The *Guidance for the Employment of the Force* is used by the Chairman of the US Joint Chiefs of Staff to develop the *Joint Strategic Capabilities Plan*, which is an implementation plan for the guidance document, but unfortunately both are classified secret (US Army War College, n.d.) and thus are unavailable for review. Whether the break in conceptual continuity regarding gender shown between the 2011 national action plan and the 2014 *Quadrennial Review* and 2015 *National Military Strategy* has continued in these important documents is unknown. Regardless, the lack of commitments to operationalising gender requirements in the *2017 National Security Strategy* is unlikely to promote efforts to bridge any gap that exists. This is not really a good example to tout as demonstrating meaningful progress.

Second, as to the work being done by the US Department of Defense regional centres, the Daniel K. Inouye Asia-Pacific Center for Security Studies (DKI APCSS) in particular has made women, peace and security considerations a key part of its work in engaging with and educating civilian and military officials throughout the Indian and Pacific Ocean areas (Yamin, 2018). For example, between 2011 and 2015, it increased participation by women in its programs from 12.5 to 21.6 percent (DKI APCSS, 2015: 5). In its Advanced Security Cooperation course in late 2016, 36 percent of the students were women (DKI APCSS, 2016: 21). From the perspective of practical scholarship, the centre also provides a publication platform that a number of its alumni have used to publish papers on the role of women and gender in security issues (DKI APCSS, 2015: 6–7). The centre has also brought its work to the field, for example, co-hosting a workshop in Mongolia in 2016 for 55 participants from ten different countries to discuss impediments to the inclusion of women, peace and security topics into security efforts (DKI APCSS, 2016: 18–19). This is a solid example of the Department of Defense fulfilling its assignments under the 2011 plan effectively.

Third, Female Engagement Teams were widely used by US and other ISAF ground forces in Afghanistan between 2009 and 2012, a period of intense combat activity prior to the full turnover of security responsibilities to Afghan forces in 2014. As the pace for combat operations dropped as international troops withdrew, the operational requirement for Female Engagement Teams in theatre decreased, and the US Marine Corps, for example, disbanded its last Female Engagement Teams in 2012 (Seck, 2015). Although the US Army essentially appears to have let its Female Engagement Team programme fade away (Prescott, 2014: 296), the US Marine Corps later recognised a continuing need for these small units, and revived its programme in 2015 (Seck, 2015). As will be discussed later in Chapter 9, 'Gender in military activities and operations', when well trained and resourced, these teams provide an important capability to interact with host nation women particularly in conservative societies that restrict contact between non-family members of different sexes. By themselves, however, unless they are supported by doctrine, education and training that familiarise commanders and planners with their roles, they represent at best a niche and ad hoc capability. These steps towards institutionalisation have largely not occurred, and the use of Female Engagement Teams is therefore not the best example of task completion under the 2011 national action plan.

On the whole, the official assessment of the Department of Defense's implementation of its tasks under the 2011 plan shows definite progress in the incorporation of women, peace and security issues into the Department of Defense's activities and operations in certain specific areas. When viewed against the vast spectrum of the Department of Defense's everyday work, it is not clear that this amount of progress is commensurate with the five years the department had to work on its tasks. This question will be examined later in this chapter when current US military doctrine is assessed in light of the time that now has elapsed since the 2016 plan.

The 2016 US national action plan

Under the 2016 *United States National Action Plan on Women, Peace, and Security*, using the sometimes critiqued 'gender mainstreaming' model (Charlesworth, 2005: 11–13), the US assigns the Department of Defense particular tasks to execute in furtherance of its five overarching goals, both internally and in its activities with other US government departments, international organisations and international military partners. These goals are: first, achieving national integration and institutionalisation of a 'gender responsive approach' in US government activities 'through interagency coordination, policy development, enhanced professional training and education, and evaluation'; second, increasing women's participation and decision-making; third, protecting women and children from abuse and sex and gender-based violence 'in conflict-affected environments'; fourth, conflict prevention; and finally, providing access to relief and recovery (The White House, 2016: 18).

With regard to the first objective, national integration and institutionalisation, the Department of Defense is tasked with incorporating the US national action plan objectives into 'appropriate DoD strategic guidance and planning documents', providing pre-deployment training and in-theatre training for all military and civilian personnel (including contractors) on women, peace and security issues, and embedding the objectives into curricula in commanders' courses as well as 'intermediate and senior service schools' (The White House, 2016: 19). The Department of Defense is required to designate action officers and senior-level officials to be responsible for the coordination and implementation of the tasks assigned to it, and to 'Develop and improve data collection mechanisms to track and report progress on WPS objectives, assess lessons learned, and identify best practices from existing programs' (20).

As to the second objective, participation in peace processes and decision-making, the Department of Defense will 'Assist partner governments in improving the recruitment and retention of women ... into government ministries and the incorporation of women's perspectives; into partner nation peace and security policy' (The White House, 2016: 21). The Department of Defense will also provide security sector training to accomplish this objective. In addition, the department will 'Leverage the participation of female U.S. military personnel to encourage and model gender integration and reach out to female and male populations in partner nations', and increase the percentage of partner nation women's participation in US training programs (21).

Interestingly, the 2016 plan focuses on the role of male military leaders personally, tasking them to 'mobilize men' in support of the plan's agenda, and to 'Utilize public diplomacy and engagement to advocate for women's leadership and participation in security-related processes and decision-making' (The White House, 2016: 21). If this were to actually become part of the basis upon which the performance of senior leaders was evaluated, this requirement might make great practical strides. The Department of Defense's final task to accomplish the second objective is to 'Assist partner nations in building their capacity to develop, implement, and enforce policies and military justice systems that promote and protect women's rights' (22).

The tasks given to the Department of Defense to accomplish the third objective, protection from violence, are ambitious and challenging. It is not clear what its working leverage would be on the bodies with which the department would need to work to meet this goal other than the large US contribution to the UN peacekeeping budget (28.43 percent in 2018) (UN Secretary General, 2015: 2). The Department of Defense has a very modest relationship with the UN in terms of providing personnel who could actually work to implement these requirements. For example, as of November 2017, US provided only a total of 46 military personnel to UN peacekeeping forces (UN, 2017). The Department of Defense is tasked, however, with advocating to the UN that peacekeeping missions have strong mandates on protecting civilians from gender-based violence – perhaps mere advocacy does not actually require boots on the ground. The Department of Defense is also required to incorporate modules on women's rights and their protection and specific needs in conflict situations into the training provided to partner nations, and to support efforts to educate US government contractors and aid workers 'on the prevention of sexual exploitation and abuse in crisis and conflict-affected environments' (The White House, 2016: 23).

In further support of this objective, the Department of Defense will address crimes committed against women and girls through its support of the development of 'effective accountability and transitional justice mechanisms', and help multilateral and international organisations in their efforts to police their own ranks against instances of sexual exploitation and assault (The White House, 2016: 25). What is particularly troubling in context is the large number of sexual assaults committed by male US service members (78 percent of the perpetrators) against their female comrades (81 percent of the victims), out of a total of 5,277 reports of in-service attacks (DoD, 2017: 8, 32). It is difficult to imagine how implementing a gender perspective in operations to better protect women and girls can really take root until these sorts of attacks are prevented or starkly reduced, to say nothing of helping other military organisations meaningfully fight sexual and gender-based violence among their ranks.

As to the fourth objective, conflict prevention, the Department of Defense is given three main tasks. First, it will use and promote the use of gender analysis 'in conflict mapping and reporting, including for mass atrocity prevention, countering violent extremism, and stabilization funding' (The White House, 2016: 27). Second, it will actively engage 'women in planning and implementing disaster and emergency preparedness and risk reduction activities, including regarding how police can better interact with women in their role as first responders' (27). The last of these tasks, supporting de-radicalisation efforts, may prove to be the most challenging to accomplish.

The Department of Defense is directed to support efforts to de-radicalise male and female extremists, and it will 'promote tolerance and pluralism in their communities, and advance stabilization and reconstruction activities' (The White House, 2016: 28). Although worthwhile, these sorts of initiatives have had an uneven record of success (Horgan and Braddock, 2010). Further, those critical of the creep of 'countering violent extremism' (CVE) into security-related UNSCR 1325 efforts have pointed out that different efforts have failed to address either violent extremism or gender considerations under UNSCR 1325 efficiently, that there is often 'a lack of appropriate gender training for practitioners engaged in CVE program delivery', and that there is a 'lack of trust between community groups and state actors' (Shepherd, 2017).

As with the fourth objective, under the fifth objective, access to relief and recovery, the department likewise has three tasks. The first task is to promote equal access to aid supplies and services irrespective of gender or age. Second, the Department of Defense will 'Ensure that U.S. Government crisis response and recovery teams have access to appropriate gender and protection expertise, such as a designated gender advisor, to integrate gender considerations in U.S. Government-supported relief and recovery efforts' (The White House, 2016: 29). This task might be difficult to accomplish in the short term, because it would appear to require the development of a field of specific expertise which is only now being created in the US military, but for which there still does not yet appear to be a permanent specific requirement in the US force structure. The third task is to support demobilisation, disarmament and reintegration programmes 'that address the distinct needs of male and female ex-combatants' (30). Given the small commitment of forces made by the US to peacekeeping efforts, it is not clear how the Department of Defense will be able to directly accomplish this task.

The tasks assigned to the Department of Defense under the 2016 national action plan are a curious mix. Many could be accomplished within its ranks with sufficient command emphasis, such as pushing male military leaders to start raising UNSCR 1325 concerns in their interactions with partner nation leaders. Others would tax even the most diligent efforts to achieve them, such as bringing about change in the way UN peacekeeping forces protect civilians from sexual and gender-based violence almost solely through budgetary means. Regardless, one thing that is missing from these tasks is a requirement to review and overhaul the vast amount of doctrine issued by the Department of Defense to reflect in a balanced way the operational significance of gender in different kinds of operations. With this in mind, it is now useful to look at the requirements of the *Women, Peace, and Security Act of 2017*.

Women, Peace, and Security Act of 2017

The *Women, Peace, and Security Act of 2017* takes an approach that is consistent with the overall approach of UNSCR 1325. The law states that it is 'the policy of the United States to promote the meaningful participation of women in all aspects of overseas conflict prevention, management, and resolution, and post-conflict relief and recovery efforts' (*WPS Act, 2017*: § 4). As a matter of policy, it then lists seven areas for US action, including the integration of 'perspectives and interests of affected women into conflict prevention activities and strategies', promotion of 'the physical safety, economic security, and dignity of women and girls', and collecting and analysing 'gender data for the purpose of developing and enhancing early warning systems of conflict and violence' (§ 4(1)-(7)).

As an action driver, the US Congress required the President to provide a single government-wide strategy on women, peace and security to it by October 2018, and then every four years thereafter, that explained how the US would meet its requirements under the Section 4 action areas. The US Congress specified that the strategy must be aligned with and supportive of other countries' plans 'to improve the meaningful participation of women in peace and security processes, conflict prevention, peace building, transitional processes, and decision-making institutions' (*WPS Act, 2017*: § 5(a)(1)). Further, to ensure accountability and efficacy, the strategy had to include measurable goals, benchmarks, metrics, timetables and monitoring and evaluation plans (§ 5(a)(2)).

Under Section 6 of the act, although much of the work and responsibility for reporting was assigned to the Department of State and the US Agency for International Development (USAID), other US government departments and agencies were assigned specific tasks under the law, particularly in terms of training. The Department of Defense was directed to ensure its personnel received training in conflict prevention and resolution processes that addressed 'the importance of meaningful participation by women', on gender considerations in international human rights law and IHL, and on 'protecting civilians from violence, exploitation, and trafficking in persons' (*WPS Act, 2017*: § 6(b)(1), (2)). Lastly, the Department of Defense will provide training to its personnel on 'Effective strategies and best practices for ensuring meaningful participation by women' (§ 6(b)(3)).

Beginning one year after the first submission of the required strategy to the US Congress, the Secretary of State, along with the USAID Administrator and the

Secretary of Defense, is required to brief the appropriate committees of the US Congress on the training that has been conducted under Section 6 of the act (*WPS Act*, 2017: § 8(a)). Within two years of the strategy's submission, the President is required to submit to the US Congress a report that addresses the strategy's implementation and its impacts, how and to what extent US government agencies coordinated in delivering the strategy, the methodology for assessing progress in the strategy's implementation and the training conducted as required by the act (*WPS Act*, 2017: § 8(b)). To help it meet these requirements, the Department of Defense has established a Women, Peace and Security Synchronization Group composed of 'representatives from the combatant commands, military services, regional centers and professional military education institutions', which meets on a monthly basis (Broadway, 2018).

In many important ways, the *Women, Peace, and Security Act of 2017* mirrors the higher-level requirements of the 2016 US national action plan, and now enshrines what had just been policy into US domestic law. What might be problematic is that the law does not mention the US national action plan at all. Thus, from a bureaucratic perspective, the possibility exists that the Department of Defense will prioritise meeting the minimum training and reporting requirements of the law, rather than the specific measures set out in the 2016 plan and its intent. Conceivably, the Department of Defense, were it so inclined, could meet its obligations under the law merely by creating and delivering the required training programmes as stand-alone items rather than undertaking any major change in the ordinary foundation of its military education and training – its doctrine. For this reason, it is worthwhile to now spend some time examining the degree to which US doctrine already addresses gender considerations in operations, and to explore the gaps that still exist.

US joint doctrine, civilian-centric operations

Because of the vast amount of US military doctrine, and the broad range of mission sets in which the US military is expected to apply it, it is useful to focus an examination of doctrinal treatment of women, peace and security on those operations that are particularly civilian-centric, both kinetic and non-kinetic. There are nine types of joint operations conducted by US forces which appear to be logical places to house the discussion of gender's operational relevance because of their high degree of interaction with civilian populations: Civil-Military, Counterinsurgency, Foreign Humanitarian Assistance, Information, Joint Urban Operations, Multinational, Non-Combatant Evacuation, Security Cooperation and Stability. As with the last chapter on UK strategy and doctrine, this section will begin with joint level US doctrine, and then move down the doctrine hierarchy to US Army doctrine.

Joint level – kinetic operations

Curiously (and this may be due to the US military joint doctrine review and revision cycle) (Chairman, JCS, 2014a: B-12, 27, 30), although many of the relevant joint doctrinal publications postdate the 2011 national action plan, gender barely registers in most of them. For example, despite the local population being the key to conducting successful counterinsurgency operations, the joint counterinsurgency doctrine set out in *Joint Publication (JP) 3–24, Counterinsurgency*, focuses on women primarily as potential recruits for insurgencies and the different combat and auxiliary roles they can perform (Chairman, JCS, 2018b: II-8–II-9). It also notes the problems former female combatants might face in reintegrating women and children associated with insurgent groups back into society once hostilities are concluded (III-33–III-34).

Gender is mentioned just a handful of times – in the context of being a factor in recruitment to insurgencies, as the last-listed item in a number of aspects in how societies might differ from one another and in the areas of study that compose the social sciences (II-9, III-7, IV-7). These are important points, but to focus solely on civilian women in the counterinsurgency environment as potential threats rather than as potential agents of stability is not really coming to terms with the requirements of the *Women, Peace, and Security Act of 2017* or the US national action plan.

Similarly, given the ever-increasing urbanisation of the world's population and the global growth of the megacity (Serena and Clark, 2016), the failure to meaningfully address women or gender in *JP 3–06, Joint Urban Operations*, is surprising. It notes only that 'culturally inappropriate interaction with women' by US soldiers might antagonise a population, and that a population analysis of an urban area should include 'delineating its primary attributes, such as age, wealth, gender, ethnicity, religion, and employment statistics' (Chairman, JCS, 2013a: III-15, A-6). *JP 3–16, Multinational Operations*, does not mention gender, and mentions women only once, as an example of a civil-military cooperation effort by a Hungarian Provincial Reconstruction Team in Afghanistan to improve the situation of local women and create job opportunities for them (Chairman, JCS, 2013b: III-36). Related doctrine on security cooperation that is very recent addresses neither women nor gender at all (Chairman, JCS, 2017b).

Importantly for the future direction of US kinetic joint doctrine, however, *JP 3–07, Stability*, which was revised in late 2016, is a quantum improvement in terms of its embrace of the goals and principles of UNSCR 1325 as captured in the US national action plan and the *Women, Peace, and Security Act of 2017*. It is worth discussing its provisions that relate to women and gender in detail, because they deal with many of the core concepts of UNSCR 1325 in sufficient depth to form the basis for education and training as well as planning and actual operations. In this sense, *JP 3–07* is a model for the inclusion of women and considerations of gender into other joint US doctrine, both kinetic and non-kinetic.

First, in discussing the need to take an integrated approach with local populations in promoting governance and participation in stability operations, *JP 3–07* notes that equity and equality are pillars of social capital development activities. These activities must advance 'equity and equality of opportunity among citizens in terms of gender, social and economic resources, political representation, ethnicity, and race. HN [host nation] sociocultural factors must be considered before advancing certain aspects of equity and equality' (Chairman, JCS, 2016, II-7).

Next, sexual and gender-based violence is identified as one of many serious threats facing civilians in building stability, which also include war crimes, genocide and crimes against humanity. *JP 3–07* notes that this type of violence, 'with its associated psychological trauma, can often be used as a tactic of war; for instance, the rape of women and girls can be perceived as an attack on the male relatives' honor by proving their inability to act as protectors' (Chairman, JCS, 2016: III-12). This doctrine also notes that 'Rape not only terrorizes and humiliates individuals, but it can also be used as a deliberate strategy to target the roles of women in society, and thus destabilize communities as an aim of war' (III-12). It also recognises the 'long lasting economic, social, and health impact on the state and surrounding region' the use of rape in armed conflict can have (III-12).

JP 3–07 highlights the importance of steps that the military can take to deter sexual and gender-based violence. It urges that 'specific attention should be paid to investments in the required infrastructure (e.g., forensic laboratories), and human resources needed for the reception of victims', and it states that 'By adopting a more interventionist approach, forces will be able to reduce impunity for war crimes affecting women' (Chairman, JCS, 2016: III-12). Further, in building transitional public security through the establishment of an interim criminal justice system and detention facilities, stability operations 'should ensure compliance with international detention standards, which require separation by gender' (III-51).

Because of the 'recognized link between the issues of peace, security, development, and gender equality', the joint stability doctrine notes that it is insufficient to promote the participation and protection of women during conflicts and in post-conflict situations. 'Stabilization and reconstruction initiatives are also needed to ensure that these actions are supported by wider development considerations, such as the promotion of economic security and opportunities, and women's access to health services and education' (Chairman, JCS, 2016: III-12). Accordingly, achieving these effects in stability operations 'requires collaborative work with international organizations and NGOs as well as the HN government' (III-12).

With regard to stabilisation planning, *JP 3–07* recognises women's role in conflict resolution is frequently neglected. 'Conflict can disrupt gender roles often on the basis that the majority of women are not involved in major conflicting parties and are therefore left to take on male-associated roles as men engage in conflict' (Chairman, JCS, 2016: IV-25). Post-conflict, it explains that women assume many important roles in their societies, including that of 'local decision makers expected to rebuild homes', 'community leaders and heads of households', caregivers 'for orphans and survivors', and other roles predominately associated with men (IV-25). Further, gender, ethnic and minority issues must be addressed in designing disarmament, demobilisation and reintegration programmes. In sum, commanders and planners are advised that

> Incorporating women into the peacebuilding process can build on societal changes that may be occurring naturally, as a result of the cultural turmoil

that ensues from conflict. Ignoring the experiences of women risks overlooking their legitimate needs and concerns in new institutions and settlements.

(IV-25)

Joint force commands are therefore directed to support 'local women's peace initiatives and local processes to ensure women's perspectives are recognized as part of an inclusive response to conflict resolution' (IV-25).

Joint level – non-kinetic operations

In contrast to the joint stability doctrine's extensive treatment of women, peace and security concerns, on the non-kinetic side of operations the absence of meaningful discussion of women's roles and gender considerations is particularly noticeable in the overarching doctrinal publications regarding information operations, or what the US formerly termed as 'psychological operations', and intelligence functions. *JP 3–13, Information Operations*, contains no information on gender or on women, although women are of course generally half of the population that such operations seek to influence in a theatre of operations (Chairman, JCS, 2014d: 3–13). *JP 2–01.3, Joint Intelligence Preparation of the Operational Environment*, mentions gender once, noting that in researching the values of certain groups or individuals, 'religious values may conflict with generational values or gender values' (Chairman, JCS, 2014b: VII-5). Women are likewise just mentioned once in this document, as having been used as human shields by insurgent forces in Somalia fighting UN troops (VII-5, B-32). Similarly, *JP 3–68, Noncombatant Evacuation Operations*, mentions women only twice, first in the sense that special search protocols for women, children and disabled or injured people need to be established, and next in recognising that pregnant women are a category of evacuees whose evacuations are to be prioritised (Chairman, JCS, 2017c: VI-7, VI-13).

The most glaring absence of meaningful discussion of women's roles and gender considerations, however, is probably in US CIMIC doctrine. Termed 'Civil-Military Operations' by the US, these operations support the entire spectrum of military activities and operations, from the most kinetic offensive operations to relatively tame post-conflict peace-building actions. The US defines Civil-Military Operations as activities 'that establish, maintain, influence, or exploit relations between military forces, indigenous populations, and institutions, by directly supporting the attainment of objectives relating to the reestablishment or maintenance of stability within a region or host nation' (Chairman, JCS, 2017a: 34). Because of Civil-Military Operations' ubiquity and their obvious focus on the civilian population, the almost complete lack of discussion of these operational gender factors is particularly troubling, and it reflects the male-normative nature of most current US doctrine.

Although *JP 3–57, Civil-Military Operations*, is 173 pages long, it does not mention gender. Women are mentioned, but sparsely. In terms of planning for Civil-Military Operations, logistics planners are reminded to 'include logistic support normally outside military logistics, such as support to the civilian populace (e.g., women, children, and the elderly)' because Civil-Military Operations often provide support to them (Chairman, JCS, 2013c: III-10). This may require medical planners to 'adjust typical personnel and logistic packages to care for women and children affected even in operations not originally of a humanitarian nature' (A-C-2). Women are mentioned once more, in the context of negotiation and mediation. Civil-Military Operations personnel are reminded that in developing a negotiation plan for a meeting, they should determine the 'appropriate construct for a meeting, consider the culture. For example, what role do women play in the society? How is status defined in the culture?' (B-14) The significance of women's role in the host nation's culture is not further discussed, and it is a fair inference from this that women are not really recognised as being required at the negotiation table.

There are two pieces of joint doctrine for non-kinetic operations, however, that provide a decent treatment of gender considerations in missions – *JP 3–29*, *Foreign Humanitarian Assistance*, and particularly the recent *JP 3–07.3*, *Peace Operations*. *JP 3–29*, *Foreign Humanitarian Assistance*, includes women among special parts of disaster-stricken populations that require special protection measures and ensured access to humanitarian assistance, and it details the roles of USAID in providing disaster relief supplies and services to women and children, and the UN Children's Fund in monitoring the situation of women and children and preventing their sexual abuse and exploitation (Chairman, JCS, 2014e: I-4, IV-8, IV-21, D-4–D-5). Importantly, commanders are informed that they must incorporate measures to prevent trafficking in women and children 'into post-conflict and humanitarian emergency assistance programs' (Chairman, JCS, 2014e: IV-21).

JP 3–29 also provides important guidance on gender issues in staging and phasing medical care, noting that in disasters, 'the first wave of patients is trauma care, and the second wave is disease control', and that in this latter stage, 'women's and children's care takes special planning to ensure the right mix of providers and medications are available' (Chairman, JCS, 2014e: E-4). Further, because of cultural sensitivities, 'the gender of both providers and interpreters should be considered' (E-4). Although *JP 3–29*'s treatment of women and gender is not extensive, it is a qualitative improvement over the other US civilian-centric joint doctrine applicable to non-kinetic operations already assessed.

JP 3–07.3, *Peace Operations*, contains an extensive discussion of the relevance of gender in providing inclusive security for an area of operations and the risks women face in conflict situations, and it states that when appropriate, peace operations are expected to include the objectives of the US national action plan in their missions (Chairman, JCS, 2018a: II-22–II-23). *JP 3–07.3* also contains an appendix that is entirely devoted to the protection of civilians (Appendix B). This appendix lodges protection of women and girls within this overarching theme of civilian protection, but it does not male-norm them as just an undifferentiated group of host nation civilians. Instead, it impresses upon planners and operators the need to understand risks civilians face in peacekeeping situations, and focuses on women and children as specific at-risk groups with different needs that need to be assessed (Appendix B, B-2 – B-4).

Further, the appendix explains in detail the different risks that women and girls face with regard to sexual and gender-based violence, and the impacts to the larger security situation in the event that they suffer this type of attack (Chairman, JCS, 2018a: Appendix B, B-7). For the US, this doctrine could be an example of a best practice in terms of relating UNSCR 1325's goals to specific US policies and executive orders so as to avoid any unnecessary and negative political reactions to non-US authorities on women, peace and security such as UNSCR 1325 itself. Although it is now possible to trace this thread of gender considerations, cast as protection of civilians, from joint doctrine to US Army doctrine as will be described next, there is a concern that these concepts are perhaps being marginalised in US doctrine. They only seem to be found in those mission areas the US forces do not emphasise in their training or operations, such as peacekeeping.

US Army doctrine

As noted in the previous chapter in the discussion of UK doctrine, ideally, service-level doctrine should nest within joint doctrine, and amplify and focus upon that information particularly relevant to the service's operations. Consistent with that, Headquarters, Department of the Army (HQDA) has set out for the US Army its own doctrine hierarchy, starting with Army Doctrinal Publications at the highest level, followed by Army Doctrine Publications (ADPs), then Field Manuals (FMs) and finally at the lowest level, Army Techniques Publications (ATPs) (HQDA, 2014a: 2-4–2-5). This hierarchy establishes 'which publication should be used when a conflict exists between publications' (2-5). Although the US Army recognises that 'Sometimes lower-level doctrine drives higher-level doctrine' (2-3), it provides no guidance for sorting out instances of doctrinal insubordination.

In the context of its treatment of women and gender, US Army doctrine largely fits the typical hierarchical pattern that currently exists in NATO and UK doctrine, but with some interesting differences. First, like the bulk of joint doctrine, the vast majority of US Army doctrine largely ignores women or gender considerations in operations, or instead treats them as threats. For example, *FM 3–0, Operations*, observes that although the threat of 'large-scale combat operations against a peer threat' became unlikely after the end of the Cold War and the US then found itself fighting counterinsurgencies in Afghanistan and Iraq, this kind of conflict now represents the most significant readiness requirement (HQDA, 2017: ix). This doctrine now 'contains fundamentals, tactics, and techniques focused on fighting and winning large scale combat operations' (xi), and it mentions neither women nor gender. When it does discuss aspects of civilian-centric operations, such as the six military tasks in stability operations, it identifies abstract roles in which US Army forces will engage with local national populations, including 'establish civil security' and 'support governance', without providing any indication that performing these roles successfully requires understanding of the operational relevance of gender (8-11–8-13).

Similarly, *FM 3–24, Insurgencies and Countering Insurgencies*, notes only that women can play an important role in an insurgent auxiliary force (HQDA, 2014g:

4-17). At the lowest-level of US Army doctrine, *ATP 2–01.3, Intelligence Preparation of the Battlefield/Battlespace*, mentions women once, in that they might be couriers for enemy forces, and gender once, in that analysis of populations includes studying gender norms and roles (but without any explanation of the purpose of this analysis) (HQDA, 2014b: 10-15). In a similar fashion, *ATP 3–55.4, Techniques for Information Collection During Operations Among Populations*, notes only that women might carry improvised explosive devices, and that 'Certain cultures prohibit interactions by people of certain genders or cultural status. In predominantly Muslim countries, cultural prohibitions may render a female interpreter ineffective in certain circumstances' (HQDA, 2016: 1-4, B-2).

Second, as with joint US doctrine, the area of US Army doctrine that one would expect to deal most fully with women and gender in non-kinetic civilian-centric operations, Civil-Military Operations, barely deals with these topics at all. Third, where US Army doctrine does deal meaningfully with women and gender in an operational sense, it does so with great specificity, as shown by US Army stability operations doctrine, for instance. The best example of this is probably the US Army doctrine on protection of civilians, which has no direct counterpart in joint doctrine although it is in the family of stability operations doctrine at the land force level.

As will be discussed shortly, the protection of civilians doctrine is remarkable for its extensive explanation of the operational relevance of women and gender and it might serve as a bridging doctrine between different families of US Army doctrine. Located as it is on the lowest tier of US Army doctrine within the stability operations family, however, it is not clear that educators, trainers, planners and operators would necessarily be consulting it regarding operations that were not specifically stability-oriented. Further, these differences in doctrinal treatment might be due to the nature of the military organisations writing the doctrine. Whilst US Army stability operations doctrine is prepared by the US Army Peacekeeping and Stability Operations Institute, *FM 3–0, Operations* was prepared by the US Army Combined Arms Doctrine Directorate (HQDA, 2017: vii).

Kinetic operations – stability

The higher level *ADP 3–07, Stability*, makes no mention of women and only mentions that the training of host nation military forces might cover topics such as human rights, child protection and the prevention of gender-based violence (HQDA, 2013a: 12). Its subordinate doctrine, however, such as *FM 3–07, Stability*, and *ATP 3–07.5, Stability Techniques*, provide extensive information regarding women and gender in operations. This discontinuity is surprising, since the US Army Peacekeeping and Stability Operations Center at the US Army War College was the 'preparing agency' for all three documents, and it has been assigned the lead role in working to operationalise gender issues related to UNSCR 1325 in the US Army (Kristine Petermann, 2017: personal communication). *FM 3–07* states plainly in the early part of the publication that 'Army leaders account for effective engagement that includes females in the population. Leaders consider how gender norms differ by culture. They plan and prepare to include female teams and interpreters, which are critical to effectiveness' (HQDA, 2014f: 1-25). Units are advised to 'be careful not to overlook or marginalize important groups such as women or minorities' in engaging the population (1-8).

Importantly, from the perspective of better understanding the risks faced by certain population groups such as women and children, *FM 3–07* points out that 'Civilians near military targets may be more vulnerable to collateral damage, and dislocated civilians who flee their homes may be more vulnerable to disease, starvation, and crime' (HQDA, 2014f: 1-9). This phrasing is very close in content to the NATO doctrinal language discussed earlier in Chapter 4, 'NATO strategy, doctrine and gender', that raises the possibility that the differentiated and disparate impacts of armed conflict upon women and girls could be considered in a commander's proportionality analysis. However, as will be discussed in greater detail in Chapter 8, 'International humanitarian law and gender', this thread is not picked up in US legal policy on IHL (Department of Defense, 2016), and it appears to be addressed neither in targeting doctrine (Chairman, JCS, 2013d) nor in targeting practice (Chairman, JCS, 2009; DoD, 2009).

FM 3–07 also reminds commanders and planners of the need to ensure equity in the distribution of relief supplies, including considerations of gender in the ease of access to supplies, and that distribution to just one gender 'can aggravate social tensions resulting in violence' (HQDA, 2014f: 1-25). Despite the troubling absence of Female Engagement Teams from other US military doctrine, the field manual identifies trained Female Engagement Teams as a vehicle for such efforts, because they can 'develop an understanding of gender and family issues, provide care to victims, and influence a significant but often inaccessible part of the population' (1-20). It further emphasises the importance of having trained Female Engagement Teams, because the Female Engagement Team 'functions will likely differ from their normal responsibilities' (1-20). To its credit, the field manual does not shy away from addressing certain problems identified with the use of Female Engagement Teams in Iraq and Afghanistan, advising that 'Female teams have been known to experience difficulties with integration into military units; therefore, commanders emphasize their importance and integration to avoid problems' (1-20).

Interestingly, the most developed portion of the field manual dealing with gender is the part that addresses disarmament, demobilisation and reintegration operations, although US Army forces have not generally been involved in such peacekeeping-related efforts. Perhaps this is because the US Army Peacekeeping and Stability Operations Institute was the preparing agency for the field manual. *FM 3–07* explains that the process requires careful planning and a significant investment in resources mindful of the significance of gender, because in camps, men, women, child soldiers and families require separate barracks and latrines. *FM 3–07* notes combatants undergoing the process include not just active fighters, but also service support personnel who are often women, who must also be disarmed, demobilised and reintegrated into society. Further, units must also 'identify women who are victims of sex slavery and forced marriage and place them in secure barracks' (HQDA, 2014f: 1-20–1-21). Reintegration programs must account 'for the specific needs of women and children associated with armed forces and groups, as well as those of civilians forced to flee their homes after violent conflict or disaster' (1-21). Not only must 'education and vocational skills training' be provided, but 'Specialized assistance, such as psychiatric counselling for traumatized combatants and devices for disabled combatants, ensures victims of conflict receive necessary assistance' (1-21).

The lower-level *ATP 3–07.5, Stability Techniques*, tracks with many of the points addressed in the field manual on women and gender, and it significantly amplifies a number of them. *ATP 3–07.5* states that in terms of understanding the area into which they have deployed, units can collect information using a tactical conflict survey which will include gender-disaggregated data (HQDA, 2012: B-4). *ATP 3–07.5* notes the relationship between sexual and gender-based violence and instability in conflict situations, and that this sort of violence 'requires focused efforts to protect the population, bring perpetrators to justice, provide specialized treatment for its victims, and eventually create a normal environment with necessary levels of security' (1-6, 4-6). Therefore, in the initial phase of a stability operation, US Army forces will monitor the situation of at-risk population groups, and recognise that although 'A host-nation constitution may articulate many rights, such as gender employment equality ... local cultural norms may not meet such standards' (4-12).

For US military personnel interacting with victims, *ATP 3–07.5* reminds them that the victims 'often fear uniformed or armed personnel' (HQDA, 2012: 1-6). *ATP 3–07.5* also advises medical personnel to be conscious of local social norms regarding medical treatment and gender, and that they might need to 'obtain permission from the patient's spouse or a respected local leader' before working with patients of a different gender (4-4). In terms of medical services, it recognises the male-normative nature of US Army forces, noting that 'These challenges arise due to population characteristics that differ significantly from military forces (such as age, gender, chronic illness vice trauma, cultural considerations, and so on)' (A-4). Further, in support to governance, US Army units will help establish interim legislative processes, such as the use of temporary councils that can deal with stability tasks, such as security, essential services and gender-related issues (5-4).

As the stability operation moves into subsequent phases, the stability techniques doctrine notes that 'Army units expand human rights protection from protection against unlawful violence to political and social areas such as gender equality and freedom of speech' (HQDA, 2012: 4-12). This also has a practical aspect, as US Army units seek to provide employment opportunities for local nationals of all ages and genders, and those with special needs (6-4). *ATP 3–07.5* also informs US military members that in these efforts, and in demobilisation, disarmament and reintegration initiatives, they need to remember that 'During violent conflict, many women become heads of large households. They have acquired critical skills to adapt to food shortages and become micro-entrepreneurs in the informal economy', and that not providing opportunities that reflect this potentially fosters destabilisation (6-4).

The depth in which the stability techniques doctrine meaningfully treats UNSCR 1325 topics is impressive. However, like *FM 3–07*, not only is this doctrine a bit of an outlier to the rest of the US Army's operations doctrine, it also located on the lowest rung of the US Army doctrinal ladder. Thus, it is not clear how familiar US Army planners or operators would be with it. However, if they were to select *ATP 3–07.5* as the guidance to be followed in effectively engaging with host nation women in a theatre of stability operations, they would likely be well-served because of its practicality and its well-developed consistency with the legal requirements set out by the US Congress in the *Women, Peace, and Security Act of 2017*.

Non-kinetic operations – civil-military operations

Unfortunately, as with joint-level US CIMIC doctrine, some US Army Civil-Military Operations doctrine simply ignores women and gender altogether. *ATP 3–57.20, Multi-Service Support Techniques for Civil Affairs Support to Foreign Humanitarian Assistance* (HQDA, 2013c), *ATP 3–57.30, Civil Affairs Support to Nation Assistance* (assisting a host nation with its own defence, security assistance 'to promote long-term stability, pluralistic government, and sound democratic institutions') (HQDA, 2014c: iii), and *ATP 3–57.70, Civil-Military Operations Center* (HQDA, 2014e) fall in this category. Other US Army CIMIC doctrine does mention women and gender, but not in a way that makes it seem like they are seen as important factors to work with in the operational environment, and perhaps instead fall in the category of 'optional'. *ATP 3–57.10, Civil Affairs Support to Populace and Resources Control* (HQDA, 2013b: A-12, A-21, A-30), and *ATP 3–57.60, Civil Affairs Planning* (HQDA, 2014d: B-12) all note that in describing host nation personnel who might be helpful in conducting operations, plans and orders should note their gender 'if applicable'. Not surprisingly, the higher-level *FM 3–57, Civil Affairs Operations*, makes no mention of gender, and only notes that the public health and welfare section of a civil affairs command should determine the capability of host nation public welfare systems to assist at-risk population groups such as women (HQDA, 2011: 2-25).

The US Army considers civil affairs units as special operations forces (SOF), along with Special Forces and Rangers. Thus, it is useful in this context to examine US Army Special Operations doctrine in general with regard to gender. As explained in *ADP 3–05, Special Operations*, because they have a deep understanding of civil considerations and sociocultural factors, 'special operations forces ... operate in small teams in friendly, politically sensitive, uncertain, or hostile environments to achieve U.S. objectives' (HQDA, 2018: 8-10). Further, special operations soldiers are 'are uniquely trained, organized, equipped, and employed in tailored operational packages', and are multilingual, 'culturally astute', 'politically nuanced', and expected to anticipate the ripple effects of their tactical actions (11–12). Ironically, given the well-developed understanding of local populations upon which the special operations soldiers base their activities, there is no mention of women or gender in the document, nor in the joint doctrine within which it nests, *JP 3–05, Special Operations* (Chairman, JCS, 2014c).

As will be discussed in Chapter 9, 'Gender in military activities and operations', however, US Army Special Operations units used female Cultural Support Teams significantly in their operations in Afghanistan, although the US Army as a whole has not institutionally embraced the concept of the Female Engagement Team. Perhaps this means though that the US Special Operations community only sees the Cultural Support Team as a niche capability, an enabler, rather than a feature of operations significant enough to warrant doctrinal discussion. In this regard, it might be relevant that the proponent responsible for Civil-Military Operations doctrine is the US Army John F. Kennedy Special Warfare Center and School. In a fairly recent survey of US Special Operations personnel concerning the integration of women into Special Operations Forces in 2015, researchers concluded that the 'opposition to opening SOF specialties to women is both deep and wide, with high levels of opposition across all SOF elements' (Szayna et al., 2015: iii).

Protection of civilians – bridging the gap between kinetic and non-kinetic operations?

As detailed as the stability field manual and the stability techniques publication are in their treatment of women and gender in stability operations, they do not approach the depth of operational considerations and practical measures concerning women and gender set out in *ATP 3–07.6*, *Protection of Civilians*, a stability operations sister document to *ATP 3–07.5*. This publication notes that US Army units should implement protection of civilians measures in operations (not just stability operations) taking account of a number of factors, including 'legal considerations, legitimacy, unity of effort, host nation ownership and capacity, and gender issues' (HQDA, 2015: 1-5). In doing so, however, it advises commanders that 'Some host-nation cultural norms (such as gender rights and acceptable levels of corruption) differ from international norms, creating challenges for actors that seek to protect civilians effectively while navigating the cultural landscape' (1-6).

ATP 3–07.6 is forthright as to the historical treatment of women and gender in doctrine. It plainly states 'Often overlooked in the past, there is now increasing awareness of sexual violence during conflicts and other unstable situations. Sexual violence is usually, but not always, directed against women and girls' (HQDA, 2015: 1-6). Further, it also notes that 'There is also greater recognition that women are vital to establishing peace and maintaining future stability' (1-6). Thus, in these environments, US Army units support the distinct needs of women and children, including ensuring safe and equitable access to humanitarian assistance (4-18).

From the protection perspective, *ATP 3–07.6* states that efforts must be made to mitigate 'harm, exploitation, discrimination, abuse, conflict-related sexual violence, and human trafficking while holding perpetrators accountable' as well as addressing access to 'humanitarian assistance, relief, and recovery and protection of human rights' (HQDA, 2015: 1-6). In this regard, *ATP 3–07.6* explains that 'Sexual discrimination, marginalization, and sexual exploitation and abuse by those in authority, including military forces or unified action partners, can be serious problems' (2-8). Importantly it also reemphasises that males can be victims of sexual violence as well, although most victims are female (2-8).

ATP 3–07.6 explains that sexual violence 'can be fostered by a culture of impunity, culturally accepted biases against women, lack of discipline in security forces, and beliefs and behaviour that violate accepted human rights standards', and that high levels of sexual violence show that civilians are not receiving adequate protection (HQDA, 2015: 2-8). Importantly, *ATP 3–07.6* plainly states the scope of sexual violence, noting that it is more than just rape, and includes 'sexual slavery, enforced prostitution, forced pregnancy, enforced sterilization, mutilation, indecent assault, human trafficking, inappropriate medical examinations, and strip searches' (4-18).

ATP 3–07.6 approaches these topics from a very practical perspective. For example, it explains that even the threat of sexual or gender-based violence has a destabilising effect, because 'potential victims may avoid necessary activities such as traveling, working, farming, obtaining water, or collecting firewood if they are vulnerable when doing so' (HQDA, 2015: 4-20). It notes that ordinary military activities can have a positive impact, such as patrols having a secondary effect of reducing 'vulnerabilities and threats related to conflict-related sexual violence' (4-23). Perhaps most importantly in this regard, this publication states that 'Leaders must emphasize the importance of eliminating conflict-related sexual violence, as well as its significance to the mission, and ensure that Soldiers receive adequate training' (4-23). It points out that such training could include scenario-based exercises that address 'conflict-related sexual violence incidents' (4-23). This might prove challenging for typical US Army training organisations to recreate properly – perhaps it would be better suited to seminar and discussion group formats rather than the field Situational Training Exercises that have proven so useful in training soldiers for deployments (Prescott and Dunlap, 2000). Finally, it informs commanders that this topic must be emphasised in their engagement with host nation key leaders, so that local authorities realise how serious the US is about this issue (4-27).

From the participation perspective, *ATP 3–07.6* notes that 'Institutions are more effective and societies are more stable when women are integrated and not marginalized', and that women's participation needs to be promoted in the host nation's 'political, economic, and security sectors and institutions' (HQDA, 2015: 1-6). Accordingly, US Army units will 'support prospects for inclusive, just, and sustainable peace by promoting and strengthening women's rights and effective leadership and substantive participation in peace processes, conflict prevention, peace building, transitional processes, and decision-making institutions in conflict-affected environments' (4-18). It also personally addresses commanders – 'Perhaps more importantly, they should encourage such gender perspectives in the host nation' (4-7). Finally, leaders are reminded that they 'must ensure that their own Soldiers are disciplined and that unit climates preclude conflict-related sexual violence and sexual exploitation and abuse' (4-7).

Just as *ATP 3–07.5, Stability Techniques*, stands out for its extensive treatment of women, peace and security issues on the kinetic side of civilian-centric operations, so too is *ATP 3–07.6, Protection of Civilians*, remarkable for the depth in which it covers important points of UNSCR 1325 principles and goals in practical ways. Importantly, the protection of civilians doctrine also recognises that 'Civilians in the proximity of military targets are more vulnerable to collateral damage', and that 'Some groups are vulnerable in certain contexts, such as women, children, or the elderly' (HQDA, 2015: 2-1). Although as noted earlier in the discussion of the US Army stability field manual, this phrasing is not as explicit as the NATO doctrine language was in specifically linking the differentiated and disparate impact of armed conflict upon women and girls with the notion of collateral damage, it is a step in the right direction.

Summary

The sheer size and complexity of the US military establishment means that any strategy and doctrinal documents must be carefully cross-walked with developments regarding women and gender in the day-to-day activities of the Department of Defense. For, example, since the first US national action plan's publication in 2011, the Department of Defense has opened all military positions to personnel irrespective of sex (Rosenberg and Phillips, 2015). This positive change should work to increase women's access to career-enhancing assignments in branches of service from which they were formerly excluded, although potentially physically qualified. This change should also assist the Department of Defense in executing its tasks to accomplish the US national action plan's second objective of increasing women's participation in peace processes and decision-making by effectively engaging military women with host nation women in security operations.

As to doctrine, although one of the designated US Joint Chiefs of Staff implementing offices, the J-7 staff section which is responsible for the development of joint doctrine (Chairman, JCS, n.d.: 'Force Development'), has begun to include operational gender considerations in revisions of joint doctrine (Elizabeth Lape, 2015: personal communication), at the time of this writing publically available versions of joint doctrine are largely silent in this regard and likely to spend years in the new doctrine review and revision cycle (Chairman, JCS, 2014a: B-30). The broad inclusion of women, peace and security considerations in US Army doctrine likely faces similar challenges. As with the potential impact of the *Women, Peace, and Security Act of 2017* on training Department of Defense personnel, the question as to whether the law will be an important external driver of doctrinal change at the joint and service levels will not likely be known for a number of years.

Further, the Joint Chiefs of Staff's J-3 section, responsible for current operations and plans (Chairman, JCS, n.d.: 'Operations'), does not appear to have received implementation tasks that would require a thorough reassessment of the use of kinetic force in the context of UNSCR 1325 and the law of armed conflict. Such a reassessment is necessary given armed conflict's different and more severe impacts on women and girls in general. Until the IHL principle of proportionality is interpreted as requiring consideration of this operational factor, any assessment by a commander of potential collateral damage will be incomplete (Prescott, 2013: 101–02, 107).

Finally, there appear to be distinct differences between US doctrinal documents depending not only on their placement in organisational review and revision cycles, but also upon which military office prepared them or served as their proponents in the drafting process. For example, in the discussion of US CIMIC doctrine's almost complete lack of gender considerations, the very male-normative nature of significant portions of the US Special Operations community was suggested as a potential explanation for this. However, women had already been serving in Civil Affairs units for almost 20 years by the time the Obama Administration rescinded the Direct Ground Combat Rule which had excluded women from many operational positions, and by 2015 the percentage of women soldiers in Civil Affairs and Military Information Support Operations was comparable to the percentage of women in the US Army overall (Vergun, 2015). The cause of this reluctance in the Civil Affairs community to deal with gender considerations in its doctrine is unknown, and perplexing.

Let's turn now to Australia. The Australian Department of Defence has strongly embraced the goals of UNSCR 1325 in its activities and operations, and there have been powerful political drivers of equality and equal treatment for women military personnel in recent years that have likely contributed to this approach by the Australian military establishment. Because of the long-standing bilateral defence relationship between Australia and the US, there is a very real possibility that US military measures to implement the UNSCR 1325 goals will be jumpstarted in part by Australian initiatives. Just as the US military is open to the idea that subordinate doctrine might drive doctrinal change in the levels above it in the doctrine hierarchy, perhaps too US practice might be driven by its work with the smaller but very capable Australian Defence Force.

References

Broadway, C (2018) 'DoD Works to Incorporate More Gender Perspective in Operations'. Department of Defense News. Available at: www.defense.gov/News/Article/Article/146185/dod-works-to-incorporate-more-gender-perspective-in-operations/ (Accessed 26 June 2018).

Chairman, Joint Chiefs of Staff (Chairman, JCS) (n.d.) 'Force Development'. Joint Chiefs of Staff. Available at: www.jcs.mil/Directorates/J7%7CJointForceDevelopment.aspx (Accessed 17 December 2017).

Chairman, Joint Chiefs of Staff (n.d.) 'Operations'. Joint Chiefs of Staff. Available at: www.jcs.mil/Directorates/J3%7COperations.aspx (Accessed 22 June 2018).

Chairman, Joint Chiefs of Staff (2009) *Chairman of the Joint Chiefs of Staff Instruction 3160.01, No-Strike And The Collateral Damage Estimation Methodology* (13 February). Washington, DC: Joint Chiefs of Staff. Available at: www.aclu.org/files/dronefoia/dod/drone_dod_3160_01.pdf (Accessed 23 June 2018).

Chairman, Joint Chiefs of Staff (2013a) *JP 3–06, Joint Urban Operations* (20 November). Washington, DC: Joint Chiefs of Staff. Available at: www.jcs.mil/Portals/36/Documents/Doctrine/pubs/jp3_06.pdf (Accessed 24 June 2018).

Chairman, Joint Chiefs of Staff (2013b) *JP 3–16, Multinational Operations* (16 July). Washington, DC: Joint Chiefs of Staff. Available at: www.jcs.mil/Portals/36/Documents/Doctrine/pubs/jp3_16.pdf (Accessed 5 July 2018).

Chairman, Joint Chiefs of Staff (2013c) *JP 3–57, Civil-Military Operations* (11 September). Washington, DC: Joint Chiefs of Staff. Available at: www.jcs.mil/Portals/36/Documents/Doctrine/pubs/jp3_57.pdf (Accessed 24 June 2018).

Chairman, Joint Chiefs of Staff (2013d) *JP 3–60, Joint Targeting* (31 January). Washington, DC: Joint Chiefs of Staff. Available at: www.cfr.org/content/publications/attachments/Joint_Chiefs_of_Staff-Joint_Targeting_31_January_2013.pdf (Accessed 5 July 2018).

Chairman, Joint Chiefs of Staff (2014a) *Chairman of the Joint Chiefs of Staff Manual 5120.01A, Joint Doctrine Development Process* (29 December). Washington, DC: Joint Chiefs of Staff. Available at: www.jcs.mil/Portals/36/Documents/Doctrine/pubs/cjcsm5120_01a.pdf (Accessed 24 June 2018).

Chairman, Joint Chiefs of Staff (2014b) *JP 2–01.3, Joint Intelligence Preparation of the Operational Environment* (21 May). Washington, DC: Joint Chiefs of Staff. Available at: https://fas.org/irp/doddir/dod/jp2-01-3.pdf (Accessed 24 June 2018).

Chairman, Joint Chiefs of Staff (2014c) *JP 3–05, Special Operations* (16 July). Washington, DC: Joint Chiefs of Staff. Available at: www.jcs.mil/Portals/36/Documents/Doctrine/pubs/jp3_05.pdf?ver=2018-03-15-111255-653 (Accessed 24 June 2018).

Chairman, Joint Chiefs of Staff (2014d) *JP 3–13, Information Operations* (20 November). Washington, DC: Joint Chiefs of Staff. Available at: www.jcs.mil/Portals/36/Documents/Doctrine/pubs/jp3_13.pdf (Accessed 24 June 2018).

Chairman, Joint Chiefs of Staff (2014e) *JP 3–29, Foreign Humanitarian Assistance* (3 January). Washington, DC: Joint Chiefs of Staff. Available at: www.jcs.mil/Portals/36/Documents/Doctrine/pubs/jp3_29.pdf (Accessed 24 June 2018).

Chairman, Joint Chiefs of Staff (2015) *The National Military Strategy of the United States of America 2015 – The United States Military's Contribution To National Security* (June). Washington, DC: Joint Chiefs of Staff. Available at: www.jcs.mil/Portals/36/Documents/Publications/2015_National_Military_Strategy.pdf (Accessed 22 June 2018).

Chairman, Joint Chiefs of Staff (2016) *JP 3–07, Stability* (3 August). Washington, DC: Joint Chiefs of Staff. Available at: www.jcs.mil/Portals/36/Documents/Doctrine/pubs/jp3_07.pdf (Accessed 24 June 2018).

Chairman, Joint Chiefs of Staff (2017a) *DOD Dictionary of Military and Associated Terms* (August). Washington, DC: Joint Chiefs of Staff. Available at: www.dtic.mil/doctrine/new_pubs/dictionary.pdf (Accessed 17 December 2017).

Chairman, Joint Chiefs of Staff (2017b) *JP 3–20, Security Cooperation* (23 May). Washington, DC: Joint Chiefs of Staff. Available at: www.jcs.mil/Portals/36/Documents/Doctrine/pubs/jp3_20_20172305.pdf (Accessed 24 June 2018).

Chairman, Joint Chiefs of Staff (2017c) *JP 3–68, Noncombatant Evacuation Operations* (14 November). Washington, DC: Joint Chiefs of Staff. Available at: www.jcs.mil/Portals/36/Documents/Doctrine/pubs/jp3_68pa.pdf (Accessed 24 June 2018).

Chairman, Joint Chiefs of Staff (2018a) *JP 3–07.3, Peace Operations* (1 March). Washington, DC: Joint Chiefs of Staff. Available at: https://pksoi.armywarcollege.edu/conferences/

psotew/documents/wg2/jp3_07_3%20Peace%20Operations%201Mar18.pdf (Accessed 24 June 2018).

Chairman, Joint Chiefs of Staff (2018b) *JP 3–24, Counterinsurgency* (25 April). Washington, DC: Joint Chiefs of Staff. Available at: www.jcs.mil/Portals/36/Documents/Doctrine/pubs/jp3_24.pdf?ver=2018-05-11-102418-000 (Accessed 24 June 2018).

Charlesworth, H (2005) 'Not Waving, but Drowning: Gender Mainstreaming and Human Rights in the United Nations'. *Harvard Human Rights Journal* 18: pp. 1–14. Available at: https://openresearch-repository.anu.edu.au/bitstream/1885/80561/2/01_Charlesworth_Not_Waving_but_Drowning:_2005.pdf (Accessed 25 June 2018).

DKI APCSS (Daniel K. Inouye Asia-Pacific Center for Security Studies) (2015) *Annual Report 2015*. Honolulu, HA: Daniel K. Inouye Asia-Pacific Center for Security Studies. Available at: https://apcss.org/wp-content/uploads/2016/06/annual_report_2015v2_final.pdf (Accessed 22 June 2018).

DKI APCSS (Daniel K. Inouye Asia-Pacific Center for Security Studies) (2016) 'Enhancing Access: Mongolia Workshop Focuses on Increasing Women's Roles'. *Currents – Inclusive Security* 28: pp. 18–21. Available at: https://apcss.org/wp-content/uploads/2016/12/CurrentsFall2016_Full-Issue.pdf (Accessed 24 June 2018).

Department of Defense (DoD) (2009) 'Joint Targeting Cycle and Collateral Damage Estimation Methodology (CDM)' [PowerPoint presentation] Washington, DC (10 November). Available at: www.aclu.org/files/dronefoia/dod/drone_dod_ACLU_DRONES_JOINT_STAFF_SLIDES_1-47.pdf (Accessed 23 June 2018).

Department of Defense (2014) *Quadrennial Defense Review 2014* (4 March). Washington, DC: Department of Defense. Available at: http://archive.defense.gov/pubs/2014_Quadrennial_Defense_Review.pdf (Accessed 22 June 2018).

Department of Defense (2016) *Law of War Manual* (December). Washington, DC: Department of Defense. Available at: www.defense.gov/Portals/1/Documents/DoD_Law_of_War_Manual-June_2015_Updated_May_2016.pdf (Accessed 5 July 2018).

Department of Defense (2017) *Department of Defense Annual Report on Sexual Assault in the Military, Fiscal Year 2017, Appendix B, Statistical Data on Sexual Assault*. Washington, DC: Department of Defense. Available at: http://sapr.mil/public/docs/reports/FY17_Annual/Appendix_B_Statistical_Data_on_Sexual_Assault.pdf (Accessed 12 June 2018).

Department of Defense (2018) *Summary of the National Defense Strategy of The United States of America*. Washington, DC: Department of Defense. Available at: www.defense.gov/Portals/1/Documents/pubs/2018-National-Defense-Strategy-Summary.pdf (Accessed 20 June 2018).

Department of State (2016) *United States Strategy To Prevent And Respond To Gender-Based Violence Globally*. Washington, DC: Department of State. Available at: www.state.gov/documents/organization/258703.pdf (Accessed 23 May 2018).

Headquarters, Department of the Army (HQDA) (2011) *Field Manual (FM) 3–57, Civil Affairs Operations* (October). Washington, DC: Department of the Army. Available at: `24 June 2018).

Headquarters, Department of the Army (2012) *Army Techniques Publication (ATP) 3–07.5, Stability Techniques* (31 August). Washington, DC: Department of the Army. Available at: https://armypubs.army.mil/epubs/DR_pubs/DR_a/pdf/web/atp3_07x5.pdf (Accessed 24 June 2018).

Headquarters, Department of the Army (2013a) *ADP 3–07, Stability* (15 February). Washington, DC: Department of the Army. Available at: https://armypubs.army.mil/epubs/DR_pubs/DR_a/pdf/web/adp3_07.pdf (Accessed 24 June 2018).

Headquarters, Department of the Army (2013b) *ATP 3–57.10, Civil Affairs Support to Populace and Resources Control* (6 August). Washington, DC: Department of the Army.

Available at: https://armypubs.army.mil/epubs/DR_pubs/DR_a/pdf/web/atp3_57x10.pdf (Accessed 24 June 2018).

Headquarters, Department of the Army (2013c) *ATP 3–57.20, Multi-Service Support Techniques for Civil Affairs Support to Foreign Humanitarian Assistance* (15 February). Washington, DC: Department of the Army. Available at: https://armypubs.army.mil/epubs/DR_pubs/DR_a/pdf/web/atp3_57x20.pdf (Accessed 24 June 2018).

Headquarters, Department of the Army (2014a) *ADP 1–01, Doctrine Primer* (2 September). Washington, DC: Department of the Army. Available at: https://armypubs.army.mil/epubs/DR_pubs/DR_a/pdf/web/adp1_01.pdf (Accessed 24 June 2018).

Headquarters, Department of the Army (2014b) *ATP 2–01.3, Intelligence Preparation of the Battlefield/Battlespace* (10 November). Washington, DC: Department of the Army. Available at: https://armypubs.army.mil/epubs/DR_pubs/DR_a/pdf/web/atp2_01x3.pdf (Accessed 24 June 2018).

Headquarters, Department of the Army (2014c) *ATP 3–57.30, Civil Affairs Support to Nation Assistance* (1 May). Washington, DC: Department of the Army. Available at: https://armypubs.army.mil/epubs/DR_pubs/DR_a/pdf/web/atp3_57x30.pdf (Accessed 24 June 2018).

Headquarters, Department of the Army (2014d) *ATP 3–57.60, Civil Affairs Planning* (27 April). Washington, DC: Department of the Army. Available at: https://armypubs.army.mil/epubs/DR_pubs/DR_a/pdf/web/atp3_57x60.pdf (Accessed 24 June 2018).

Headquarters, Department of the Army (2014e) *ATP 3–57.70, Civil-Military Operations Center* (5 May). Washington, DC: Department of the Army. Available at: https://armypubs.army.mil/epubs/DR_pubs/DR_a/pdf/web/atp3_57x70.pdf (Accessed 24 June 2018).

Headquarters, Department of the Army (2014f) *FM 3–07, Stability* (2 June). Washington, DC: Department of the Army. Available at: https://armypubs.army.mil/epubs/DR_pubs/DR_a/pdf/web/fm3_07.pdf (Accessed 24 June 2018).

Headquarters, Department of the Army (2014g) *FM 3–24, Insurgencies and Countering Insurgencies* (13 May). Washington, DC: Department of the Army. Available at: https://armypubs.army.mil/epubs/DR_pubs/DR_a/pdf/web/fm3_24.pdf (Accessed 24 June 2018).

Headquarters, Department of the Army (2015) *ATP 3–07.6, Protection of Civilians* (29 October). Washington, DC: Department of the Army. Available at: https://armypubs.army.mil/epubs/DR_pubs/DR_a/pdf/web/atp3_07x6.pdf (Accessed 24 June 2018).

Headquarters, Department of the Army (2016) *ATP 3–55.4, Techniques for Information Collection During Operations Among Populations* (5 April). Washington, DC: Department of the Army. Available at: https://armypubs.army.mil/epubs/DR_pubs/DR_a/pdf/web/atp3_55x4.pdf (Accessed 24 June 2018).

Headquarters, Department of the Army (2017) *FM 3–0, Operations* (6 December). Washington, DC: Department of the Army. Available at: https://armypubs.army.mil/epubs/DR_pubs/DR_a/pdf/web/ARN6687_FM%203-0%20C1%20Inc%20FINAL%20WEB.pdf (Accessed 24 June 2018).

Headquarters, Department of the Army (2018) *ADP 3–05, Special Operations* (January). Washington, DC: Department of the Army. https://armypubs.army.mil/epubs/DR_pubs/DR_a/pdf/web/ARN10343_ADP%203-05%20FINAL%20WEB.pdf (Accessed 24 June 2018).

Horgan, J and Braddock, K (2010) 'Rehabilitating the Terrorists? Challenges in Assessing the Effectiveness of De-radicalization Programs'. *Terrorism and Political Violence* 22: pp. 267–91. DOI: 10.1080/09546551003594748.

Prescott, JM (2013) 'NATO Gender Mainstreaming and the Feminist Critique of the Law of Armed Conflict'. *Georgetown Journal of Gender and the Law* 14(1): pp. 83–131. Available at: https://papers.ssrn.com/sol3/papers.cfm?abstract_id=2246000## (Accessed 25 June 2018).

Prescott, JM (2014) 'Climate Change, Gender, and Rethinking Military Operations'. *Vermont Journal of Environmental Law* 15(4): pp. 766–802. Available at: http://vjel.vermontlaw.edu/files/2014/06/Prescott_ForPrint.pdf (Accessed 22 June 2018).

Prescott, JM and Dunlap, J (2000) 'Law of War and Rules of Engagement Training for the Objective Force: A Proposed Methodology for Training Role-Players'. *The Army Lawyer* (September): pp. 43–47. Available at: www.loc.gov/rr/frd/Military_Law/pdf/09-2000.pdf (Accessed 24 June 2018).

Rosenberg, M and Phillips, D (2015) 'All Combat Roles Now Open to Women, Defense Secretary Says'. *New York Times* (3 December). Available at: www.nytimes.com/2015/12/04/us/politics/combat-military-women-ash-carter.html?mcubz=3 (Accessed 22 June 2018).

Seck, HH (2015) 'Marine Corps Revives Female Engagement Team Mission'. *Marine Corps Times* (5 August). Available at: www.marinecorpstimes.com/news/your-marine-corps/2015/08/05/marine-corps-revives-female-engagement-team-mission/ (Accessed 3 February 2018).

Serena, C and Clark, C (2016) 'A New Kind of Battlefield Awaits the U.S Military – Megacities'. *Reuters* (22 July). Available at: www.reuters.com/article/usa-commentary-megacities-idUSKCN0X52B2 (Accessed 22 July 2018).

Shepherd, LJ (2017) 'The Role of the WPS Agenda in Countering Violent Extremism'. *The Strategist* [online] (8 March). Available at: www.aspistrategist.org.au/role-wps-agenda-countering-violent-extremism/ (Accessed 3 April 2018).

Szayna, TS; Larson, EV; O'Mahoney, A; Robson, S; Schaefer, AG and Matthews, M (2015) 'Considerations for Integrating Women into Closed Occupations in the U.S. Special Operations Forces'. RAND National Defense Research Institute, Forces and Resources Center. Available at: www.defense.gov/Portals/1/Documents/wisr-studies/SOCOM%20-%20Considerations%20for%20Integrating%20Women%20into%20Closed%20Occupations%20in%20the%20US%20Special%20Operations%20Forces.pdf (Accessed 23 December 2017).

UN Peacekeeping (2017) 'Contributors to UN Peacekeeping Operations by Country and Post'. UN (30 November). Available at: http://peacekeeping.un.org/sites/default/files/contributor.pdf (Accessed 22 June 2018).

UN Secretary General (2015) *Implementation of General Assembly Resolutions 55/235 and 55/236*. UN Doc. A/70/331/Add. 1. UN. Available at: www.un.org/en/ga/search/view_doc.asp?symbol=A/70/331/Add.1 (Accessed 22 May 2018).

US Army War College (n.d.) 'Joint Strategic Capabilities Plan (JSCP)'. US Army War College. Available at: https://ssl.armywarcollege.edu/dde/documents/jsps/terms/jscp.cfm (Accessed 22 June 2018).

Vergun, D (2015) 'Directive Opens 4,100 Special OPS Positions to Women'. *US Army.mil* (13 March). Available at: www.army.mil/article/144515/directive_opens_4100_special_ops_positions_to_women (Accessed 25 June 2018).

The White House (2011) *United States National Action Plan on Women, Peace, and Security* (December). Washington, DC: The White House. Available at: https://obamawhitehouse.archives.gov/sites/default/files/email-files/US_National_Action_Plan_on_Women_Peace_and_Security.pdf (Accessed 22 June 2018)

The White House (2015) *National Security Strategy* (February). Washington, DC: The White House. Available at: http://nssarchive.us/wp-content/uploads/2015/02/2015.pdf (Accessed 22 June 2018).

The White House (2016) *The United States National Action Plan on Women, Peace, and Security* (June). Washington, DC: The White House. Available at: www.usaid.gov/sites/default/files/documents/1868/National%20Action%20Plan%20on%20Women%2C%20Peace%2C%20and%20Security.pdf (Accessed 22 June 2018).

The White House (2017) *National Security Strategy of the United States of America* (December). Washington, DC: The White House. Available at: www.whitehouse.gov/wp-content/uploads/2017/12/NSS-Final-12-18-2017-0905.pdf (Accessed 22 June 2018).

Women, Peace, and Security Act of 2017, Public Law 115-68 (6 October). Available at: www.congress.gov/115/plaws/publ68/PLAW-115publ68.pdf (Accessed 24 June 2018).

Yamin, S (2018) 'Advancing Security through a Gender Lens: Building Capacity of Asia Pacific Security practitioners Through Executive Education'. In: *'Promoting Global Leadership', 2018 Women, Peace and Security Conference*. Providence, RI: Brown University (31 May–1 June) (Copy on file with author).

7
AUSTRALIAN STRATEGY, DOCTRINE AND GENDER

The Australian Defence Force (ADF) has strongly engaged with the tasks it has received under the Australian national action plan on women, peace and security, and it has begun to make significant headway in the operationalisation of gender issues in its activities and missions. But where the UK's focus on gender is driven primarily by foreign policy and it targets a specific set of countries in which it is already engaged, and the US efforts will perhaps be geared towards meeting legislative requirements in its activities and operations overseas, Australia's drivers of change are different. As will be discussed later, it is important to consider the Australian Defence Force's efforts in operationalising gender as part of a larger effort to achieve gender equality within the Australian government itself (Australian Human Rights Commission, 2011, 2012).

Reflecting this expression of political will in the domestic arena, the Australian defence establishment has instituted a set of mechanisms within its governance structure that continue to push the incorporation of gender considerations forward in its work. This sort of institutionalisation can serve as a consistent engine of change, long after the original decision-makers who drafted the policy being implemented have left the stage. Thus, before considering the Australian national and defence strategies, Australia's national action plans and Australian Defence Force doctrine, it is worthwhile to first examine the way these policies are being implemented on an institutional basis.

Implementation infrastructure

Evidencing a very high-level commitment to operationalising gender, since 2014 the Australian Defence Force has appointed a GENAD to advise the Chief of Defence Force, who is a civilian outside of the force (Department of Defence, 2014). The Department of Defence has also established a Gender Equality Advisory

Board, which is a direction-setting advisory body that drives and shapes the direction of the 'Secretary of Defence's and Chief of Defence Force's gender equality priorities within the broader Defence cultural reform agenda' (Department of the Prime Minister and Cabinet, 2014: 72). The board is jointly chaired by the Secretary of Defence and the Chief of Defence Forces, indicating institutional championing of this effort (Jennifer Wittwer, 2015: personal communication). Board members include 'Defence officials, ADF Women's advisors, senior private sector and civil society representatives and a Special Advisor (Sex Discrimination Commissioner, Australian Human Rights Commission)' (Department of Prime Minister and Cabinet, 2014: 72). The board meets quarterly, and, since December 2013, it has included women, peace and security and national action plan implementation as standing agenda items. The activities of the Gender Equality Advisory Board are complemented by the Defence National Action Plan Implementation Plan Working Group, consisting of representatives from the different services and other groups within the department, which is tasked with facilitating progress on the national action plan (Department of the Prime Minister and Cabinet, 2014: 72–73).

This organisational structure is buttressed by required annual reports on the status of women in the Australian Defence Force. The 2017 annual report is 153 pages long, and it analyses a wide spectrum of data related to women's participation in the force and assesses outcomes against metrics approved by the Chiefs of Service Committee (Department of Defence, 2017: v). For example, Chapter 4 of this report focuses on performance, skills and career management of women, and it looks at aspects of a successful military career that are deeper and more relevant than merely counting numbers of female military students or commanders of units. It drills down into more subtle, but very significant trends, such as how women personnel are encouraged to stay and progress in the Australian Defence Force through mentoring, networking and sponsorship, and how women deal with cultural barriers to reaching senior leadership positions (29–48). It is difficult to overemphasise the importance of reliable data collected against well-conceived metrics to the meaningful operation of any organisational body tasked with accomplishing transformational change in a national military force. Against this backdrop of institutional commitment, the national-level strategies of Australia can be better placed in the context of Australia's efforts to implement the goals and principles of UNSCR 1325.

National level security-related strategy

National security strategy

The most recent document on national-level security strategy for Australia dates from 2013, *Strong and Secure: A Strategy for Australia's National Security* (Department of the Prime Minister and Cabinet, 2013). A fairly concise document, it mentions women only once, in providing the example of Australia's whole-of-government partnership with Afghanistan to build a girls' school in Uruzgan Province (37). Gender is likewise only noted once, in describing the range of issues that affect security, stability and economic prosperity in certain Pacific island countries (38). In his 2017 national security statement to Parliament, Prime Minister Turnbull discussed a number of strategies in specific areas that had been recently implemented and actions Australia had taken, in the area of cybersecurity for example, but did not mention efforts regarding women or gender (Turnbull, 2017). It is difficult to pick up the threads of UNSCR 1325 in Australian security strategy and trace them through subordinate levels of government security-related documents. Instead, this line of thought registers elsewhere at the national level.

National-level strategy impacting the Department of Defence

The 2016 Australian government strategy published by the Australian Public Service Commission, *Balancing the Future: The Australian Public Service Gender Equality Strategy 2016–19*, is intended to help resolve the gender imbalance in the Australian government agencies by requiring them to undertake efforts in five specific areas. These areas include fostering a supportive and enabling workplace culture, achieving gender equality in agency leadership, working innovatively to embed gender equality in employment practices, increasing the use of flexible work-arrangements by both men and women and developing methods to measure and assess whether the first four actions are working (Australian Public Service Commission, 2016: 5). Although these measures are not specifically directed at operationalising gender in the Australian Defence Force, from an infrastructural perspective they could prove complementary to such efforts in the long term.

Against the backdrop of earlier high-level assessments of the lack of gender equality and the discriminatory treatment of women in the Australian Defence Force (Australian Human Rights Commission, 2012), the Department of Defence has developed its own Defence Action Plan to implement the Public Service strategy. Although the focus of this plan is similarly on internal gender equality, it might also have a significant complementary effect on the Australian Defence Force's operationalisation of gender if its goals are fully met. For example, with regard to the third strategic action of Australian Public Service Commission's plan, embedding gender equality in employment practices, the Department of Defence has committed itself to ensuring that half of personnel selection panels are women. Further, to build more female leaders, the Department of Defence would establish gender equality targets for positions in leadership, and develop an exchange programme for senior women leaders (Department of Defence, n.d.: 'APS Gender Equality Strategy 2016–2019 – Defence Action Plan 2016').

The 2016 defence white paper

Australia's national commitment to supporting the incorporation of gender considerations into its military operations registers significantly in the annual white papers published by the Department of Defence on its policies, programmes and initiatives in the security sector. The 2016 *Defence White Paper* stated that the Department of Defence 'will provide expertise to support the Australian National Action on Women, Peace and Security 2012–2018' that implements UNSCR 1325 (Department of Defence, 2016: 137). It noted the many actions under the Australian national action plan for which the Department of Defence was responsible for implementing, and it described them as concrete measures by which Australia will 'integrate a gender perspective into its peace and security efforts, protect women and girls' human rights and promote their participation in conflict prevention, management and resolution' (137). The 2016 *Defence White Paper* also recognised that under the Australian national action plan, women in the Australian Defence Force were playing a more prominent and influential role in operations conducted by it, for example, in serving as GENADs to the NATO-led military mission in Afghanistan and providing guidance on the construction of proper facilities and provision of proper equipment to female Afghan military members. Further, training for both sexes on prevention of gender-based violence and harassment was being given to all personnel (137).

Within the Australian Defence Force itself, the 2016 *Defence White Paper* noted that the military is increasing its efforts to recruit more women into its ranks, including more female Indigenous Australians, through programs that include 'targeted recruitment initiatives, retention measures and career support' (Department of Defence, 2016: 151). In addition, all combat roles in the Australian Defence Force were opened to women in 2011 (152). Importantly, these efforts towards gender equality are seen as part of the larger initiative to bring about a change in culture in the Australian Defence Force, and to 'give further attention to career opportunities for women, people from culturally and linguistically diverse backgrounds and people with disabilities' (155).

The Australian national action plan 2012–2018

The first Australian national action plan was published in 2012 and was intended to cover Australia's women, peace and security efforts through 2018 (Australian Government, 2012). It was extended in December 2017 until 30 June 2019 (Australian Government, 2017). Under the plan, the Australian government intended to improve outcomes for women and girls in conflict-affected regions in five specific thematic areas: first, conflict prevention; second, increased participation of women in political processes related to conflict; third, greater protection during all phases of armed conflict; fourth, heightened consideration in the implementation of relief and recovery efforts afterwards, and finally, the promotion of normative measures. The national action plan sets out strategies to improve outcomes, and lists a number of actions that benchmark how these strategies will be implemented (Australian Government, 2012). The Australian Defence Force has a role in the implementation of 17 of the 24 separate actions for Australian government agencies set out in the Australian national action plan (Department of Defence, n.d.: 'Women, Peace and Security in Defence').

Of the strategies set out in the national action plan in which there is a military role, three in particular appear to be relevant to its implementation of UNSCR 1325 in an operational sense – integrating 'a gender perspective into Australia's policies on peace and security', promoting implementation internationally, and taking a coordinated and holistic approach domestically and internationally to women, peace and security (Australian Government, 2012: 19). As to the use of a gender perspective, the Australian Defence Force is required to 'develop guidelines for the protection of civilians, including women and girls' (19). Accordingly, any review of the force's incorporation of a gender perspective in its activities and operations must recognise that it is occurring alongside very significant government actions to ensure equal treatment of women service members and to improve gender diversity in the military from a capability perspective. For example, the Australian Government's *2014 Progress Report Australian National Action Plan on Women, Peace and Security 2012–2018* specifically notes that reviews conducted by the Australian Human Rights Commission into sex discrimination in the Australian Defence Force intersect with the national action plan, 'which is designed to enhance military capability and contributions to peace and security efforts through increased participation of women in the force, on deployments, and in senior decision-making roles' (Department of the Prime Minister and Cabinet, 2014: 16).

Regarding Australian Defence Force efforts in conflict-affected areas, the 2012 national action plan tasked the force with promoting 'opportunities for women's leadership and participation in decision-making' at the country level, to consider the use of specific capabilities such as 'female engagement teams and the use of gender advisors', and to 'promote women's involvement in the development of institutions, including national judiciary, security and governance structures' (Australian Government, 2012: 23). The national action plan set out a number of activity benchmarks to gauge the progress of the Australian Defence Force in accomplishing its tasks, and the force has established a Defence Implementation Plan to further develop the specific actions that must be taken to meet the plan's requirements. The Defence Implementation Plan is described as a flexible matrix that provides for the inclusion of new tasks as they arise in the course of the Australian Defence Force's implementation of its tasks under the national action plan, and it is reviewed on a quarterly basis by an implementation working group (Department of Defence, n.d.: 'Defence Implementation Plan')

Appendix C of the 2012 national action plan is titled 'Development of International Human, Rights, Humanitarian and Criminal Law in Relation to Gender-based Violence in Armed Conflict' (Australian Government, 2012: 56–58). It pointed out that women and girls are disproportionally affected by such violence in armed conflict, and that although the term 'gender-based violence' includes rape, its compass is much broader and includes other sexually violent acts such as sexual slavery and forced prostitution (56). It listed the different legal protections afforded women and girls during armed conflict under IHL against sexual and gender-based violence in some detail (57–58), but it did not address violence against women and girls in the context of proportionality in kinetic operations. Unfortunately, this tends to reinforce a view that the portion of IHL that is amenable to gender perspectives is that part which intersects with international human rights law and international criminal law.

Assessment of the 2012–2018 Australian national action plan

External review

Similar to the UK, Australia sought the services of an external civil society organisation evaluator to assess its work under the 2012 national action plan. In its interim review, the Humanitarian Advisory Group found that the plan's actions were relevant to implementing the women, peace and security goals and concerns because of their positive alignment with the steps all nations were expected to take under UNSCR 1325 and subsequent resolutions. However, monitoring and evaluation measures set out in the plan were insufficient to properly track progress on these actions (Humanitarian Advisory Group, 2015: 7). It further found that government agencies had unevenly implemented the Australian plan's requirements which resulted in 'limited institutionalisation and Australian NAP awareness in some agencies', perhaps as the result of uneven funding and resourcing for this effort amongst them (8). The Humanitarian Advisory Group's review also noted concern with the quantitative approach taken in implementing the plan, and for example, recommended supplementing the metric of number of personnel trained on women, peace and security issues prior to deployment with a qualitative assessment of 'post-deployment perceptions about the impact and applicability of WPS training for day-to-day operations' (10).

Internal review

As mentioned previously regarding the importance of gender equality initiatives in the whole of Australian government to national action plan implementation, Australia has also conducted its own internal review of its progress in meeting the national action plan's requirements. For example, the plan recognises the importance of embedding gender-perspective considerations in doctrine in order to accomplish their incorporation across the spectrum of Australian Defence Force activities and operations. Accordingly, it identifies as an implementation metric 'the number, title and description of relevant policy and guidance documents that contain reference to the Women, Peace and Security agenda or resolutions 1325, 1820, 1888, 1889 and 1960' (Australian Government, 2012: 28).

The Australian government's 2014 progress report noted that these references were being included 'in key strategic guidance documents', including the 2014 version of the 'Defence Corporate Plan, the 2014 Defence Annual Plan, the Defence International Engagement Strategy and the Defence Regional Engagement Strategy' (Department of the Prime Minister and Cabinet, 2014: 15). Further, it noted that 'Operational guidance on Women, Peace and Security will be included in the Chief of the Defence Force Planning Directives which inform strategic direction and planning for operations' (15). Although the inclusion of these references is an easily verifiable metric, as the Humanitarian Advisory Group pointed out in its evaluation, focusing on this measure without correlating it with a qualitative assessment might be of limited value.

However, the Department of Defence does appear to have made continued and significant progress in meeting the benchmarks it set for itself in achieving the incorporation of gender considerations into its activities and operations. The Australian government's 2016 progress report noted that as of the end of 2015, 23 key Department of Defence policy and guidance documents had been amended to reference women, peace and security goals and concerns, including the Defence Corporate Plan (the department's principal planning document), the Defence Business Plan (provides implementation guidance on the Defence Corporate Plan), Australian Defence Force operational planning documents for both operations and exercises, and the Royal Australian Air Force's gender equality strategy. Further, and very importantly from an awareness and capability-building perspective in the land force, the Australian Army had issued a directive requiring all personnel to receive annual training on Women, Peace and Security (Department of the Prime Minister and Cabinet, 2017: 139–40).

Joint doctrine

Structure and the doctrine revision process

There are two main types of Australian joint doctrine, Australian Defence Doctrine Publications (ADDPs), 'pitched at the philosophical and high-application level', and Australian Defence Force Publications (ADFPs), intended to be used at the 'application and procedural level' (Chief of Joint Operations, 2006: iii). Because the bulk of current Australian Defence Force joint doctrine is not generally available to the public, it is not possible to examine it as thoroughly as the NATO, UK or US joint doctrine. However, a review of older Australian joint doctrine and of different reports regarding recent changes to it provide a useful understanding of the doctrinal baseline regarding women, peace and security goals and concerns in joint doctrine prior to the revision process, and an idea as to those documents that are being amended to bring them in line with the requirements of the national action plan.

As the Australian Defence Force representative at the 2015 annual Quinquepartite Combined Joint Warfare Conference (QCJWC) noted, Australian joint doctrine is continuously reassessed and amended over a four-year review cycle (QCJWC, 2015: B-3). As of 2015, the Australian Defence Force's Joint Doctrine Centre had incorporated Women, Peace and Security and Protection of Civilians concepts into its ordinary review criteria for all joint doctrine (B-5). The Joint Doctrine Centre identified three publications that would include specific content on these concepts: *ADDP 3.8, Peace Operations*; *ADDP 3.20, The Military Contribution to Humanitarian*

Operations; and *ADFP 5.0.1, The Joint Military Appreciation Process* (B-5). Importantly, *ADFP 5.0.1, Joint Military Appreciation Process*, would include a discussion of the roles and responsibilities of gender advisors, and their place in the command group (B-5).

Although *ADDP 00.3, Multinational Operations* (Chief of Joint Operations, 2011), was later noted by the Joint Doctrine Centre as being in the queue for revision at the 2016 Quinquepartite Combined Joint Warfare Conference (QCJWC, 2016: E1–E3), and it would appear to be a document for which the treatment of women, peace and security goals and concerns would be most appropriate, it was not singled out in this regard. This is perhaps unfortunate, since the older version mentioned women only twice, both times in the context of the composition of different armed forces (Chief of Joint Operations, 2011: 1-5, 3-18). There was no mention of gender or UNSCR 1325. Similarly, the 2013 information operations doctrine, *ADDP 3.13*, contained no mention of women or gender, even though its purpose was to 'provide guidance for the planning and conduct of Information Activities in the joint and multinational environments' (Chief of Joint Operations, 2013: iii).

The 2016 national action plan progress report confirmed that *ADDP 3.20, The Military Contribution to Humanitarian Operations* and *ADDP 3.8 Peace Operations*, had been updated to explain the five thematic areas of the Australian national action plan (Prevention, Protection, Participation, Relief and Recovery, and Normative) and to highlight protections for civilians and the needs of women and children (Department of the Prime Minister and Cabinet, 2017: 140–41). *ADDP 3.8, Peace Operations*, now reflects extensive revisions in this regard (Christopher Ross, 2016: personal communication). This represents a significant shift in the thinking of the Australian Defence Force with regard to these very civilian-centric operations. Previously, *ADDP 3.8* had only noted that Australia was a party to the UN Convention against Transnational Organised Crime's People Trafficking Protocol (Chief of Joint Operations, 2009a: 2-14).

The 2015 version of *ADDP 3.8* extensively explains the meaning and the significance of using a gender perspective in peace operations. It notes the gender-differentiated impact of armed conflict, and provides examples of the harms suffered by women and girls as a result of being the victims of sexual violence during armed conflict (Chief of Joint Operations, 2015: 4-14). *ADDP 3.8* provides a succinct but useful description of the UN Security Council resolutions that followed UNSCR 1325, and cross-references the reader to *ADDP 3.20, The Military Contribution Humanitarian Operations*, for additional guidance (4-14–4-15). It also contains thorough guidance for planners on how to incorporate a gender perspective into their work and identify civilians who may be at risk, and to develop courses of action to mitigate these risks in the short, medium and long term (5-12–5-13). Finally, *ADDP 3.80* contains a detailed explanation of how Australian units and personnel can take practical measures in the field to align their actions with the goals and requirements of the national action plan (5-13–5-14).

Australian Defence Force civil-military cooperation doctrine has been under review and was intended to be revised in late 2018 to include content relevant to gender and the protection of civilians (Christopher Ross, 2018: personal communication). To understand the scope of the force's undertaking in this regard, it is useful to review the prior version of the doctrine, *ADDP 3.11*, *Civil-Military Operations*, from 2009. In this document, the Australian Defence Force defined civil-military cooperation as 'any measures, activities, or planning undertaken by the military which both facilitates the conduct of military operations, and builds support, legitimacy and consent, within the civil population in furtherance of the mission' (Chief of Joint Operations, 2009b: 1–2).

Although the 2009 version of *ADDP 3.11* was very detailed and comprehensive in terms of providing guidance in working with host nation populations, it addressed women only twice. It first noted that they might be an at-risk population in host nations, and second, when detained in the course of an armed conflict, they 'are to be quartered separately from men and supervised by women' (2B-2). Whilst the idea of separate quarters for detained women being supervised by women might strike most as obvious, as will be discussed in the next chapter on IHL and gender, this was not actually required by the third Geneva Convention of 1949 that covered prisoners of war. *ADDP 3.11* mentions gender once, in that humanitarian assistance was to be rendered impartially irrespective of gender (1-11). To be useful from the perspective of UNSCR 1325 and the Australian national action plan, the revision of *ADDP 3.11* will need to be rather extensive.

Australian guidelines for the protection of civilians

Although technically not official doctrine, the Australian Guidelines for the Protection of Civilians (Guidelines) occupy a joint doctrine-like space, and they are similar thematically to the US Army protection of civilians doctrine discussed in the previous chapter. Created through collaboration between the Australian Civil-Military Centre (ACMC), government agencies and civil society organisations, the Guidelines 'provide a whole of government perspective on protecting civilians in armed conflict and other violent situations' and were required under the national action plan (ACMC, 2015: 4). They are centred on Australian Defence Force and Australian Federal Police operations (5). Whilst the US Army doctrine on protection of civilians is quite detailed, however, the Guidelines are a bit more general.

There are seven separate principles listed in the Guidelines. Of these, two seem aspirational rather than substantive: protection of civilians 'is integral to Australia's contribution to international peace and security', and 'Across all operations Australia will support, and not detract, from' protection of civilians (ACMC, 2015: 9). Two others seem a bit obvious – protection of civilians strategies reflect Australia's international legal obligations 'where applicable', and 'Australian agencies will act in coordination with all protection actors including military, police and civilian components' (9). The three remaining principles, however, are substantive, and would likely have a significant impact from both planning and operations perspectives for the Australian Defence Force.

First, when host nations 'are unable or unwilling to protect their own citizens, or when government forces themselves pose a threat to civilians', Australian protection of civilian strategies will 'support the host state's protection efforts or inform actions to protect civilians' (ACMC, 2015: 9). Second, Australian agencies will plan and consult 'with civil society organisations and local communities members with a view to creating a sustainable impact' (9) – that is, the planning and consultation must be geared to creating effects that last, rather than just 'checking a box'. Third, and most importantly for this chapter, these strategies are to address the different needs of a broad set of at-risk groups of civilians, 'including women and girls, men and boys, the frail and the wounded, people with disabilities and ethnic minorities, refugees and internally displaced persons, and professionals at risk such as medical personnel' (9). This is a significant step towards differentiating between various types of civilians to assess their actual needs, rather than just viewing the amorphous mass of the 'population' as something to be protected.

Consistent with UN peacekeeping policy, the Guidelines identify three focus areas for the protection of civilians: increasing protection through dialogue and engagement, providing physical security when necessary, and establishing a protective environment (ACMC, 2015: 9). The focus area most likely to require the Australian Defence Force to engage operationally with the kinetic aspects of the operationalisation of gender is that of physical protection. The Guidelines broadly describe this focus area as including 'the show or use of force to prevent, deter, pre-empt and respond to situations in which civilians are under threat of physical violence' (9). Significantly, the Guidelines give the example of 'Responding to violent attacks, and sexual and gender-based violence (SGBV) with all necessary means, including the use of force, where permissible' (9). As noted in Chapter 4, 'NATO strategy, doctrine and gender', in its discussion of NATO ROE and the construction of ROE to protect certain civilians and perhaps their property, this has important implications not just for planners, but for the education and training of Australian Defence Force service members so they will to be able to apply force appropriately in these situations as well.

Army doctrine

From a planning perspective, *Land Warfare Doctrine (LWD) 5–1–4, The Military Appreciation Process*, has been recently revised to include points on gender perspectives in the planning process (Department of the Prime Minister and Cabinet, 2017: 144). Apparently due to its classification it is not available to the public, and therefore cannot be assessed here to determine whether it incorporates sufficient linkages between the Guidelines and the land-oriented military planning process to effectively serve as a tool for including gender perspectives. In general, however, other higher-level Australian Army doctrine related to planning does not yet fully reference the goals and principles of UNSCR 1325 that are to be implemented under the national action plan. For example, the recent *LWD 1, The Fundamentals of Land Power*, includes no mention of gender, and mentions women only in the sense of the Australian Army having both male and female personnel 'dedicated to the service of the nation' (Australian Army, 2017a: 41). Neither *LWD 2–0, Intelligence*

(Australian Army, 2018a) nor *LWD 4–0, Logistics* (Australian Army, 2018c) makes any mention of women or gender at all.

From an operations perspective, there has also been recent significant change in the high-level doctrine employed by the Australian Army that has yet to be reflected fully in subordinate operational doctrine. For example, the 2015 version of *LWD 3–0, Operations*, did not mention women or gender at all (Australian Army, 2015). The 2018 version, however, shows marked improvements in this regard. First, it describes UNSCR 1325 and related resolutions, and then explains how the national action plan requires operational planners to incorporate the 'whole-of-government implementation measures' into their work (Australian Army, 2018b: 57). The role of the GENAD in the land force headquarters is succinctly and usefully described from a functional perspective as being threefold: to understand the human terrain and cultural framework in an area of operations through the lens of gender, to develop plans to protect at-risk populations from violence, and to report on 'conflict-related sexual and gender-based violence within the area of operations' (57).

This gender perspective has not yet been reflected in lower-level operational doctrine. For example, *Land Warfare Procedures – General 3–9–6, Operations in Urban Environments*, notes only that detainees are to be 'treated humanely and with respect for their personal life, gender, race and religion' (Australian Army, 2012: 2-57). Another subordinate document, *LWD 3–0–1, Counterinsurgency*, mentions gender only once, in the sense of understanding the importance of insurgent propaganda by understanding among other things, 'the gender relationship in the dissemination of information' (Chief of Army, 2009: 7-13). From a historical perspective, it notes Australian counterinsurgency efforts in the Vietnam War, and describes a simple set of rules adopted by the Australian Army to relate to the civilian population better, including the simple but important Regulation 7, 'Don't molest or rape women' (6-14).

Recent revisions to important non-kinetic doctrine do reflect the trend seen in the new *LWD 3–0*. For example, a significant difference can now be seen in the treatment of women and gender in the latest version of *LWD 3–8–6, Civil-Military Cooperation*. The 2010 version noted only that international organisations and non-governmental organisations 'may have particular knowledge and skills in one or more "sectors"', and that one of these sectoral groupings could be gender issues (Commander Training Command – Army, 2010: 2-11–2-12). Women were only mentioned in the sense that female detainees 'are to be quartered separately from men and supervised by women' (6-4).

The 2017 version of *LWD 3–8–6, Civil-Military Cooperation*, however, takes a different approach. First, it emphasises the importance of protection of civilians in Australian international engagement, both in the area of diplomacy and operations (Australian Army, 2017b: 38–39). Next, it devotes a fair amount of space to the discussion of incorporating a gender perspective in operations under the Australian national action plan and UNSCR 1325. In keeping with the tendency to view UNSCR 1325 as being primarily concerned with preventing sexual assaults, it notes that UNSCR 1325 'recognises that the experiences and needs of women and girls differ from those of men and boys' in armed conflict situations, 'particularly in relation to human rights violations such as sexual and gender-based violence' (40). Almost as an aside, however, it mentions the 'role of women in conflict prevention, management and resolution' (40) without really developing this topic further.

Regardless, the 2017 version takes a very important step forward in operationalising gender issues by providing practical points and examples that could be used by troops in the field. Regarding humanitarian operations, *LWD 3–8–6* now also notes the particular risks that women and children face in a 'complex emergency', which it defines as 'a multifaceted humanitarian crisis ... where there is a total or considerable breakdown of authority resulting from internal or external conflict' (Australian Army, 2017b: 41). As to principles that guide the use of economic quick impact projects by Australian Defence Force units in operational areas, the 2017 version notes that it is important to 'Engage with, and listen to, the local population so that it feels able to contribute. Women sometimes find it more difficult than men to contribute' (86). It also establishes an information formatting example, setting out a template for a status report from a civil-military cooperation unit to the headquarters S-9 staff section (CIMIC), which includes addressing the status of women in the host nation, as well as local leaders who support host nation women (129). Finally, in terms of constructing an analysis of an operational area, it provides women's associations as an example of local organisations to be considered (195).

Summary

Australia has made significant progress so far in the incorporation of relevant parts of the women, peace and security goals and concerns set out in its national action plan into its military activities and operations, even though its overarching national security strategy does not really address it. Forthright engagement with the national action plan by the Department of Defence and the Australian Defence Force has resulted in important high-level actions such as the directives issued by senior military leadership specifying the actions to be taken in force service components to implement the national action plan into service activities and operations (Department of the Prime Minister and Cabinet, 2017: 141–43). Other important achievements include the training of all deploying military personnel on women, peace and security, and incorporating these goals and concepts into curricula at the Australian Command and Staff Course, the Defence and Strategic Studies Course, and for cadets at the Australian Defence Force Academy (Department of the Prime Minister and Cabinet, 2017: 21–22). In the field, Australia has also been very reliable in providing GENADs for the continuing multinational mission in Afghanistan and for Australian Joint Task Force 633 in the Middle East (Jennifer Wittwer, 2015: personal communication).

From a doctrinal perspective, the Australian Defence Force appears to be meaningfully embedding gender considerations into the review criteria for its joint doctrine, and although classification concerns do not allow a detailed assessment of the products of these review efforts, qualitative reports and certain revised doctrine suggest that these inclusions have been productive. As an example of the Australian Defence Force's commitment to this effort, in the 2015 Qinquepartite Combined Joint Warfare Conference report, Australia was the only Five Eyes country to explain how it was incorporating gender into its joint doctrine (QCJWC, 2015: Enclosure B). The same was true at the 2016 conference (QCJWC, 2016: Enclosure E). Similarly, at the Australian Army level, newly revised land force doctrinal documents, such as the 2017 civil-military cooperation doctrine and the 2018 operations doctrine, show real promise of meaningful and practical inclusion of gender considerations at the tactical level of operations.

From a different perspective, Wadham et al. (2016) argue that irrespective of measures taken to incorporate the goals of the national action plan into the activities and operations of the Australian Defence Force, the chances of it becoming a truly gender-inclusive workplace are limited by its 'preoccupation' with maintaining military effectiveness – that is, framing 'women's participation in the ADF as an issue of capacity rather than rights' (3). From an operational perspective, the notion that a military organisation should not be highly concerned with its effectiveness is a bit of a surprise. Likewise, the idea that women's participation in militaries might not actually lead to an increase in capabilities in certain respects, particularly as gender is being operationalised, appears to ignore some of the most important points of UNSCR 1325. What is the best way to choose the criteria by which gender mainstreaming efforts in any military organisation are assessed and evaluated?

The point of this book is that separate and apart from the changes necessary to ensure gender equality within the ordinary functioning of a military organisation, such as equal opportunity for women and men to compete for training opportunities based on individual merit, the inclusion of gender considerations in the planning and execution of missions should be premised on operational relevance, and relevant gender considerations will not be the same from operation to operation. Further, depending on the specific operation, they might not be relevant at all – but commanders and planners will not know this unless they conduct a proper gender analysis of the operational area. Over time, the changes the Australian Defence Force is making by building its GENAD capability, as will be discussed later in Chapter 9, 'Gender in military activities and operations', and requiring consideration of gender perspectives in operations as has already been discussed, should create the demand among planners and operators for sex and gender-disaggregated information and analysis to support their mission activities (Prescott, 2013: 59). This information will be necessary to conduct a proper gender analysis, and to identify the risks of not considering gender in operations.

Given Australia's engagement with its allies in the Pacific area on operational gender issues, including Japan (Fielding, 2017) and the US as will be discussed in Chapter 9 in the context of the *Talisman Sabre* exercises, the Australian Defence Force appears poised to play a constructive and influential role in realising the goals of UNSCR 1325 within the international security community. Before looking at *Talisman Sabre* and other some useful examples of the gender-related work relevant to military activities and operations being conducted around the world in

Chapter 9, however, the next chapter will take us on a short but necessary detour into the relationship between IHL and gender in operations. IHL actually is a gendered body of law, but unless the prevention and prosecution of sexual and gender-based violence is on the table, it frankly tends to be ignored in discussions of operationalising gender.

References

Australian Army (2009) *Land Warfare Doctrine (LWD) 3-0-1, Counterinsurgency* (16 December). Australian Army. Available at: www.army.gov.au/sites/g/files/net1846/f/lwd_3-0-1_counterinsurgency_full_0.pdf (Accessed 23 June 2018).

Australian Army (2012) *Land Warfare Procedures – General 3-9-6, Operations in Urban Environments* (24 May). Australian Army. Available at: www.army.gov.au/sites/g/files/net1846/f/lwp-g_3-9-6_operations_in_urban_environments_full.pdf (Accessed 23 June 2018).

Australian Army (2015) *LWD 3–0, Operations* (9 November). Australian Army (Copy on file with author).

Australian Army (2017a) *LWD 1, The Fundamentals of Land Power*. Australian Army. Available at: www.army.gov.au/sites/g/files/net1846/f/lwd_1_the_fundamentals_of_land_power_full_july_2017.pdf (Accessed 23 June 2018).

Australian Army (2017b) *LWD 3-8-6, Civil-Military Cooperation*. Australian Army. Available at: www.army.gov.au/sites/g/files/net1846/f/lwd_3-8-6_civil-military_cooperation_interim.pdf (Accessed 23 June 2018).

Australian Army (2018a) *LWD 2–0, Intelligence*. Australian Army. Available at: www.army.gov.au/sites/g/files/net1846/f/lwd_2-0_intelligence_full.pdf (Accessed 23 June 2018).

Australian Army (2018b) *LWD 3–0, Operations*. Australian Army. Available at: www.army.gov.au/sites/g/files/net1846/f/lwd_3-0_operations_full.pdf (Accessed 23 June 2018).

Australian Army (2018c) *LWD 4–0, Logistics*. Australian Army. Available at: www.army.gov.au/sites/g/files/net1846/f/lwd_4-0_logistics_full.pdf (Accessed 23 June 2018).

Australian Civil-Military Centre (2015) 'Australian Guidelines for the Protection of Civilians'. Australian Civil-Military Centre (December). Available at: www.acmc.gov.au/wp-content/uploads/2016/01/160128_POC-Guidelines.pdf (Accessed 23 June 2018).

Australian Government (2012) *Australian National Action Plan on Women, Peace and Security 2012–2018*. Canberra: Department of Families, Housing, Community Services and Indigenous Affairs. Available at: www.pmc.gov.au/sites/default/files/publications/national-action-plan_on_women%20peace-security-%202012-18.pdf (Accessed 6 July 2018).

Australian Government (2017) 'Addendum to the Australian National Action Plan on Women, Peace and Security 2012–2018'. Australian Government (7 December). Available at: www.pmc.gov.au/sites/default/files/publications/addendum-australian-national-action-plan.pdf (Accessed 23 June 2018).

Australian Human Rights Commission (2011) *Report on the Review into the Treatment of Women at the Australian Defence Force Academy, Phase 1 of the Review into the Treatment of Women in the Australian Defence Force*. Sydney: Australian Human Rights Commission. Available at: http://defencereview.humanrights.gov.au/sites/default/files/ADFA_2011.pdf (Accessed 23 June 2018).

Australian Human Rights Commission (2012) *Review into the Treatment of Women in the Australian Defence Force, Phase 2 Report*. Sydney: Australian Human Rights Commission. Available at: http://defencereview.humanrights.gov.au/sites/default/files/adf-complete.pdf (Accessed 23 June 2018).

Australian Public Service Commission (2016) *Balancing the Future: The Australian Public Service Gender Equality Strategy 2016–19*. Canberra: Australian Public Service Commission.

Available at: www.apsc.gov.au/sites/g/files/net4441/f/FINAL-Balancing-the-future-the-Australian-Public-Service-gender-equality-strategy-2016–19.pdf (Accessed 25 June 2018).

Chief of Army (2009) *LWD 3-0-1, Counterinsurgency* (16 December). Puckapanyal, VIC: Land Warfare Development Centre. Available at: www.army.gov.au/sites/g/files/net1846/f/lwd_3-0-1_counterinsurgency_full_0.pdf (Accessed 23 June 2018).

Chief of Joint Operations (2006) *ADDP 7.0, Doctrine and Training* (January). Williamtown, NSW: Joint Warfare Doctrine and Training Centre. Available at: http://indianstrategicknowledgeonline.com/web/AUSSI%20-Training.pdf (Accessed 23 June 2018).

Chief of Joint Operations (2009a) *ADDP 3.8, Peace Operations*, 2nd Ed (14 December). Williamtown, NSW: Joint Warfare Doctrine and Training Centre. Available at: http://indianstrategicknowledgeonline.com/web/AUSSIE%20PEACE%20OPS.pdf (Accessed 23 June 2018).

Chief of Joint Operations (2009b) *ADDP 3.11, Civil-Military Operations* (1 April). Canberra: Australian Defence Force Warfare Centre. Available at: www.defence.gov.au/adfwc/Documents/DoctrineLibrary/ADDP/ADDP3.11-Civil-MilitaryOperations.pdf (Accessed 23 June 2018).

Chief of Joint Operations (2011) *ADDP 00.3, Multinational Operations* (October). Williamtown, NSW: Joint Warfare Training and Doctrine Centre. Available at: www.defence.gov.au/adfwc/Documents/DoctrineLibrary/ADDP/ADDP00.3-MultinationalOperations.pdf (Accessed 23 June 2018).

Chief of Joint Operations (2013) *ADDP 3.13, Information Activities*, Ed. 3 (6 November). Canberra: Headquarters Joint Operations Command. Available at: www.defence.gov.au/FOI/Docs/Disclosures/330_1314_Document.pdf (Accessed 25 June 2018).

Chief of Joint Operations (2015) *ADDP 3.8, Peace Operations* (27 February). Canberra: Joint Doctrine Centre (Copy on file with author).

Commander Training Command – Army (2010) *LWD 3-8-6, Civil-Military Cooperation* (24 May). Puckapanyal, VIC: Land Doctrine Centre. Available at: *https://tgsacu.files.wordpress.com/2017/06/lwd_3-8-6_civil-military_cooperation_full.pdf* (Accessed 23 June 2018).

Department of Defence (n.d.) 'APS Gender Equality Strategy 2016–2019 – Defence Action Plan 2016'. Department of Defence. Available at: www.defence.gov.au/Diversity/_Master/docs/APS-Gender-Equality-Action-Plan-20170424.pdf (Accessed 6 July 2018).

Department of Defence (n.d.) 'Defence Implementation Plan'. Department of Defence. Available at: www.defence.gov.au/Women/NAP/ImplementationPlan.asp (Accessed 23 June 2018) (The Defence Implementation Plan itself is not available to the public at this point).

Department of Defence (n.d.) 'Women, Peace and Security in Defence'. Department of Defence. Available at: www.defence.gov.au/jcg/Women_Peace_Security/ (Accessed 23 June 2018).

Department of Defence (2014) 'Chief of Defence Force appoints Gender Adviser'. Defence News and Media [press release] (26 March). Available at: https://news.defence.gov.au/media/media-releases/chief-defence-force-appoints-gender-adviser (Accessed 23 June 2018).

Department of Defence (2016) *Defence White Paper 2016*. Canberra: Department of Defence. Available at: www.defence.gov.au/WhitePaper/Docs/2016-Defence-White-Paper.pdf (Accessed 23 June 2018).

Department of Defence (2017) *Women in the ADF Report 2016–17, A Supplement to the Defence Annual Report 2016–17*. Canberra: Department of Defence. Available at: www.defence.gov.au/annualreports/16-17/Downloads/WomenInTheADFReport2016-17.pdf (Accessed 23 June 2018).

Department of the Prime Minister and Cabinet (2013) *Strong and Secure: A Strategy for Australia's National Security*. Canberra: Australian Government. Available at: www.files.

ethz.ch/isn/167267/Australia%20A%20Strategy%20for%20National%20Securit.pdf (Accessed 23 June 2018).

Department of the Prime Minister and Cabinet (2014) *2014 Progress Report – Australian National Action Plan on Women, Peace and Security 2012–2018*. Canberra: Australian Government. Available at: https://pmc.gov.au/resource-centre/office-women/2014-progress-report-australian-national-action-plan-women-peace-and-security-2012-2018 (Accessed 6 July 2018).

Department of the Prime Minister and Cabinet (2017) *2016 Progress Report on the Australian National Action Plan on Women, Peace and Security 2012–2018*. Canberra: Australian Government. Available at: www.pmc.gov.au/sites/default/files/publications/progress-report-nap-women-peace-security.pdf (Accessed 6 July 2018).

Fielding, A (2017) 'WPS in the Japan-Australia Bilateral Agenda – Integrating a Gender Perspective into Military Operations' [PowerPoint Presentation]. In: *'The Next Decade: Amplifying the Women, Peace and Security Agenda', Women, Peace and Security Conference 2017*. Newport, RI: US Naval War College (10–11 August) (Copy on file with author).

Humanitarian Advisory Group (2015) *Independent Interim Review of the Australian National Action Plan on Women, Peace and Security 2012–2018* (30 October). Melbourne, VIC: Humanitarian Advisory Group. Available at: https://humanitarianadvisorygroup.org/wp-content/uploads/2016/06/nap-interim-review-report.pdf (Accessed 23 June 2018).

Prescott, JM (2013) 'NATO Gender Mainstreaming: A New Approach to War Amongst the People?' *Royal United Services Institute Journal* 158(5): pp. 56–62. DOI: 10.1080/03071847.2013.847722.

Quinquepartite Combined Joint Warfare Conference (QCJWC) (2015) *Annual Report 2014–2015* (15 April). Washington, DC: Joint Education and Doctrine Directorate. Available at: https://wss.apan.org/2411/QCJWC%20Document%20Library/QCJWC%202015%20(UK)/Annual%20Report%202014-15.pdf (Accessed 6 July 2018).

Quinquepartite Combined Joint Warfare Conference (2016) *Annual Report 2015–2016* (14 March). Shrivenham, UK: Development, Concepts and Doctrine Centre. Available at: https://wss.apan.org/2411/QCJWC%20Document%20Library/QCJWC%202016%20(AS)/QCJWC_Annual_Report_Final_2015-16.pdf (Accessed 25 June 2018).

Turnbull, M (2017) 'National Security Statement' [transcript of statement to House of Representatives, Parliament House, Canberra, Australia]. Office of the Prime Minister (13 June). Available at: www.pm.gov.au/media/national-security-statement (Accessed 23 June 2018).

Wadham, B; Bridges, D; Mundkur, A and Connor, J (2016) '"War-Fighting and Left-Wing Feminist Agendas": Gender and Change in the Australian Defence Force'. *Critical Military Studies* [online]: pp. 1–17. DOI: 10.1080/23337486.2016.126837.

Wittwer, JA (2013) 'The Gender Agenda: Women, Peace and Security in the Conduct of NATO-Led Operations and Missions'. *Australian Defence Journal* 191 (July/August): pp. 59–65. Available at: www.defence.gov.au/adc/adfj/Documents/issue_191/191_2013_Jul_Aug.pdf (Accessed 25 June 2018).

8
INTERNATIONAL HUMANITARIAN LAW AND GENDER

Having just completed an examination of the relationships between armed conflict, women and climate change, and a review of how NATO and certain member states of the Fives Eyes community approach gender in the areas of strategy, policy and military doctrine, it might seem puzzling to spend time discussing the relationship between IHL and the protection of women. Rather than exploring a body of law that is almost universally accepted by the international community, at least with regard to the applicable treaty law, it might appear more useful to just begin the survey of how military organisations are actually working to operationalise gender in their activities and operations today. Additionally, the pronouncements and writings of the International Committee of the Red Cross (ICRC), the recognised guardian and arbiter of IHL (Henckaerts and Doswald-Beck, 2006: xxvi), would suggest that such a discussion is not really necessary. From the International Committee of the Red Cross' perspective, modern IHL has been truly ground-breaking in that it requires that all people be treated equally irrespective of their sex, and it actually includes numerous provisions that specifically address women and their needs during armed conflict (ICRC, n.d.: 'H. Paragraph 4: Treatment of Women'). What is required, it has stated until recently, is not new law, but rather better compliance with the protective law that already exists (Lindsey-Curtet, Holst-Roness and Anderson, 2004: 9; Durham and O'Byrne, 2010: 34–37).

This argument has a solid grounding, and there is little, if any, evidence to suggest that the militaries of stable democracies today are seeking to implement IHL in their work in a way that deliberately and intentionally discriminates against women. What this argument does not address, however, is the pervasive male-normative context within which this law was founded, and the effect that this continues to have on the way IHL is understood and applied. This is important, because its long-standing norms undergird the writing of doctrine, professional military education, realistic training, thorough planning and the careful conduct of modern military operations.

The priority between IHL and international human rights law as to which body of law applies in situations of armed conflict when their applicable provisions are themselves in conflict is unsettled. Writers such as Paust (2015) ably argue that IHL should not be regarded as *lex specialis* that will automatically trump international human rights law in these situations. Paust argues that human rights law continues to apply during wartime, and that neither treaty nor customary international law supports the idea that IHL completely displaces human rights law. In fact, he states that human rights should be understood as having 'a recognized and unavoidable primacy in any social context', because of their 'status as rights guaranteed by the U.N. Charter, their additional peremptory status under customary law as rights *jus cogens*, and their nonderogable status under certain treaties' (510). Even Paust recognises, however, that when the taking of enemy combatants' lives occurs, the applicable body of law to determine whether the taking is lawful is not international human rights law, but IHL (532–36). This is not merely a difference in standards – it is fundamental to the way military organisations educate, train, and plan to carry out missions involving the use of armed force.

This chapter begins with an examination of an area in which international human rights law and IHL strongly overlap – the prevention and prosecution of sexual violence against women and girls in war – and then turns to an area where they diverge almost completely, the standards applicable to the use of lethal force. Next, it looks at the feminist critique of IHL, and the development of IHL particularly with regard to its treatment of gender. Then, it briefly reviews the way that Australian, UK and US legal doctrine reflect the traditional male-normative approach of IHL, and likely function as a conservative force acting against the critical review of ILH necessary to fully effect the goals of UNSCR 1325. Finally, it assesses different factors that likely have a role to play in understanding why IHL and militaries' understandings and applications of it have been slow to incorporate the operational relevance of gender.

UNSCR 1325, international human rights law and IHL

A convergence of norms across the spectrum of violence

There are certain areas in military operations in which there is an overlap between international human rights law and IHL standards. For example, Crawford (2010) has argued persuasively that there has been a convergence of norms regarding detention standards across the spectrum of violence in types of armed conflict, and advocates cogently for the application of universal standards for the detention of combatants regardless of whether the conflict is considered international or non-international (23–48, 78–80). An important example of convergence in the area of sexual violence in armed conflict is the Rome Statute of the International Criminal Court (ICC), which establishes the court's jurisdiction over the crime of genocide, crimes against humanity and war crimes (2002: art. 5). As defined in the statute and the elements of the crimes under the court's jurisdiction (arts. 6–8), each of these crimes includes acts of sexual violence. Although its trial judgement was later overturned on appeal (ICC, 2018), the International Criminal Court has convicted a defendant for both human rights violations and IHL violations for the same substantive acts of sexual violence (*Prosecutor v. Jean-Pierre Bemba Gombo*, 2016), following the path set out by the International Criminal Tribunal for Rwanda (*Prosecutor v. Jean Paul Akayesu*, 1998), and the International Criminal Tribunal for the Former Yugoslavia *(Prosecutor v. Anto Furundzija, 1998)*.

Barrow (2010) identifies this judicial recognition as a welcome reorientation from the traditional concept of rape as a crime against a woman's honour to a focus instead on the serious bodily and psychological harm that results from the crime (226–28). She also notes, however, that some writers are concerned that the focus on women as victims of sexual violence 'undermines women's agency as active participants and distorts discourses by failing to consider women's broader experience of armed conflict' (222). This concern is well placed.

Although UNSCR 1325 is built broadly enough to cover the range of women's experiences before, during and after armed conflict, UN Security Council action since its promulgation in 2000 has largely been focused on the prevention and remediation of sexual violence (Prescott, 2013a: 85). This effort is undoubtedly necessary, and these subsequent resolutions are important landmarks in the fight to reduce the impacts of armed conflict on women. The attention placed on this area of clear overlap between IHL and international human rights law, however, risks marginalising efforts to increase the protection of women in areas where the two bodies of law have little in common. To fully meet the requirements of UNSCR 1325, it is just as important to address those areas where there really is no overlap between the two, especially in the standards applied in use of armed force.

The convergence of violence around norms

The two bodies of law differ markedly in the standards that must be met before human life might be taken legally. Under the International Covenant on Civil and Political Rights, a person's life can be taken when certain due process requirements are met in judicial proceedings (1966: arts. 6, 14). This level of due process is not found on battlefields. Similarly, although it is generally not thought of as a human right, and it is perhaps better described as an individual right, a person, such as a law enforcement officer, might lawfully take another's life when that person responds immediately and proportionately in self-defence upon perceiving the likely suffering of grievous bodily harm or death (Hessbruegge, 2017: 3, 75–90, 92–189).

Similar but less restrictive conditions are found in the *jus ad bellum* category of international law regarding the exercise of self-defence by nations (Taft, 2004: 305–06). As to individuals under IHL, from the perspective of *jus in bello*, however, a person's life can be taken if another combatant is reasonably certain that the person is taking a direct part in hostilities, under the circumstances at the time of engagement (Melzer, 2009: 70, 76). As set out in Additional Protocol I to the 1949 Geneva Conventions (AP I), a completely innocent person's life can also be taken, for example, if a commander decides that the direct military advantage to be gained by the application of military force is not outweighed by excessive harm to civilians or their property, under the IHL definition of proportionality (AP I, 1977: arts. 51(5)(b), 57(2)(a)(iii)).

As a matter of policy in certain civilian-centric military operations, many nations and defence organisations have promulgated ROE which often limit uses of armed force below IHL legal ceilings. For example, in the NATO operation in Libya that led to the downfall of Muammar Gaddafi, air strikes on targets were aborted if the NATO forces became aware that there were civilians in the weapons' effects areas (Prescott, 2013b: 355–56). Another example would be different tactical directives put in place by successive ISAF commanders to reduce the number of civilian casualties, which were jeopardising the command's relationship with the Afghan government (Prescott, 2013b: 354). A third example of possible convergence between human rights and IHL norms might be the hybridisation of IHL detention standards with national processes with greater due process protections used by the US in the later years of the conflict in Afghanistan during the Obama Administration (Prescott, 2010: 83). To some, this could give the appearance of movement towards the incorporation of international human rights law approaches in the conduct of armed conflict. It is probably more accurate to say that this is a conscious tempering of the use of force to meet political and operational concerns.

Some suggest that practices of this sort might give rise to new IHL implementation norms that would significantly restrict the use of force by the US in armed conflict (Frederick and Johnson, 2015: 88). UK experts have assessed that similar trends have already led to a conflation of police-style legal constraints with combat operations realities, with resulting confusion as to what the applicable legal standards actually are (Clarke, 2013). Because the principles of proportionality and distinction are so fundamental to the use of force in armed conflict, instead of looking for a legal convergence of IHL and international human rights norms in these areas, we should instead view them from a functional perspective and understand that there is also a convergence of violence around these IHL norms in military operations quite unlike that experienced in ordinary police situations, and that this further reinforces their essential differences from international human rights norms.

For example, feedback loops appeared to exist between the application of force on the basis of the perception of direct participation in hostilities in combat and combat-related detention in Afghanistan during the time the US military was holding Afghan detainees suspected of being insurgents. Many detainees reported that they had become involved in the insurgency because someone they knew had been shot, presumably by Afghan National Security Forces or ISAF soldiers. In many instances, soldiers and police had mistakenly opened fire on approaching vehicles at checkpoints, because they believed they were under attack. The US Army recognised that the unacceptably high rate of civilian Afghan deaths resulting from these checkpoint incidents contributed to the impetus to join the insurgency, that is, it led Afghans to decide to take a direct part in hostilities and therefore become targetable by ISAF forces (Prescott, 2016: 66). If captured, these now-combatants were then detained in some form of military or security force custody.

Given this dynamic relationship, the work that has been done so far by IHL practitioners and international human rights advocates in tackling the issue of sexual violence in armed conflict, particularly against women, might represent a plateau in development along this path. Of course, effective implementation and enforcement in this area will continue to require very significant collaboration and effort on the parts of both militaries and international human rights advocates. It is difficult to see, however, how cooperation in this area would translate meaningfully to areas in which the two bodies of law do not overlap in terms of their long-established norms, such as in the use of lethal force. Further, despite worthwhile developments in addressing sexual violence against women in armed conflict, the traditional formulation of the rationale behind it – questions of honour, modesty and stereotyped victimhood – appears to have become subtly embedded into understandings and applications of the use of other types of military force against military and civilian women under IHL. These understandings will not be dislodged by increased emphasis upon rape in armed conflict as a war crime in judicial proceedings – something more is required.

Further, the latitude afforded commanders under IHL in deciding when and how to use lethal force is much less justiciable (Prescott, 2018: 329) than sexual and gender-based violence. The latter will always be a crime – the former would be devilishly difficult to prosecute, and under current understandings and applications of IHL considerations of gender are largely irrelevant. Thus, despite IHL being textually committed to the complete equality of all persons, and if anything, affording even greater protections to women, it continues to reflect its male-normative heritage in ways that conflict with the international community's practical appreciation of armed conflict's differentiated and disparate effects upon women and girls. Currently, the very structure of its most salient norms provides no space for consideration of these factors (Prescott, 2013a: 93–94).

The feminist critique of IHL

To better appreciate the nature of this problem, it is useful to first assess IHL treaty and customary law regarding the treatment of women during armed conflict as they have developed over the last 75 years as a result of treaty negotiations, state practice, judicial decisions and international law scholarship. Although the feminist critique of IHL has some practical limitations given the general development of state practice since the 1949 Geneva Conventions, its most important points are not based on academic applications of international human rights law principles to IHL. They are instead based on a straight forward review of the history of IHL treaties and a textual analysis of their provisions as they relate to the protection of women, beginning with the 1949 Geneva Conventions.

The feminist critique sees IHL as constructed to reflect states' interests and the integral role combatants play in the furtherance of those interests – combatants, of course, who even today, are generally men (Gardam, 1989: 277). In this sense, Oosterveld (2009) has described IHL 'as a reflection of masculine assumptions that do not take into account global systematic gender inequality' (387). Protections given to the sick, wounded, prisoners of war and civilians under IHL are also seen as male-normative, and any special and specific protections given to women under IHL perpetuate this underlying bias (Gardam and Charlesworth, 2000: 161).

Further, IHL fails to recognise that women's and girls' greater risk to the effects of armed conflict results from pre-existing gender discrimination (Lijnzaad, 2005: 500), and it does not recognise that non-Western civilian women who are most at risk are those whose needs are least acknowledged in its application (Gardam and Jarvis, 2001: 253). Finally, in its provisions governing the actual used of armed force, such as article 51.5(b) of Additional Protocol I which sets out the general parameters of the principle of proportionality, it fails to account for the differentiated and more severe impacts of armed conflict upon women and girls than upon men. Commanders therefore are not required to include in their proportionality analysis the disproportionate effect upon women and girls that damage to their homes and other infrastructure would cause (Gardam and Jarvis, 2001: 126), for example.

The Geneva Conventions of 1949

The conference of government experts

Negotiated in the aftermath of World War II's widespread mistreatment of civilians and prisoners of war and the staggering loss of life amongst them (ICRC, 1949b: 9; ICRC, 2014), the 1949 Geneva Conventions (GCs) were the result of a systematic overhaul of existing treaties dealing with the protection of victims of armed conflict. In April 1947, Abplanalp (1995) recounts how the International Committee of the Red Cross convened a Conference of Government Experts to consider revisions to existing IHL treaties (533). The experts were divided into three different commissions to conduct this work, which was completed over a two-week period (ICRC, 1947: 1). In reviewing the experts' consideration of how women were to be better protected, it is important to note the different approaches taken by the different commissions, and how opportunities were missed to base the eventual protections for women on their equality with men rather than on what was viewed as their inherent weakness or their roles as wives, daughters and mothers.

The First Commission revisited the Geneva Convention of 27 July 1929 for the Relief of Wounded and Sick in Armies in the Field and the Hague Convention of 18 October 1907, for the Adaptation to Maritime Warfare of the Principles of the Geneva Convention of July 6, 1906. On the basis of the International Committee of the Red Cross' position on the way in which women should be afforded protection, the commission rejected language that would have required equal treatment of persons 'without any distinction' (ICRC, 1947: 12) of sex. It 'did not consider it opportune to include the words "of sex"', because it 'thought that the spirit of the convention itself indicates that women in particular should be granted preferential treatment' (12). It instead proposed that language from article 3 of the 1929 Geneva prisoner of war convention should be used: 'Women shall be treated with all consideration due their sex' (ICRC, 1947: 12–13). The First Commission also considered the protection of wounded and sick civilians, and with regard to pregnant women it discussed whether they should be expressly mentioned in any revision, in addition to the sick and wounded. In the end, the First Commission 'considered that such women were assimilated to the sick in general' (ICRC, 1947: 69).

The Second Commission considered revisions to the 1929 Geneva prisoner of war convention itself. In contrast to the First Commission, the Second Commission believed that the language in proposed article 3 (women 'shall be treated with all consideration due to their sex') was insufficient to protect women (ICRC, 1947: 119). On 'the contrary, it considered that, as women in many countries were still placed on an inferior footing and received less consideration than men', it proposed to add language to article 3 that established that the treatment afforded women prisoners of war 'shall in no case be inferior to that accorded men' (ICRC, 1947: 119). This formulation would have made great progress in considering women as equal to men before the law whilst still considering their different needs and requirements.

The Third Commission's task was quite different from those assigned to other commissions. Rather than reviewing existing treaty law, the Third Commission considered draft provisions for an entirely new treaty that would provide for the protection of civilians in time of war (ICRC, 1947: 269). The International Commiteee of the Red Cross proposed that very specific measures to protect pregnant women and women with young children should be part of this treaty, including that they should not be 'enlisted in armed forces', 'employed on any work or activity connected with the war', or 'compelled to do work for which they [were] physically unfit' (ICRC, 1947: 301). In the end, the Third Commission recommended that these measures should be studied further (ICRC, 1947: 301), and approved language that stated women 'shall be treated with all consideration due to their sex, and children with all consideration due to their age and helpless condition' (ICRC, 1947: 275).

The diplomatic conference

The work done by the Conference of Government Experts was included in the four draft conventions brought forward to a drafting conference held in Stockholm in 1948. The drafts were reviewed by the conference's Legal Commission, which made a few amendments and approved them. Abplanalp (1995) reports that the approved drafts were then submitted to the diplomatic conference in Geneva which began in April 1949 (533). As the government experts had done earlier, the delegates worked in committees dedicated to particular conventions. Committee I focused on the wounded and sick, and maritime warfare. By its 33rd meeting, it had drafted an article setting the scope of protection under what became the first and second Geneva Conventions that forbid 'any adverse distinction . . . founded on sex, race, nationality, religion, political opinions or any other similar criteria' and provided that women 'shall be treated with all consideration due to their sex' (ICRC, 1949b: 156, 159). This language would appear to have eliminated the concerns the First

Commission of Government Experts had earlier as to whether equal treatment on the basis of sex needed to be explicitly stated, but it does not necessarily reflect a change in the understanding that being pregnant was essentially akin to being sick.

Committee II worked on what became GC III, the prisoner of war convention. The delegates discussed camp arrangements for women, and their exchanges focused largely on the need for separate dormitories and 'separate conveniences for women' (ICRC, 1949b: 256, 259). The scope of medical examinations were also discussed, generally in the context of whether and how often prisoners of war should be x-rayed as a means of diagnosing contagious disease (ICRC 1949b, 260). Initially, the language establishing that women were to be given special treatment with regard to their sex was seen by most of the delegates, almost all of whom were men (ICRC 1949a, 158–70), as sufficient for the protection of women (ICRC, 1949b: 489). The British delegation, however, which did include women (ICRC, 1949a: 170), noted that 'the position of women was a special one and in the last war all armies had women soldiers' (ICRC 1949b, 490). The UK was persistent, and in the end it successfully supported specific language establishing protections for women prisoners of war who were being punished for penal or disciplinary offences (ICRC, 1949b: 489–90, 494, 498, 502).

Committee III worked on the establishment of the new convention for the protection of civilian persons in time of war, what became the fourth Geneva Convention. As to article 11 of the draft convention, setting forth the general scope of protections to be accorded to civilians, the British delegate stated that 'there was a danger in providing for treatment according to sex; as in some countries this might unfortunately mean that women were treated worse than men' (ICRC, 1949b: 625). The UK delegate was likely mindful of the horrific treatment endured by civilian women and children at the hands of imperial Japanese forces in World War II (Wolff, 2014). This specific concern was not addressed by the other delegates at the time (ICRC, 1949b: 625–26), nor at the draft article's second reading in committee prior to the committee's adoption of language that did not include it (ICRC, 1949b: 785–86). Finally, the discussion of equal protection irrespective of sex was not was not specifically mentioned in the committee's report to the diplomatic conference on its work (ICRC, 1949b: 816)

Committee III then turned to draft article 27, which concerned protections afforded to women and children (ICRC, 1949b: 643). The approved draft which had been submitted to the diplomatic conference stated that women 'shall be specially protected against any attacks on their honour or dignity' (ICRC, 1949a: 118). The International Committee of the Red Cross expressed its hope that language proposed earlier by recognised observers at the diplomatic conference, the International Abolitionist Federation and the International Council of Women, would be adopted instead. Under this formulation, women would be 'specially protected against any attacks on their honour, in particular against rape, enforced prostitution and any form of indecent assault' (ICRC, 1949b: 643). The UK delegate concurred with the proposal, and discussion moved on to other parts of the proposed article (ICRC, 1949b: 643).

These groups' proposed language marked an improvement over the draft article in the protection of women in the sense that it was more specific about the acts that constituted offences against women's 'honour'. However, these offences were in fact still cast in terms of violations of 'honour' – a conservative social construction that would seem to exclude women who had been characterised as 'dishonoured' – rather than on the violation of women's human rights as persons. Thus, the women's rights advocates who appear to have gained the most traction in the formulation of the Geneva Conventions' treatment of women also appear to have reinforced a traditional essentialist view of women in what was already a largely male and male-normative undertaking.

At the next committee meeting, the Indian delegate concurred with this language favoured by the International Committee of the Red Cross, but the Italian delegate suggested

> that in view of the extreme gravity of offences against the honour and dignity of women, a specific reference should be made to the responsibility of the Commander of the armed forces, as in the similar provisions in Article 51 of the Hague Convention.
>
> *(ICRC, 1949b: 644)*

Had this sort of enforcement provision been accepted, the Italian delegate's proposal might have had a significant impact on how sexual assaults against civilian women were handled. Article 51 of the Hague Regulations of 1907 prohibits forces occupying enemy territory from levying cash contributions on the civilian population for purposes other than defraying the costs of the occupation, and only then 'on the responsibility of a commander-in-chief' (Convention (IV) Respecting the Laws and Customs of War on Land and its Annex, 1907: art. 51). Had instances of sexual violence against women by occupying troops been specifically viewed as matters of command responsibility in the fourth Geneva Convention, this could have established the seriousness with which the international community handles these crimes decades before the international tribunal decisions in *Akayesu* and *Furundzija*.

Common Article 3, Geneva Conventions I-IV

Common Article 3 of the Geneva Conventions of 1949 was an important innovation in the development of IHL, and it is important to review its provisions before examining the separate conventions in detail. Elder (1979) reports that as early as the 1946 Preliminary Conference of the National Red Cross Societies, national parties had indicated a strong interest in extending basic IHL protections to victims of conflicts of other than an international nature, and the subsequent experts' conference recommended to the international committee a partial application of these principles irrespective of the parties' legal status (42–43). After significant negotiation over the text of the proposed article, delegates to the diplomatic conference finally agreed to language which clarified that the article would apply only in instances of actual armed conflict (43–52).

As set out in *Geneva Convention for the Amelioration of the Condition of the Wounded and Sick in Armed Forces in the Field* (1949) (GC I), for example, the common article established that as a minimum requirement in international and non-international armed conflict, wounded and sick armed forces personnel, prisoners of war and civilians under conditions of occupation or internment are to be treated 'humanely, without any adverse distinction founded on race, colour, religion or faith, sex, birth or wealth, or any other similar criteria' (GC I, 1949: art. 3(1)). In particular, acts that do 'violence to life and person, in particular murder of all kinds, mutilation, cruel treatment and torture' (GC I, 1949: art. 3(1)(a)), as well as 'outrages on personal dignity, in particular humiliating and degrading treatment', are 'prohibited at any time and in any place whatsoever' (GC I, 1949: art. 3(1)(c)). Textually, Common Article 3 is premised on equal treatment of all persons, and women would appear to receive a measure of differentiated and presumably favourable treatment under its terms.

Consideration of Common Article 3 in a larger context, however, suggests a different picture. Women are not to be accorded any differentiated protection given their special needs on the basis of equality between the sexes, they are instead to be protected according to subordinating social notions of modesty and chastity, and their relationships to others, such as children. Turning to the actual conventions within which the common article is embedded, the same picture is presented.

Geneva Conventions I and II

The diplomatic conference finally concluded on 12 August 1949, and the parties all agreed to the language of the four different conventions that had been negotiated (Abplanalp, 1995: 533). In addition to Common Article 3, the first Geneva Convention and the *Geneva Convention for the Amelioration of the Condition of Wounded, Sick and Shipwrecked Members of the Armed Forces at Sea* (1949) (GC II), both covering wounded and sick armed forces personnel, also have a Common Article 12, which repeats the Common Article 3 requirement that members of the armed forces are 'to be treated humanely ... without any adverse distinction founded on ... sex' (GC I and GC II, 1949: arts. 12(2)). Article 12 also adds a provision that women 'shall be treated with all consideration due to their sex' (GC I and GC II, 1949: art. 12(4)).

Although the Commentaries to the Geneva Conventions are not official texts (Pictet, 1952: 7), they are considered authoritative interpretations of the Conventions' provisions. Their weight of authority is in large part justified by experiences and work of the primary editor, Jean S. Pictet, the International Committee of the Red Cross' Director for General Affairs, and the other authors, both during World War II and in the different conferences that culminated in the 1949 conventions. The first Geneva Convention's Commentary explains that article 12's purpose was to not exclude 'distinctions made in favour of enemy wounded or sick and in order to take their physical attributes into account' (Pictet, 1952: 137–38). In this context, the consideration due to women was seen as the 'consideration which is accorded in every country to beings who are weaker than oneself and whose honour and modesty call for respect' (Pictet, 1952: 140). The second Geneva Convention's Commentary on article 12 matches the first Geneva Convention explanation (Pictet, 1960a: 92). Thus, the protection afforded to women as women under these treaties is based on the traditional social inferiority of women rather than equality (Gardam and Jarvis, 2001: 10–11).

Geneva Convention III

Protections for women

The *Geneva Convention Relative to the Treatment of Prisoners of War* (1949) (GC III) establishes more specific protections for women in addition to the Common Article 3 requirements. Prisoners of war (POWs) 'are entitled in all circumstances to respect for their persons and their honour', and women 'shall be treated with all the regard due to their sex and shall in all cases benefit by treatment as favourable as that granted to men' (GC III, 1949: art. 14). Female POWs will have separate sleeping quarters from male POWs (GC III, 1949: art. 25), and separate hygienic facilities (GC III, 1949: art. 29). Pregnant women POWs and those who suffer from obstetrical disorders may be accommodated in neutral countries if possible – if not, they are to be repatriated directly back to their own country (GC III, 1949: art. 110; Annex I, A(3)(*f*)). Women POWs who are 'mothers with infants and small children' may also be accommodated in neutral countries, but the detaining country is not required to repatriate them (GC III, 1949: Annex I, B(7)).

Women POWs convicted of penal offences or camp disciplinary infractions are to receive punishments no more severe 'than a woman member of the armed forces of the Detaining Power dealt with for a similar offense' (GC III, 1949: art. 88). Female offenders will serve their sentences 'in separate quarters and be under the supervision of women', in the 'same establishments and under the same conditions as in the case of members of the armed forces of the Detaining Power' (GC III, 1949: art. 108). Similarly, female POWs serving punishments for disciplinary infractions are to be 'confined in separate quarters from male prisoners of war and be under the immediate supervision of women' (GC III, 1949: art. 97).

Deficiencies in the protections afforded by Geneva Convention III

These are significant protections, but when examined in context the limited nature of their scope, their grounding in a discriminatory male-normative approach to the protection of POWs becomes evident. The same rationale for affording women 'due consideration' in the first and second Geneva Conventions also applies to the third. The third Geneva Convention Commentary notes that the Conference of Government Experts had opined that the convention should provide for treatment of women prisoners of war that was at least equal to that afforded men, and that to make proper sense of article 14, for example, its second paragraph should be switched about so that treatment of women prisoners of war was to be based on equivalent treatment instead of 'the rather vague idea of "regard"' (Pictet, 1960b: 146).

The third Geneva Convention's Commentary notes that the 'regard' due to women POWs is not easy to define, but that is essentially based on three factors: women's weakness; their 'honour and modesty' and the health challenges of pregnancy and child-birth (Pictet, 1960b: 147). 'Honour and modesty' are functionally defined as protection against 'rape, forced prostitution and any form of indecent assault' (Pictet, 1960b: 147). Realistically, mindful of the experience of so-called 'comfort women' enslaved into prostitution by imperial Japanese forces in World War II (Hicks, 1997), the difference between 'enforced' prostitution versus a 'voluntary' choice in an austere POW camp setting is likely one of little distinction.

The approach of seeing women as a separate sub-class of POWs requiring special protection but then neglecting their specific needs and risks is exacerbated by the lack of women's representation in camps under the third Geneva Convention. In POW camps with only enlisted personnel, the POWs elect representatives by secret ballot to represent them to the detaining country and external organisations (GC III, 1949: art. 79). In camps with mixed officer and enlisted populations, the senior POW officer assumes the leadership role, and the enlisted are allowed to elect representatives to act as the senior officer's assistants. POWs of each nationality are allowed to have their own representatives as well (GC III, 1949: art. 79). Women POWs, however, will likely constitute a minority of any detained military population, and if the election of representatives is based on winning a simple majority of votes, as in the case with camps housing only enlisted personnel, they could find their specific requirements and voices marginalised. In this case, the spatial separation between men and women POWs, designed for the women's protection, might work against them in certain respects unless they are allowed their own representatives.

As to the health risks that women POWs face as a result of being pregnant, they are at least in part created by the deficiencies in the third Geneva Convention regarding reproductive health. The separate hygienic facilities provided to women are defined as 'adequate' infirmaries (GC III, 1949: art. 30); baths and showers, soap and water and latrines (GC III, 1949: art. 29). All prisoners are required to undergo at least a monthly medical inspection, which focuses on health, nutrition, cleanliness and detecting contagious diseases, such as venereal disease (GC III, 1949: art. 31). It is difficult to argue that the third Geneva Convention truly contemplates adequate hygienic facilities and services for women POWs when there is no mention of contraception or the need for women's sanitary supplies. Further, as to violent or coerced sexual relations between women POWs and men, whether fellow POWs or detaining personnel, and any resulting pregnancies, these risks to women POWs are unfortunately built into the treaty. For example, even though women POWs in punishment facilities are to be supervised by women, there is no such requirement for regular dormitories.

Tellingly, the third Geneva Convention's Commentary notes that the particular regard afforded by the treaty to women POWs who become pregnant whilst detained is premised on a failure of 'the precautions taken' (Pictet, 1960b: 148) – which apparently amount only to separate showers, latrines and dormitories. Unfortunately, this is not just a historical case of women soldiers' needs having been overlooked in the middle of the 20th century – it is a continuing problem. It was not until fairly recently that the US Army, for example, recognised that it had not been adequately addressing female soldiers' particular health needs in operational settings, and that 'basic improvements' were 'needed to help women avoid higher rates of urinary tract or vaginal infections', and 'stress-related menstrual difficulties' (Zoroya, 2012). Examples of useful solutions to some of these problems include the greater awareness and use of the Female Urinary Diversion Device, and the development of medical protocols for male medics and health care providers with limited experience with women patients, for such things as evaluation and treatment of abnormal uterine bleeding and emergency contraception (Wissemann, 2015).

Geneva Convention IV

The *Geneva Convention Relative to the Protection of Civilian Persons in Time of War* (1949) (GC IV), which sets out the protections accorded to civilians during armed conflicts, contains the greatest number of provisions dealing with women among all the conventions. There are two important limitations on the scope of its application, however. First, other than the minimum protections accorded under Common Article 3 even in non-international armed conflicts, it is focused on situations where one country occupies another, or when the nationals of one country are located in the territory of a different country, with which their home country is now engaged in armed conflict. Second, irrespective of Common Article 3 and article 27 the fourth Geneva Convention's coverage seems limited from a gender perspective – it allows 'no adverse distinction based, in particular, on race, nationality, religion, or political opinion' (GC IV, 1949: art. 13), but not sex.

In general, civilians in occupied countries are to be respected in 'their persons, their honour, their family rights . . . and to be humanely treated' (GC IV, 1949: art. 27). There are considerable protections for the female populations of occupied countries (GC IV, 1949: arts. 16–18, 20–23). However, the rationale for these protections is the same as that which underlies special protections for women in the other three conventions – one based on weakness and an inferior social position (Gardam and Jarvis, 2001: 108), rather than equal treatment based on sex-differentiated needs. For example, there is no general obligation that women will receive equitable distribution of food or medical supplies. Similarly, there is no differentiation in the protection given to girls as compared to male children, even though they might in fact be one of the most at-risk population cohorts.

In terms of potential numbers, internees are likely to be a smaller group of people affected by international armed conflict, but they too receive specific protections, which are to a degree sex-differentiated. For example, interned families will be housed together, and women internees without families will be given separate sleeping quarters and hygienic facilities (GC IV, 1949: arts. 49, 85). Only women will search female internees (GC IV, 1949: art. 97), and any disciplinary punishments they receive must take account of their sex (GC IV, 1949: art. 119). Further, the fourth Geneva Convention recognises that women who are pregnant or caring for young children are a particularly at-risk group, and they receive special protections. Pregnant and nursing mothers will be given additional food as their conditions require (GC IV, 1949: art. 89), and interned women delivering babies 'must be admitted to any institution where adequate treatment can be given and shall receive care not inferior to that provided for the general population' (GC IV, 1949: art. 91). Although these are important protections, the effect of these provisions is to essentialise women in the IHL treaty law context as mothers and caregivers, rather than as being equal to men (Barrow, 2010: 224). This is particularly highlighted by the emphasis upon 'honour' in the Commentary to this treaty as it relates to women.

'Honour' was defined as 'a moral and social quality', and the 'right to respect for his honour is a right invested in man because he is endowed with reason and a conscience' (Pictet, 1958: 202). Even assuming that the definition of 'honour' would have been properly understood as applying to all persons at that time irrespective of the use of masculine pronoun, as formulated it sets out a different sense of 'honour' for women than it does for men. For example, the fourth Geneva Convention's Commentary notes that 'respect for family rights', intended to protect 'the marriage ties and the community of parents and children', is 'covered by the clause prohibiting rape and other attacks on women's honour' (Pictet, 1958: 202). Similarly, the prohibition of these offences was intended to protect women's 'absolute right to respect for their honour and their modesty, in short, for their dignity as women' (Pictet, 1958: 206). Modesty is left undefined, but 'immodest' women are arguably not included in the rationale for the protection of women in the sense of the convention (Crowe, 2016: 16).

The 1977 additional protocols

Beginning with the first diplomatic conference session held in 1974 (Junod, 1983: 32), the 1977 Additional Protocols to the 1949 Geneva Conventions were negotiated in a very different international security setting. Barrass (2009) notes that at this time, in the aftermath of the US retreat and eventual defeat in Vietnam, the Cold War between the US and the Soviet Union had begun showing serious signs of heating back up (186–211). The vast majority of European colonies had achieved independence by this time, and new states had often been born out of situations of non-international armed conflict (Betts, 2004: 23, 27, 31, 32, 40, 60–62, 112–13). A number of conflicts were still ongoing at the time, for example insurgencies in Namibia, Zimbabwe and South Africa against white apartheid rule (Turner, 1998: 16–99).

With new states at the negotiating table, with different experiences and political alignments, Additional Protocol I (AP I) took a different approach to IHL than that set by the Geneva Conventions, and it effectively merged the Geneva strand of IHL regarding victims of international armed conflict with the Hague strand of IHL concerning the conduct of international armed conflict law into one treaty

(Henckaerts and Doswald-Beck, 2006: ix). Additional Protocol II (AP II) sought to accomplish something similar with non-international armed conflicts, but once the language of Additional Protocol I, article 1 was approved extending the definition of international armed conflict to wars against colonial or racist regimes, a majority of the states party to the negotiations seemed leery about expanding the coverage of the Additional Protocol II too far into internal domestic affairs. Thus, the final version of the protocol did not include any provisions appearing to recognise the legitimacy of insurgencies, and most proposed rules regulating means and methods of warfare were stripped out (Junod, 1983: 32–34).

Additional Protocol I

Consistent with the 1949 Geneva Conventions, Additional Protocol I provides certain enhanced protections to women and girls, but many of these provisions continue the Geneva Conventions' rationale of women's status of caregivers and weakness rather than equality as the basis for protection. Other protections are indirect in nature, and although in certain circumstances they might actually mitigate the disproportionate effects of armed conflict upon women and girls, this is not noted as a basis for the protections. This might be the result of Additional Protocol I's significant focus on the principle of distinction, that is, distinguishing between persons who are not engaged in combat and are therefore to be protected, and those who are engaged in combat and therefore might be lawfully targeted with armed force (AP I, 1979: arts. 8–20, 35, 39, 43–51, 58, 67 and 79).

For example, whilst Additional Protocol I, article 76, establishes that pregnant women should not receive death sentences, and that pregnant women or mothers with dependent infants will not be executed, the intent of this provision appears to have been to enhance the protection of foetuses and born children (Sandoz, Swinarski and Zimmerman, 1987: 895–96) – not the mothers. Consistent with the fourth Geneva Convention, women who are pregnant or nursing will be given priority in the distribution of relief supplies (AP I, 1977: art. 70), and when subject to allegations of wrongdoing, pregnant detainees or those with dependent infants will have their penal 'cases considered with the utmost priority' (AP I, 1977: art. 76(2)). The rationale for these protections is the same used in the 1949 Geneva Conventions – the Commentary to Additional Protocol I states that the prohibition is still based on the traditional notion of 'honour' (Sandoz, Swinarski and Zimmerman, 1987: 892) rather than just an unqualified proscription of offences against women as equals to men in personhood.

Additional Protocol I's differentiation between civilians on the basis of sex also occurs to a degree in the indirect use of armed force. For instance, the intent of article 75's prohibition of rape, enforced prostitution and indecent assault against women was to extend the protections afforded to women and girls by GC IV in situations of occupation and internment to all women in the territories of parties to the conflict (Sandoz, Swinarski and Zimmerman, 1987: 892). In this sense, the differentiation is not based on the use of armed force to secure a military objective, but indirectly on the prohibition of the use of force against protected persons who by definition cannot be military targets.

Ironically, Additional Protocol I's liberalisation of the legal standard used to determine whether a person is a combatant so as to increase the protections available to captured irregular force personnel by detaining forces under GC III might actually work to the detriment of women and children. Under GC III, a combatant could not receive POW status unless she had a commander who was responsible for her subordinates, the combatant wore a fixed and distinctive sign visible at a distance identifying her as a combatant, she carried her weapons in the open, and the military operations in which she was involved were conducted in accordance with IHL (GC III, 1949: art. 4.A(2)).

Under Additional Protocol I, however, captured irregular combatants are accorded prisoner of war status so long as they carried their weapons in the open before they attacked, and they are not required to wear distinctive insignia and uniforms (AP I, 1977: art. 44(3)(b)). As seen in instances of Taliban fighters using civilians as cover and taking civilians hostage as they conduct attacks (BBC, 2010), Gardam (1989) argues that this has the unintended effect of legitimising 'the use of women and children as shields' just before an attack is launched (265). This complicates applying the principle of distinction to protect these female civilians when the force that is attacked responds with armed force itself (265).

Additional Protocol II

Primarily dealing with the protection of victims of non-international armed conflict, Additional Protocol II increases the minimum standard of humane treatment provided under Common Article 3. Adverse distinctions among civilians on the basis of sex are specifically prohibited (AP II, 1977: art. 2), as are 'outrages upon personal dignity, in particular humiliating and degrading treatment, rape, enforced prostitution, and any form of indecent assault' (AP II 1977, art. 4.2(*e*)). Consistent with the provisions of the fourth Geneva Convention in situations of international armed conflict, women internees without families will be held in facilities separate from men, and under the supervision of women (AP II, 1977: art. 5.2(*a*)). Further, and particularly important in light of the Islamic State of Iraq and the Levant's barbaric violations of women's and girls' persons and property (Otten, 2017), slavery (AP II, 1977: art. 4.2(*f*)) and pillage (AP II, 1977: art. 4.2(*g*)) are prohibited. Unfortunately, the rationale underlying these provisions is the same concern for women's weakness and status as caregivers found in the 1949 Geneva Conventions and Additional Protocol I (ICRC, 1987: 1375).

Customary IHL

In terms of identifying the applicable norms that need to be factored into military activities and operations, it is not enough to look only at treaty law, we must also examine the role played by the development of customary IHL. In their comprehensive study of customary IHL for the International Committee of the Red Cross, Henckaerts and Doswald-Beck (2006) note that customary international law, established by state acceptance and practice of treaty law is binding on all states regardless of whether they are parties to the underlying treaties that establish the underlying legal norms (xlii–xlv). They take the position that the norms set out in the 1949 Geneva Conventions and the Additional Protocols are now part of customary IHL (xxvii). Henckaerts and Doswald-Beck find three basic rules regarding the treatment of women in armed conflict as reflecting customary international law: the prohibition of sexual violence against them, the requirement to separate women who are detained from detained men and the consideration of women's specific protection, health and assistance needs (323–27, 431–33, 475–79).

Henckaerts and Doswald-Beck (2006) sometimes make certain generalisations about these protections that if correct would lead them to have a broader scope of application, but these generalisations are not necessarily borne out by provisions within the treaties that they rely upon. For example, as to the separation of women and men deprived of their liberties, they state that women detainees must be under the immediate supervision of other women, citing the third Geneva Convention, article 97, for support (431, n. 22). Article 97, however, only affords this protection to women who are in disciplinary confinement, rather than generally to women prisoners of war. Further, they state that the special protections afforded women under the 1949 Geneva Conventions have become customary IHL through the operation of terminology such as 'all consideration due their sex' and women being the 'object of special respect' (475–76). Given the outmoded and discriminatory social construct regarding women that gave rise to these phrases, it does not follow that this language serving as the vehicle by which women's protections have been made part of customary IHL is completely a positive thing in terms of realising the goals of UNSCR 1325.

Challenges in correcting the deficiencies

The feminist critique's failure to gain traction

The feminist critique of IHL has been set out cogently in scholarly writings for more than a quarter century (Chinkin, 1989; Gardam, 1989), but it is difficult to trace any impact it might have had so far on how militaries educate and train their personnel or conduct their planning and operations. There are a number of possible explanations for this lack of traction. Some feminist critics of militaries in general argue that that 'monolithic' nature of military organisations reinforces their qualities of 'patriarchal hegemony' and their emphasis on 'hyper-masculinity'. Authoritarian and male-dominated militaries therefore reject the fundamental revolution in the ways in which they are organised and operate that would be necessary to achieve true equality between women and men (Runyan and Peterson, 2014: 151–55; Duncanson, 2015: 234–36, 240–41). That line of argument is valid in certain respects, but it has three weaknesses that prevent it from being a particularly useful approach in operationalising gender.

First, it does not appear to fully appreciate that militaries are very conservative by nature and mission. It is difficult to find examples of militaries eagerly embracing any revolution, especially one with a vocabulary that translates so awkwardly in dealing with the crush of daily operations and established procedures. Second, militaries by their nature will be hierarchies – that is the nature of command – in which conservatism acts to mitigate the risks and the costs of failure.

Third, what would a military 'matriarchy' look like? For the sake of argument, let's say that it would be an operational military unit that is composed either mostly of women or at least equal numbers of women and men, generally commanded by women. This definition would exclude traditional service support units that would otherwise be included, such as nursing units.

There are only a few modern examples of formal units that would meet this definition. The female units formed by different Kurdish authorities in Syria (*The Economist*, 2018) and Iraq (McLaughlin, 2016) discussed in Chapter 1, 'Armed conflict and gender', might be good examples. The formed female police units deployed by India and Bangladesh in UN peacekeeping operations and the mixed-gender Israeli light infantry battalions discussed in the next chapter likely come the closest. On the basis of these limited examples, a military matriarchy might be characterised by a lower degree of indiscipline in the ranks (Anderholt, 2012: 14–15) and a commensurately higher degree of morale (Kraft, 2018) than many male units. One thing that would not likely be different is the dedication that commanders and soldiers of such units have shown to their missions, their comrades and their families (Raddatz, 2016; Gross, 2018; Kraft, 2018).

Problems in practice and in perception

Other possible explanations are less theoretical. Barrow (2010) suggests that UNSCR 1325 itself might be at fault. Because it did not move forward far enough regarding the status of women and girls under IHL, 'constructions of gender within international humanitarian law are thus constrained without any radical form of redress' (234). Another explanation could be the relative scarcity of writings on the feminist critique of IHL in feminist literature in general, which, to the extent it considers international law, tends to focus more on human rights law (Gardam and Charlesworth, 2000: 150) instead of IHL. This factor might be compounded by discussions of armed conflict's impacts upon women taking place in locations and spaces not generally frequented by those who train, plan and operate in an IHL context. The annual conferences held under the auspices of the US Naval War College on Women, Peace and Security are an important exception to this tendency (US Naval War College, 2018), but there are a limited number of examples of such work.

Perhaps stakeholders and participants in this traditionally male-influenced area of law concerning the use of force conflate this critique with human-rights influenced pacifism, and cursorily dismiss it without examining it in any depth (Prescott, 2013a: 104). Or, perhaps they see it as an attempt to incorporate international human rights law into IHL, which could dilute the *lex specialis* status that IHL traditionally has been seen as having in situations of armed conflict (Hathaway et al., 2012: 1888). This factor might also be compounded by women still having to prove themselves as capable as men in military situations, especially combat, to be accepted as equal team mates. Further, some might see arguing for a gender-differentiated approach to operations as being inconsistent with a nominally gender-blind military (Prescott, 2013a: 104), and women military personnel in these situations may feel a need to be less critical of the male-normative military culture in order to be accepted (Valenius, 2007: 41). A related factor might be simply a scarcity of women in the right places, and not just in command slots. As Gardam and Jarvis (2001) note, IHL does not require that there be 'adequate sex composition amongst personnel who are trained to facilitate' its application (105).

Official interpretations and applications of IHL

Finally, on the military side, the role of legal doctrine cannot be ignored. None of the doctrinal documents on IHL and its application from Australia (Chief of Defence Force, 2006), NATO (NATO Standardization Agency, 2013), the UK (Ministry of Defence, 2004) or the US discuss the application of UNSCR 1325 to operations conducted under IHL. The US Department of Defense *Law of War Manual* (2016) is a staggering 1,193 pages long. It contains no mention of UNSCR 1325, and its discussion of protections afforded to women and girls essentially repeats the traditional formulations found in IHL treaty law and the treaty commentaries (Department of Defense, 2016: 98, 99–100, 165–66, 256, 447, 515, 517, 520–21, 531, 546, 564–65, 619, 622, 630, 667, 670, 676–78, 683, 691, etc.). For example, in discussing the rights afforded to women internees, the *Law of War Manual* states that they 'shall be provided separate sanitary conveniences for their use' (695). This sounds potentially progressive, until one looks at the footnotes that explain these 'conveniences', and realises that they consist only of latrine, laundry and bathing facilities and supplies (695, nn. 194–97). As a final example, it states that although the third Geneva Convention 'does not specifically provide for this, female POWs should be under the immediate supervision of women' (551). 'Should' does not mean 'must'.

Militaries of nation states are not the only important actors in determining the interpretations of IHL that will be applied in armed conflict. Rather than seeking to modify existing treaty law, the International Committee of the Red Cross has apparently decided to redefine 'honour' as it has embarked on an effort to rewrite the commentaries on the 1949 Geneva Conventions and the 1977 Additional Protocols. For example, in a very practical and comprehensive guidance document designed for use by policy-makers and staff in the field to address the needs of women impacted by armed conflict, the term 'honour' is used exclusively in the context of conservative societal mores that discriminate against women by requiring them to meet certain conditions of chastity to be socially accepted (Lindsey-Curtet et al., 2004: 12, 22, 29–30, 33, 121, 141, 155, 159).

In addition, the new commentary on the first Geneva Convention now includes significant references to gender mainstreaming, that is, recognising that men, women, boys and girls are affected by armed conflict in different ways, and that therefore sensitivity 'to the individual's inherent status, capacities and needs, including how these differ among men and women due to social, economic cultural and political structures in society, contributes to the understanding of humane treatment under common Article 3' (ICRC, 2016a: para. 553). It notes that these factors should be considered in providing appropriate medical treatment to individuals (paras. 577, 578, 766). The new commentary also discusses gender-based humiliation as prohibited cruel treatment (para. 621) as well as sexual violence, sexual slavery and forced marriages (para. 672), including sexual violence against men and boys (para. 700) under the common article.

Similarly, in the new commentary specifically on the first Geneva Convention's article 12, concerning protection and care of the wounded and sick, appropriate humane treatment is seen as being predicated on gender in part (ICRC, 2016b: para. 1373), and that differentiating between sexes in providing proper medical care is based on 'social and international legal developments in relation to equality of the sexes' (para. 1427). As to the language of the original commentary referencing women's weakness, honour and modesty as the basis for their humane treatment, the new commentary notes that it 'would no longer be considered appropriate' (n. 161). Instead, the new basis for providing differentiated treatment to women is not that 'they have less resilience, agency or capacity within the armed forces or as civilians, but rather acknowledges that women have a distinct set of needs and may face particular physical and psychological risks' (para. 1427).

Whilst the language in the new commentary at least recognises that the solution to achieve greater protection for women and girls in armed conflict is more than just increased compliance with IHL, the International Committee of the Red Cross's new interpretation of social and legal developments regarding gender relations is not sufficient to erase the original language of the treaty, nor the understandings based on this language that continue to influence its application today. The legal sources cited in the footnotes justifying the new approach would suggest to one unacquainted with the negotiation history of the 1949 Geneva Conventions that these legal concepts, such as equality of the sexes, were only recognised long after the treaties were negotiated and the original commentaries written.

The history shows that instead, opportunities to ground the treaty provisions in notions of sexual equality were presented by different experts and delegates, but these understandings were not taken on board, sometimes at the urging of the International Committee of the Red Cross representatives. And, no disrespect intended, but the authors of the new commentary are not in the position of a Jean Pictet, or those who worked with him on the original commentaries. What was written then has not been unwritten by later commentary that might bear the authoritative imprimatur of the International Committee of the Red Cross, but in substance does not address the root problem of gender in treaty IHL and instead essentially states only that times have changed.

Summary

IHL provides protection to women in three primary ways. First, through its requirements of equal treatment of protected persons regardless of sex; second, its special protections for pregnant women and mothers of small children; and third, its prohibition of sexual violence against women (Oosterveld, 2009: 387). International efforts to reduce the occurrence of sexual violence in armed conflict situations and to eliminate the impunity of its perpetrators represent real and meaningful progress in implementing important goals set out for the international community under UNSR 1325. At the same time, however, IHL treaty law has continued to reflect male-normative notions of the rationales for the protection of women in situations of armed conflict. These notions were largely premised on perceptions of women's inherent weakness vis-à-vis men, and of a status defined largely by women's relationships to others and their own chastity. It was not originally based on an equal status for women that reflects their special needs and the socially rooted risks they endure to the impacts of armed conflict.

Further, in the areas of IHL that are fundamental to its application in armed conflict, such as the standards governing the use of lethal force, there is little evidence of any meaningful reassessment of IHL in light of the recognised differentiated and more severe impacts of armed conflict on women and girls. The conservative role played by legal doctrine that uncritically reflects these notions and perspectives should not be underestimated, and it likely tends to reinforce these concepts in the activities and operations of military organisations, which are by their nature, irrespective of gender issues, generally averse to new ideas. Mindful of this, let's finally turn to the ongoing efforts different militaries across the world have engaged in to begin to operationalise gender's relevance in military activities and operations.

References

Abplanalp, P (1995) 'The International Conferences of the Red Cross as a Factor for the Development of International Humanitarian Law and the Cohesion of the International Red Cross and Red Crescent Movement'. *International Review of the Red Cross* 380: pp. 520–49. Available at: www.icrc.org/eng/resources/documents/article/other/57jmr9.htm (Accessed 27 June 2018).

Anderholt, C (2012) 'Female Participation in Formed Police Units – A Report on the Integration of Women in Formed Police Units of Peacekeeping Operations'. *US Army Peacekeeping & Stability Operations Institute Papers* (September): pp. 1–60. Available at: http://pksoi.armywarcollege.edu/default/assets/File/Formed_Police_Units.pdf (Accessed 11 June 2018).

AP I (Protocol Additional to the Geneva Conventions of 12 August 1949, and Relating to the Protection of Victims of International Armed Conflicts (adopted 8 June 1977, entered into force 7 December 1978) 1125 UNTS 3.

AP II (Protocol Additional to the Geneva Conventions of 12 August 1949, and Relating to the Protection of Victims of Non-International Armed Conflict (adopted 8 June 1977, entered into force 7 December 1978) 1125 UNTS 609.

Barrass, GS (2009) *The Great Cold War: A Journey through the Hall of Mirrors*. Stanford: Stanford University Press.

Barrow, A (2010) 'UN Security Council Resolutions 1325 and 1820: Construing Gender in Armed Conflict and International Humanitarian Law'. *International Review of the Red Cross* 92(877): pp. 221–34. Available at: www.icrc.org/eng/assets/files/other/irrc-877-barrow.pdf (Accessed 27 June 2018).

BBC Staff (2010) 'Afghanistan Taliban "using Human Shields" – General'. *BBC News* (17 February). Available at: http://news.bbc.co.uk/2/hi/south_asia/8519507.stm (Accessed 21 June 2018).

Betts, RF (2004) *Decolonization*, 2nd Ed. New York: Routledge.

Chief of Defence Force (2006) *ADDP 06.4, Law of Armed Conflict* (11 May). Canberra: Australian Defence Headquarters. Available at: www.defence.gov.au/adfwc/documents/doctrinelibrary/addp/addp06.4-lawofarmedconflict.pdf (Accessed 21 June 2018).

Chinkin, C (1989) 'A Gendered Perspective to the International Use of Force'. *Australian Year Book of International Law* pp. 279–93. Available at: http://www5.austlii.edu.au/au/journals/AUYrBkIntLaw/1989/13.html (Accessed 27 June 2018).

Clarke, M (2013) 'UK Armed Forces Personnel and the Legal Framework for Future Operations'. www.parliament.uk (11 December). Available at: https://publications.parliament.uk/pa/cm201314/cmselect/cmdfence/writev/futureops/law18.htm (Accessed 21 June 2018).

Convention (IV) Respecting the Laws and Customs of War on Land and its Annex: Regulations Concerning the Laws and Customs of War on Land (adopted 18 October 1907, entered into force 26 January 1910) (Hague Regulations). Available at: file:///C:/Users/01jpresc/Downloads/IHL-19-EN.pdf (Accessed 6 July 2018).

Crawford, E (2010) *The Treatment of Combatants and Insurgents under the Law of Armed Conflict*. Oxford: Oxford University Press.

Crowe, A (2016) '"All the regard due to their sex": Women in the Geneva Conventions of 1949'. *Harvard Research Working Paper Series*, HRP 16–001 (December). Harvard Rights Program. Available at: http://hrp.law.harvard.edu/wp-content/uploads/2016/12/Anna-Crowe_HRP-16_001.pdf (Accessed 21 June 2018).

Department of Defense (US) (2016) *Law of War Manual*. Washington, DC: Office of General Counsel. Available at: www.defense.gov/Portals/1/Documents/DoD_Law_of_War_Manual-June_2015_Updated_May_2016.pdf (Accessed 6 July 2018).

Duncanson, C (2015) 'Hegemonic Masculinity and the Possibility of Change in Gender Relations'. *Men and Masculinities* 18(2): pp. 231–45. DOI: 10.1177/1097184x15584912.

Durham, H and O'Byrne, K (2010) 'The Dialogue of Difference: Gender Perspectives on International Humanitarian Law'. *International Review of the Red Cross* 92(877): pp. 31–52.

The Economist Staff (2018) 'Where the Fighters Are Female – A Short History of Kurdish Women on the Front Lines'. *The Economist* (22 March). Available at: www.economist.com/middle-east-and-africa/2018/03/22/a-short-history-of-kurdish-women-on-the-front-lines (Accessed 29 June 2018).

Elder, DA (1979) 'The Historical Background of Common Article 3 of the Geneva Convention of 1949'. *Case Western Reserve Journal of International Law* 11(1): pp. 37–69. Available at: https://scholarlycommons.law.case.edu/cgi/viewcontent.cgi?article=1940&context=jil (Accessed 27 June 2018).

Frederick, B and Johnson, DE (2015) *The Continued Evolution of U.S. Law of Armed Conflict Implementation*. Santa Monica, CA: Rand Corporation. Available at: file:///C:/Users/jmprescott/Downloads/RAND_RR1122.pdf (Accessed 27 June 2018).

Gardam, J (1989) 'A Feminist Analysis of Certain Aspects of International Humanitarian Law'. *Australian Year Book of International Law* 12: pp. 265–78. Available at: www.austlii.edu.au/au/journals/AUYrBkIntLaw/1989/12.pdf (Accessed 27 June 2018).

Gardam, J and Charlesworth, H (2000) 'The Need for New Directions in the Protection of Women in Armed Conflict'. *Human Rights Quarterly* 22(1): pp. 148–66. Available at: www.jstor.org/stable/4489270 (Accessed 27 June 2018).

Gardam, JG and Jarvis, MJ (2001) *Women, Armed Conflict and International Law*. The Hague: Kluwer Law International.

GC I (Geneva Convention for the Amelioration of the Condition of the Wounded and Sick in Armed Forces in the Field (adopted 12 August 1949, entered into force 21 October 1950)) 75 UNTS 31.

GC II (Geneva Convention for the Amelioration of the Condition of Wounded, Sick and Shipwrecked Members of the Armed Forces at Sea (adopted 12 August 1949, entered into force 21 October 1950)) 75 UNTS 85.

GC III (Geneva Convention Relative to the Treatment of Prisoners of War (adopted 12 August 1949, entered into force 21 October 1950)) 75 UNTS 135.

GC IV (Geneva Convention Relative to the Protection of Civilian Persons in Time of War (adopted 12 August 1949, entered into force 21 October 1950)) 75 UNTS 287.

Gross, JA (2018) 'Arguments Rage Over IDF's Inclusion of Women, But Co-ed "Lions" Have a Job to Do'. *The Times of Israel* (11 March). Available at: www.timesofisrarel.com/arguments-rage-over-idfs-inclusion-of-women-but-co-ed-lions-have-a-job-to-do/ (Accessed 10 June 2018).

Hathaway, O; Crootof, R; Levitz, P; Nix, H; Perdue, W and Purvis, C (2012) 'Which Law Governs During Armed Conflict? The Relationship Between International Humanitarian Law and Human Rights Law'. *Minnesota Law Review* 96: pp. 1883–943. Available at: http://digitalcommons.law.yale.edu/cgi/viewcontent.cgi?article=5734&context=fss_papers (Accessed 27 June 2018).

Henckaerts, J-M and Doswald-Beck, L (2006) *Customary International Humanitarian Law, Vol. 1, Rules*. Geneva: ICRC.

Hessbruegge, JA (2017) *Human Rights and Personal Self-Defense in International Law*. Oxford: Oxford University Press.

Hicks, G (1997) *The Comfort Women: Japan's Brutal Regime of Enforced Prostitution in the Second World War*. New York: WW Norton & Company.

International Committee of the Red Cross (ICRC) (n.d.) 'H. Paragraph 4: Treatment of Women'. *ICRC*. Available at: www.icrc.org/applic/ihl/ihl.nsf/Comment.xsp?action=openDocument&documentId=CECD58D1E2A2AF30C1257F15004A7CB9#_Toc437265696 (Accessed 21 June 2018).

International Committee of the Red Cross (1947) *Report on the Work of the Conference of Government Experts for the Study of the Conventions for the Protection of War Victims*. Geneva: ICRC.

International Committee of the Red Cross (1949a) *Final Record of the Diplomatic Conference of Geneva of 1949, Vol. I*, Berne: Swiss Federal Political Department.

International Committee of the Red Cross (1949b) *Final Record of the Diplomatic Conference of Geneva of 1949, Vol. II-A*. Berne: Swiss Federal Political Department.

International Committee of the Red Cross (1987) *Commentary on the Additional Protocols of 8 June 1977 to the Geneva Conventions of 12 August 1949*. Geneva: Martinus Nijhoff.

International Committee of the Red Cross (2014) 'A History of Humanity'. *ICRC* [video] (26 May). Available at: www.youtube.com/watch?v=lBdQ-Rr4BuQ (Accessed 21 June 2018).

International Committee of the Red Cross (2016a) 'Commentary of 2016 – Article 3: Conflicts Not of an International Character'. *ICRC*. Available at: https://ihl-databases.icrc.org/applic/ihl/ihl.nsf/Comment.xsp?action=openDocument&documentId=59F6CDFA490736C1C1257F7D004BA0EC (Accessed 21 June 2018).

International Committee of the Red Cross (2016b) 'Commentary of 2016 – Article 12: Protection and Care of the Wounded and Sick'. *ICRC*. Available at: https://ihl-databases.

icrc.org/applic/ihl/ihl.nsf/Comment.xsp?action=openDocument&documentId=CECD58D1E2A2AF30C1257F15004A7CB9#159 (Accessed 21 June 2018).

International Covenant on Civil and Political Rights (adopted 16 December 1966, entered into force 23 March 1976) 999 UNTS 171.

International Criminal Court (2018) 'Case Information Sheet – *The Prosecutor v. Jean-Pierre Bemba Gombo*, ICC-01/05–01/18'. *ICC*. Available at: www.icc-cpi.int/car/bemba/Documents/BembaEng.pdf (Accessed 27 June 2018).

Junod, S-S (1983) 'The American Red Cross – Washington College of Law Conference: International Humanitarian and Human Rights Law In Non-International Armed Conflicts, April 12–13, 1983: Additional Protocol II: History And Scope'. *American University Law Review* 33: pp. 29–40. Available on Lexis™ (Accessed 17 September 2017).

Kraft, D (2018) 'Why Mixed-Gender Combat Units Are on the Rise in Israel'. *Christian Science Monitor* (23 March). Available at: www.csmonitor.com/World/Middle-East/2018/0208/Why-mixed-gender-combat-units-are-on-the-rise-in-Israel (Accessed 25 June 2018).

Lijnzaad, L (2005) 'Book Review: J.G. Gardam; M.J. Jarvis, *Women, Armed Conflict and International Law*'. *Netherlands International Law Review* 52(3): pp. 496–500. DOI: 10.1017/S0165070X05234897.

Lindsey-Curtet, C; Holst-Roness, FT and Anderson, L (2004) *Addressing the Needs of Women Affected by Armed Conflict: An ICRC Guidance Document*. Geneva: ICRC. Available at: www.icrc.org/eng/assets/files/other/icrc_002_0840_women_guidance.pdf (Accessed 27 June 2018).

McLaughlin, E (2016) 'Women at War: Meet the Female Peshmerga Fighters Taking on ISIS'. *ABC News* (16 May). Available at: https://abcnews.go.com/International/women-war-meet-female-peshmerga-fighters-taking-isis/story?id=39142160 (Accessed 29 June 2018).

Melzer, N (2009) *Interpretive Guidance on the Notion of Direct Participation in Hostilities Under International Humanitarian Law*. Geneva: ICRC.

Ministry of Defence (UK) (2004) *JSP 383, The Joint Service Manual of the Law of Armed Conflict*. Shrivenham, UK: Joint Doctrine and Concepts Centre. Available at: www.gov.uk/government/uploads/system/uploads/attachment_data/file/27874/JSP3832004Edition.pdf (Accessed 14 March 2017).

NATO Standardization Agency (2013) *ATrainP-2, Training in the Law of Armed Conflict*, Ed. A.Ver. 1 (March). Brussels: NATO. Available at: www.act.nato.int/images/stories/budfin/rfp016046.pdf (Accessed 27 June 2018) (found in RFP-ACT-SACT-16–46, pp. 171–318).

Oosterveld, V (2009) 'Feminist Debates on Civilian Women and International Humanitarian Law'. *Windsor Yearbook of Access to Justice* 27: pp. 385–402. Available at: https://ir.lib.uwo.ca/cgi/viewcontent.cgi?article=1105&context=lawpub (Accessed 27 June 2018).

Otten, C (2017) 'Life After ISIS Slavery For Yazidi Women And Children'. *The New Yorker* (31 August). Available at: www.newyorker.com/news/news-desk/life-after-isis-slavery-for-yazidi-women-and-children (Accessed 2 February 2018).

Paust, JJ (2015) 'Human Rights on the Battlefield'. *The George Washington International Law Review* 47(3): pp. 509–61.

Pictet, JS (1952) *The Geneva Conventions of 12 August 1949, Volume I, Commentary, Geneva Convention for the Amelioration of the Condition of the Wounded and Sick in Armed Forces in the Field*. Geneva: ICRC.

Pictet, JS (1958) *The Geneva Conventions of 12 August 1949, Volume IV, Commentary, Geneva Convention Relative to the Protection of Civilian Persons in Time of War*. Geneva: ICRC.

Pictet, JS (1960a) *The Geneva Conventions of 12 August 1949, Volume II, Commentary, Geneva Convention for the Amelioration of the Condition of the Wounded, Sick and Shipwrecked Members of Armed Forces at Sea*. Geneva: ICRC.

Pictet, JS (1960b) *The Geneva Conventions of 12 August 1949, Volume III, Commentary, Geneva Convention Relative to the Treatment of Prisoners of War*. Geneva: ICRC.
Prescott, JM (2010) 'Detention Status Review Process in Transnational Armed Conflict: Al Maqaleh v. Gates and the Parwan Detention Facility'. *University of Massachusetts Law Review* 5(1): pp. 34–88. Available at: https://scholarship.law.umassd.edu/cgi/viewcontent.cgi?article=1085&context=umlr (Accessed 27 June 2018).
Prescott, JM (2013a) 'NATO Gender Mainstreaming and the Feminist Critique of the Law of Armed Conflict'. *Georgetown University Journal on Gender and the Law* 14(1): pp. 83–131. Available at: https://papers.ssrn.com/sol3/papers.cfm?abstract_id=2246000 (Accessed 27 June 2018).
Prescott, JM (2013b) 'The North Atlantic Treaty Organization'. In: Zyberi, G (ed.) *An Institutional Approach to the Responsibility to Protect*, Cambridge: Cambridge University Press, pp. 338–61.
Prescott, JM (2016) 'The Convergence of Violence around a Norm: Direct Participation in Hostilities and Its Significance for Detention Standards in Non-International Armed Conflict'. In: Rose, G and Oswald, B (eds.) *Detention of Non-State Actors Engaged in Hostilities – The Future Law*. Leiden: Brill Nijhoff, pp. 65–92.
Prescott, JM (2018) 'Claims'. In: Fleck, D (ed.) *The Handbook of the Law of Visiting Forces*, 2nd Ed. Oxford: Oxford University Press, pp. 275–338.
Prosecutor v. *Anto Furundzija* (judgment) IT-95–17/1-T (10 December 1998).
Prosecutor v. Jean Paul Akayesu (judgment) ICTR-96–4-T (2 September 1998).
Prosecutor v. Jean-Pierre Bemba Gombo (judgment) ICC-01/05–01/08 (21 June 2016).
Raddatz, M (2016) 'Meet the Female Peshmerga Fighters Taking on ISIS'. *ABC News* [video] (16 May). Available at: http://abcn.ws/1TdzEcm (Accessed 29 June 2018).
Rome Statute of the International Criminal Court (adopted 17 July 1998, entered into force 1 July 2002) 2187 UNTS 3 (Rome Statute).
Runyan, AS and Peterson, VS (2014) *Global Gender Issues: In the New Millennium*, 4th Ed, Boulder, CO: Westview Press.
Sandoz, Y; Swinarski, C and Zimmerman, B (1987) *Commentary on the Additional Protocols of 8 June 1977 to the Geneva Conventions of 12 August 1949*. Geneva: Martinus Nijhoff.
Taft, WH (2004) 'Self-Defense and the Oil Platforms Decision'. *Yale Journal of International Law* 29(2): pp. 295–306. Available at: http://digitalcommons.law.yale.edu/cgi/viewcontent.cgi?article=1232&context=yjil (Accessed 27 June 2018).
Turner, JW (1998) *Continent Ablaze: The Insurgency Wars in Africa 1960 to the Present*. London: Arms and Armour Press.
US Naval War College (2018) 'Women, Peace and Security Conference Series: 4th Conference'. US Naval War College [press release]. Available at: https://usnwc.edu/News-and-Events/Events/Women-Peace-and-Security-Conference-Series-4th-Conference (Accessed 27 June 2018).
Valenius, J (2007) *Gender Mainstreaming in ESDP Missions*, Chaillot Paper No. 101. Paris: Institute for Security Studies. Available at: www.iss.europa.eu/sites/default/files/EUISSFiles/cp101.pdf (Accessed 27 June 2018).
Wissemann, M (2015) 'The Women's Health Task Force'. *US Army* (28 May). Available at: www.army.mil/article/149407/the_womens_health_task_force (Accessed 3 August 2015).
Wolff, I (2014) 'The Forgotten Women of the "war in the East"'. *BBC Magazine* (19 October). Available at: www.bbc.com/news/magazine-29665232 (Accessed 13 December 2017).
Zoroya, G (2012) 'Army Task Force: Females Troops Need Better Health Care'. *USA Today* (6 June). Available at: http://wwwusatoday.com/news/military/story/2012-06-06/female-soldiers-need-better-health-care (Accessed 3 August 2015).

9
GENDER IN MILITARY ACTIVITIES AND OPERATIONS

Strategy, policy, doctrine and understandings and applications of IHL comprise a significant portion of the intellectual infrastructure militaries rely upon to develop curricula for military education, training objectives and evaluation criteria for exercises, standard operating procedures for planning operations, and guidance for operations. As examined in some detail in the prior chapters on NATO, the UK, the US and Australia, there are significant areas within these militaries' intellectual infrastructure where the incorporation of gender considerations is evident, and there are many meaningful efforts under way to promote further integration of gender into these militaries' activities and operations. At this time, however, it is difficult sometimes to continuously trace the understanding of gender's operational relevance from high-level strategy all the way down to service-level doctrine. These gaps will need to be closed to ensure that gender considerations are operationalised to the fullest extent possible.

Despite these gaps, military organisations are not standing still on operational gender issues, and different countries and international as well as non-governmental organisations have undertaken tangible efforts across the spectrum of activities and missions to address gender's operational relevance. This trend is also observable in certain police establishments that are active in peacekeeping efforts. Up until this point, this book has primarily focused on the work being done by NATO and two of its English-speaking members, and their close ally Australia. This chapter will undertake a more representative survey of how gender is actually being incorporated into the daily work of military activities and operations across the international community. To set the stage for reviewing examples in the broad functional areas of education, training, planning and operations, this chapter will first consider two very different non-NATO countries, Ireland and Japan, and their approaches to mainstreaming gender in their defence establishments.

Ireland: a quick transformation

Ireland's national action plans

Ireland has been a consistent contributor to UN peacekeeping efforts (Defence Forces Ireland, n.d.: 'Ireland's involvement with the UN'), especially the UN Interim Force In Lebanon (UNIFIL) (Defence Forces Ireland, n.d.: 'United Nations Interim Force Lebanon'), which will be discussed in greater detail later in this chapter. Ireland's first national action plan was issued in 2011, and the second in 2015. *Ireland's Second National Action Plan on Women, Peace and Security 2015–2018* carries forward much of the first plan, and it commits the Irish government to strengthening 'women's leadership and participation in decision-making in conflict and post-conflict situations', incorporating a gender perspective into its peacekeeping and post-conflict activities overseas, and working to prevent gender-based violence (Government of Ireland, 2015: 6). The plan gives Defence Forces Ireland a number of specific tasks to implement these commitments, as set out in Annex I, 'Monitoring Framework' (14–21). These tasks are not just evaluated by measures of performance, such as 'Number of Irish and international military and civilian deployees and potential deployees trained on Women, Peace and Security issues' for training, but by measures of effectiveness as well, as in this case, 'Regular, quality, up-to-date training on issues relating to Women, Peace and Security' (14).

The Defence Forces action plans

Pursuant to the first national action plan, Defence Forces Ireland promulgated a detailed and thorough implementation plan to carry out its objectives. Signed by the Chief of Staff for Defence Forces Ireland, it demonstrated commitment to incorporating gender considerations into the military's work by its senior leadership (Defence Forces Ireland, 2013: 4). Further, it emphasised the significant role that the government had carved out for Defence Forces Ireland in implementing the plan, noting that the military was tasked in six of the 12 objective categories established by the plan, and that this required 'integrating UNSCR 1325 into policy, training and operations at home and overseas' (10).

Importantly, in terms of making its requirements clear to military leaders and personnel, before it listed the specific actions that Defence Forces Ireland would take undertake in its implementation programme, the Defence Forces Action Plan translated the broad taskings of the national action plan into specific action areas, such as 'the training of specialist Gender Advisers, Gender Field advisers and Gender Focal Points so that a Gender perspective can be achieved at all levels within the Defence Forces' (12). Just as the national action plan set out an annex specifying the tasks to the Irish government agencies, the Defence Forces Action Plan contained a similar chart that listed the task, described the actions necessary to accomplish the task in detail, the staff section or office responsible for the task, indicators of both performance and effectiveness, and importantly, set a time frame within which the work was to occur. For example, the J-7 (Training) staff section was required to audit existing training materials to ensure they were in compliance 'with best practice and permissible mandates and responsibilities of Irish Defence Forces' as they related to UNSCR 1325 – not just that training materials now had to mention UNSCR 1325 – by the end of September 2013 (19).

Defence Forces Ireland issued an updated implementation plan in 2016. Despite the short time between the implementation plans, the second plan indicates a significant maturation in the implementation of gender considerations in Defence Forces Ireland. For example, with regard to training, training materials were to be audited and amended on a regular bi-annual basis by the UN Training School Ireland to ensure that they were current. Further, Defence Forces Ireland was moving to conduct post-deployment surveys of personnel to generate data to be fed into the audit process (Defence Forces Ireland, 2016: 5). Other examples of this maturation are the detailed descriptions provided of the roles of GENADs in Defence Forces Ireland and in the defence forces' joint staff sections and down to the brigade level (9–10). It also set the terms of reference for the duties of Gender Focal Points, including that their gender mainstreaming duties should constitute about 5 percent of their workload and their role whilst deployed (11–12).

In an internal update from March 2018, Defence Forces Ireland noted that GENADs had been appointed down to the brigade level, and that a course for Gender Focal Points had been developed and was in the process of being conducted across the force. Further, all deploying troop units were required to have at least 25 Gender Focal Points in their ranks. To meet these requirements, in 2017, for example, six GENADs and 122 Gender Focal Points had been trained. In addition, four officers had attended the UN's women officers' course (Defence Forces Ireland, 2018: 2–7) which will be described in greater detail later in this chapter.

Some might argue that it is easier for a country with a small military that has a relatively small mission set such as Ireland (approximately 9,000 personnel, 6 percent women) (Barry, 2017) to more quickly incorporate gender considerations throughout its armed forces than a country such as the US. This argument is not without merit. Chapter 6, 'US strategy, doctrine and gender', was designed in part to give the reader an impression of the ponderous nature of the US military establishment and the slow pace of change within it. The difference between the two national militaries in this regard, however, is not merely one of size and organisational complexity. Ireland's political leadership prioritised the operationalisation of gender in its defence forces – the US has not (at least not prior to the passage of the *Women, Peace, and Security Act of 2017* described in Chapter 6), even though the percentage of women in the US military is roughly three times greater than in Defence Forces Ireland.

As part of their efforts to implement the national action plan, Defence Forces Ireland has invested in recruiting efforts to bring more women into the forces, although they have faced significant challenges in doing so (Barry, 2017). In 2017, however, the number of female inductees had increased to 10 percent, roughly doubling 2016's numbers (Defence Forces Ireland, 2018: 8). Recruitment, of course, is just the first part of making space for women in armed forces – retention is another problematic area as well. Like Ireland, Japan is grappling with these issues now as well.

Japan: personnel recruitment and retention in the Japan Self-Defense Forces

Like Ireland, Japan has been a strong supporter of UN peacekeeping efforts (Ministry of Foreign Affairs, 2014b). But whilst Ireland's population is just under 5 million, Japan has about 127 million people (Worldometers, 'Ireland' and 'Japan', 2018a, b), and in 2017 Japan was ranked as the world's third largest economy (Gray, 2017). Commensurate with its economic strength, in 2016 Japan's military expenditure was the eighth largest in the world (about $46 billion) (Tian et al., 2017: 2), although technically the Japan Self-Defense Forces (JSDF) are not a military organisation (Prime Minister of Japan and His Cabinet, Constitution of Japan, 1946: art. 9).

Women in the JSDF

In response to negative demographic pressures and the efforts of the international community to improve gender equality, the administration of Prime Minister Abe began a national programme to increase women's participation in Japan's economic and political life in 2013 (Ministry of Foreign Affairs, 2013). As assessed by different metrics, Japanese society is male-normative and male-dominated in terms of economic and political power. For example, although Japanese women comprised 77 percent of the part-time and temporary workforce in 2012, in 2011 only 1 percent were senior, executive-committee-level managers (*The Economist*, 2014). The Abe Administration included efforts by the JSDF to recruit more women into this programme (Spitzer, 2013), to provide them more meaningful career opportunities and to provide conditions of service more accommodating to family needs (Ministry of Defense, 2014a). Japan's national action plan, officially announced in 2015, seeks to increase participation and opportunities for women in the JSDF, and to formalise aspects of gender considerations in the JSDF's mission set, which includes operations with non-JSDF units and personnel in peacekeeping and disaster relief operations (Prescott, Iwata and Pincus, 2015: 5).

Women have been formally part of the JSDF since 1952 when 57 female nurses were given regular service status in the ground formations of the then-National Safety Forces (Sato, 2012: 2). Their numbers had remained low over time – in 1995, they constituted 4.23 percent of the JSDF (Ministry of Defense, 2017b: 1), and by 2007 this number had risen only to 4.9 percent (Ministry of Defense, 2017a: 406). Further, women tended to be clumped in service support positions, such as medical services (Prescott, Iwata and Pincus, 2015: 33–34). Interestingly, unlike the situations in Australia, the UK and the US, Japanese women had not campaigned politically for gender integration in the JSDF. As Sato has noted, by 'contrast, Japanese women activists, deeply concerned about the threat of resurgent militarism in Japan, have not prioritized support for expanded opportunities in the SDF' (Sato, 2012: 8).

In 2011, women in the JSDF played a significant and visible role in the recovery efforts after the Great East Japan Earthquake (Simpson, 2011). Around the time that the Abe Administration launched its programme to promote the role of women in Japan, female naval officers had also begun to command training vessels (Spitzer, 2013). Soon after, the Ministry of Defense began to publicise the role that female personnel played in Japan's humanitarian deployments to Timor and South Sudan (Ministry of Defense, 2014a, 2014b) and the first assignment of a female officer as a gender advisor to NATO (Ministry of Foreign Affairs, 2014a). It is not clear, however, that the favourable attention drawn to these women's work resulted in significant changes in JSDF female recruitment and retention.

JSDF Female Personnel Empowerment Initiative

In April 2017, the Ministry of Defense announced its JSDF Female Personnel Empowerment Initiative, which included a goal of doubling the number of women in the JSDF and giving them more active roles, and lifting recruitment quotas based on gender. Further, almost all JSDF positions would be opened to women 'after a thorough review of existing restrictions', except for those in submarines, tunnelling units and certain parts of the unit that handles defence against nuclear, biological and chemical threats, because of applicable labour law and equipment space limitations (Ministry of Defense, 2017c). This initiative also contained measures that specifically targeted women service members' ability to progress in their careers to positions of greater responsibility. For example, steps were to be taken to boost the number of women promoted to field grade position (major or lieutenant commander and above) and increase the promotion pool by ensuring that women were assigned as commanders or assistant commanders at the company-grade level (captain or lieutenant and below) (Ministry of Defense, 2017a: 407).

The 2017 Ministry of Defense *White Paper* included a special feature on the role of women in the ministry and the JSDF, which emphasised the rewarding roles they played and their accomplishments. Importantly, this document also highlighted the role that men and boys play in supporting women so they could achieve this success and fulfilment, including as parents taking paternity leave, grandparents providing child care when female personnel deploy, children who are proud of their mothers' service and bosses who are sensitive to the need to accommodate parents with children in the pace of work (Ministry of Defense, 2017b: 15–17). Finally, this part of the 2017 *White Paper* also identified Ministry of Defense and JSDF efforts to provide flex-time employment opportunities and child-care facilities as benefits for personnel, and to educate senior officers to raise their awareness of work-style reform (Ministry of Defense, 2017b: 14–16).

Japan's efforts to recruit and retain more women in the ranks of the JSDF might have already begun paying dividends. Even before the JSDF's efforts had begun in earnest, many women had already been attracted to service in the JSDF because it was seen as less sexist than the Japanese corporate world, with more interesting and rewarding work. These positive aspects were often counterbalanced, however, by the requirements of military life in general and gender expectations which caused many women to leave service once they had their own families (Frühstück, 2014: 2–7). By 2017, the number of women serving with the JSDF had increased slightly to 6.1 percent (Ministry of Defense, 2017a: 406), up from 5.6 percent in 2014 (Ministry of Defense, 2014a: 351). Further, women were being assigned to roles with greater responsibility and wider visibility. For example, a female officer was appointed to command a naval squadron that included Japan's new helicopter carrier, the *JS Izumo* (Reuters, 2018).

Although Japan has a unique culture in many ways, the concerns that women serving in the JSDF have regarding equality of opportunity and finding a decent balance between life and work needs and requirements are shared with military women of other countries. The Ministry of Defense has committed itself to addressing these concerns using a number of methods, such as providing child-care facilities on bases and ensuring women receive command postings at the company-grade level to make them more competitive for field grade promotion. These are important steps, but even if the Ministry of Defense is successful in its ambitious goal to raise the number of women in the JSDF to about 9 percent, the question remains whether this significant increase will have a positive effect on increasing incorporation of gender considerations into JSDF activities and operations. After all, the US military has about 15 percent women in its ranks (CNN, 2013), and as we have seen through the examination of US strategy, policy and doctrine, the treatment of gender in its intellectual security infrastructure is at best uneven.

Education

With these realistic factors regarding women's participation in militaries and the inclusion of gender considerations in operations as background, let's now explore the wide range of military and peacekeeping-related programs, projects and resources that have been developed by different nations and organisation that further the goals of UNSCR 1325 irrespective of any unevenness in the development of relevant high-level strategy, policy and doctrine. The first functional area we will look at is education, followed by training. Given that militaries are functionally geared towards problem-solving, the line between education and training is not as crisp as it would be in the civilian world. This chapter will use a somewhat arbitrary definition of education as activities related to imparting information that can be used to develop intellectual skills, either in a residential setting or online.

In her 2017 assessment of Professional Military Education institutions in the US, Johnson-Freese found that there was little evidence of women, peace and security issues being included in the core curricula of these schools. Similarly, whilst there were a couple of courses offered at different schools on the history of women in combat and in the US military, only one elective course was identified that dealt directly with UNSCR 1325 (Johnson-Freese, 2017). There are a number of US civilian universities for example that offer courses directly related to UNSCR 1325, such as Brandeis (Brandeis University, n.d.), Georgetown (Georgetown University, 2018) and Tufts University (Tufts University, n.d.), but US military personnel do not likely have convenient access to them. Instances of educational offerings that cover the operationalisation of gender should be considered against this backdrop.

Resident instruction

US Marine Corps University and cultural competence

The US Marine Corps (US Marine Corps University, 2016) has published a strategy that specifically addresses the need for individual Marines to develop familiarity with a specific language and region, to include general cultural concepts and skills. The Center for Advanced Operational Culture Learning was established in 2006 by the US Marine Corps University as one of the programmes to meet the strategy's aims (US Marine Corps University, n.d.). Although the strategy makes no mention of women or gender specifically, a new *Culture General Guidebook for Military Professionals* produced by the centre offers concepts and skills for learning about culture as one way of broadening students' understanding about gender discrimination, gender roles and socialisation, gender identity and women's economic roles in developed and less developed countries (Fosher et al., 2018: 24–25, 27, 35–38, 41, 52, 55, 65, 67, 78). The guidebook is currently used in diverse military instructional contexts, including distance as well as resident education, by both the US Marine Corps and the US Army (Lauren Mackenzie, 2018: personal communication). Importantly, from the perspective of interoperability, this textbook addresses many of the points raised in the UK's *JDN 4/13, Culture and Human Terrain*, discussed earlier in Chapter 5, 'UK strategy, doctrine and gender'.

International Peace Support Training Centre and the Female Military Officers Course

The International Peace Support Training Centre (IPSTC) was established in 2001 as a part of the Kenyan Defense Staff College, and it originally was geared towards training Kenyan personnel for peacekeeping missions (IPSTC, n.d.: 'About Us'). Today, its focus is international, and its partner nations include Australia, Belgium, Denmark, France and Japan (IPSTC, n.d.: 'IPSTC Partners'). In partnership with the Peace Operations Training Institute based in the US, it offers many free online educational courses, including a number related to gender (POTI, n.d.: 'POTI Online Classroom'). This gives the centre important international reach in this study area, but what is likely as important is the large number of courses it teaches on site at its campus outside Nairobi, including courses related to gender issues in operations, such as 'Gender in Peace Support Operations', 'Women Police Peace Support Operations', and 'Sexual Gender Based Violence' (IPSTC, 2018).

In 2016, the International Peace Support Training Centre hosted the eighth iteration of the Female Military Officers Course, in conjunction with UN Women (UN Kenya, n.d.). The purpose of the course is to enable women officers to perform more effectively in UN missions whether as armed staff officers or unarmed military experts on mission (Department of the Prime Minister and Cabinet, 2017, 64). As of May 2018, a total of 135 female officers had been trained in the different iterations of the course, 104 of them military, and 31 police, from 50 different countries (Jade Mali Mizutani, 2018: personal communication). Many of these women already had deployment experiences, but a number had not had the opportunity to participate in courses that were focused on their roles and challenges as female officers (UN Women, 2016). The course has been successful in building gendered deployment capacity – the course organisers estimate that approximately 75 percent of the some 200 students taught so far have deployed (UN Kenya, n.d.), several in staff leadership positions in UN peacekeeping missions (Jade Mali Mizutani, 2018: personal communication). The intent of the course organisers is that it becomes a standard for national peacekeeping centres in all countries that contribute troops to UN missions (Department of the Prime Minister and Cabinet, 2017: 64).

Online instruction – 'I Know Gender 1–2–3'

The use of the Internet as a platform for the training of gender issues provides the means to reach large, distributed training audiences in a relatively inexpensive way. It also offers flexibility in scheduling that is a huge logistical advantage in finding ways to fit operational gender training into busy pre-deployment training schedules. An example of a quality online instruction programme that is available to the public at no cost is the 'I Know Gender 1–2–3' set of courses offered by the UN Women Training Centre. Located in Santo Domingo, Dominican Republic, the UN Women Training Centre offers an eLearning Campus which is funded by Denmark, Estonia and Switzerland that could be used for basic gender issues education for deploying peacekeepers. Many of the courses it offers are free of charge, and they are conducted in Arabic, English, French and Spanish (UN Women Training Centre, n.d.).

To access the free online courses, prospective students must create an account, and then take a basic set of three modules designed to provide a basic understanding of gender concepts as a prerequisite to taking the complementary courses (UN Women, n.d.). When one takes the 'I Know Gender 1–2–3' courses, one finds that the basic modules take about two and one-half hours to complete. The first module, 'Gender Concepts to Get Started', covers the difference between gender and sex, how gender is socialised and gender discrimination. The second module, 'International Frameworks of Gender Equality', focuses on human rights, international agreements, and the relationship between gender equality and the UN Sustainable Development Goals. The final basic module, 'Promoting Gender Equality Throughout the UN System', explains gender mainstreaming and the requirement for equal representation of women and men in UN activities.

The quality of the training platform and the substantive materials is very high. The lessons are well organised, contain interesting videos and interactive features for students, and require students to achieve passing scores on quizzes before moving on to subsequent sections. The complementary module 'Women, Peace and Security' shares these features, and takes about 45 minutes to complete. It provides a sound introduction to the impacts of armed conflict and natural disasters on women and girls, explains the different UN Security Council resolutions on women, peace and security, and provides examples of gender perspectives being incorporated into operations in the field. Its material is not redundant with the basic modules, and it provides a very useful overview of this topic area.

Teaching the educators – 'Teaching Gender in the Military – A Handbook'

The Geneva Centre for Democratic Control of the Armed Forces (DCAF) and the Partnership for Peace Consortium of Defence Academies and Security Studies Institutes have jointly created *Teaching Gender in the Military – A Handbook* (DCAF, 2016). It has a number of features which make it an excellent tool for instructors from different nations to determine how best to approach education on gender issues in their respective countries. First, there is significant national diversity amongst its authors, most of whom are women, but who are from many different countries including Bosnia-Herzegovina, Bulgaria, Finland, Germany, Georgia, The Netherlands, Norway and Ukraine. Second, it is designed to take instructors stepwise from the theoretical, such as a chapter on the legal and policy framework relevant to gender (27–62), to the practical, such as a chapter on addressing gender dynamics in the classroom (113–30). Because it translates general gender concepts and considerations into material directly geared towards military instruction, it can be very useful for new instructors who might find themselves teaching courses on gender – courses which themselves might be completely new in their respective educational institutions. Providing quality instruction on a new topic is challenging even for experienced teachers, and delivering the instruction effectively from the start would help build credibility for these courses among fellow faculty and the students.

Training

Whilst this chapter defines education in the military context to be primarily geared towards imparting information to students and fostering the development of general intellectual skills regarding its use, it sees training as being focused on specific skill development in the context of specific mission situations. There have been training events focused on gender held across the world, often with a regional composition, such as the workshop hosted by the Argentine Center of Training for Peace Operations for Argentine and Uruguayan peacekeepers in 2012 (UN Women, 2012). Another example is the Five Eyes Plus Gender Conference hosted by Canada in 2017, that brought together representatives from the Five Eyes community with representatives from the UN, NATO, Finland, Germany, The Netherlands, and Spain 'to share best practices and lessons learned regarding the integration of gender in operations' (Shoemaker, 2017: 1). Importantly from the perspective of capacity building, 'train the trainer' programmes have also been being held, such as the course for military and police peacekeepers on conflict-related sexual violence supported by the UN Department of Peacekeeping Operations and the Department of Field Support (UN DPKO and DFS) and conducted in Entebbe in 2015 (UN DPKO and DFS, 2015).

Training resources – 'Gender Education and Training Package'

NATO's Allied Command Transformation and Sweden's Nordic Centre for Gender in Military Operations have produced a gender perspective training package that is designed to convey essential information about NATO's efforts in gender mainstreaming and to serve as a template into which national directives and policies could be included during pre-deployment training (HQ ACT Office of the Gender Advisor, n.d.). The materials can be downloaded as a compressed file, and they contain modules for training at the strategic/operational and the tactical levels. The pre-deployment training is composed of three full lecture modules, and an alternative short lecture which synopsises the three modules (Office of the Gender Advisor, 2017: 4). It is a fairly well-developed resource package, but it is overly optimistic as to how long it would take to deliver the training to a quality standard. For example, for the first module, which deals with the strategic-operational impacts of using a gender perspective, the instructor's guidance estimates that about three hours of in-class time would be required for lesson delivery, but this does not appear to include time for meaningful interaction between the students or evaluation of what they had learned (Office of the Gender Advisor, 2017: 12–15). Inexperienced military instructors might struggle to make instruction using these modules other than just talking through a series of PowerPoint slides unless they were to receive the recommended gender instructor training (Office of the Gender Advisor, 2017: 9) given by the Nordic Centre for Gender in Military Operations (NCGM) (Forsvarsmakten, n.d.: 'GToT'). Optimistic schedules are also a feature of an even better developed training resource packet, the UN's 'Core Pre-deployment Training Material 2017'.

Training resources – UN 'Core Pre-deployment Training Material 2017'

In 2017 the UN published online its 'Core Pre-deployment Training Materials for United Nations Peacekeeping Missions, CPTM 2017' (CPTM 2017; UN DPKO and DFS, 2017a). Despite one external evaluation of the instructional resource packet that found that the CPTM 2017 programme of instruction was too short (Razakamaharavo, Ryan and Sherwood, 2018), this particularly well-developed set of training resources probably suffers instead from the opposite problem. Although it is a very high quality set of products, a thorough review of its materials suggests that the estimates of time given to conduct the training to standard are not reliable. This would make it difficult for busy training schedulers and instructors to fit it into what are often abbreviated training opportunities for different units.

CPTM 2017 can be found on the online UN Peacekeeping Resource Hub. The UN Peacekeeping Resource Hub has a section devoted to peacekeeping training, and organised under this section is a user-friendly information hierarchy that includes links to lists of peacekeeping training institutions, training materials that can be downloaded, specialised materials on conflict-related sexual violence and protection of children and online resources such as the UN Police Gender Toolkit. The 'Training Materials' section links further to standard operating procedures to set up Mobile Training Support Teams and how to train personnel who will be trainers themselves. Additional links are provided to the CPTM 2017 materials, as well as to supporting general and specialised training materials. To complete its coverage of the training cycle, this section is also linked to guidelines and manuals, including the Practical Guide to Peacekeeping Training Evaluation (UN DPKO and DFS, n.d.: 'UN Peacekeeping Resource Hub').

The CPTM 2017 consists of three training modules, and a number of videos that can be used to support the training on the modules. CPTM 2017 is intended to cover 'the essential knowledge required by all peacekeeping personnel – military, police and civilians – to function effectively in a UN peacekeeping mission' (UN DPKO and DFS, 2017a: 11), and to provide Specialised Training Materials for different categories of personnel and particular subject areas that complement the core training for any mission. From the UN's perspective, together these materials 'constitute the mandatory minimum requirements for UN pre-deployment training' (UN DPKO and DFS, 2017a: 12).

The UN has promulgated guidance for instructors to accompany these materials that shows real andragogic depth. The UN distinguishes the new training materials as providing a stronger emphasis on learning than the previous materials, and in this context, the instructor's guidance explains that the instructional support materials are geared to address three main kinds of learning: knowledge; skills, that is, the ability to apply the knowledge to a task; and impacts on students' attitudes and values (UN DPKO and DFS, 2017b: 1). Logistically, the instructor guidance recommends that lessons should be about 45 minutes long, and the lessons should be divided equally between lecture and student interaction and participation in learning activities. An example of the latter is group discussion facilitated by the instructor using 'key questions' (UN DPKO and DFS, 2017b: 2).

Module 1 provides an overview of UN peacekeeping operations, Module 2 covers the mandated tasks of peacekeeping operations and Module 3 addresses the specific requirements expected of individual peacekeeping personnel (UN DPKO and DFS, 2017a: 5–6). Modules 2 and 3 contain the lessons that are most relevant to gender considerations in operations. Among the tasks of operations covered by Module 2, Lessons 4 through 7, Women, Peace and Security; Protection of Civilians; Conflict Related Sexual Violence; and Child Protection, respectively, are the lessons that provide specific instruction on these tasks from a gender perspective (UN DPKO and DFS, 2017a: 5–6).

Lesson 2.4, Module 2, is titled 'Women, Peace and Security'. Like each of the other lessons for the modules, it contains a textual description of the lesson itself, lesson slides, resources to support the learning activities promoted in the Instructor Guidance, slides to support those activities and different options to evaluate learning (UN Peacekeeping Resource Hub, 2017a). The recommended duration for Lesson 2.4 is 60 minutes, with 16 slides. The lesson curriculum contains four different learning activities, covering the differences between sex and gender, the impact on conflict on women and girls, gender mainstreaming and a film, 'Women in Peacekeeping: The Power to Empower' (UN Peacekeeping Resource Hub, 2017b).

For the learning activity on the differences between sex and gender using small group discussion, Lesson 2.4, Module 2, provides a short option of five minutes, with three minutes for group work and 2 minutes for discussion, and a longer option that goes for 10–15 minutes. During this time, the students are expected to define 'sex' and 'gender', give examples of gender stereotypes, discuss how gender results in inequalities and whether treatment of genders should be same or different (UN Peacekeeping Resource Hub, 2017b: Lesson 2.4, 4). The time budgeted for the three other learning activities is about the same, with the film itself being eight minutes long (10, 16, 30).

One problem with these well-crafted materials is that they are simply too much for most pre-deployment training schedules. Module 2 has seven lessons, which are estimated to take a total of six in-class hours to conduct. The topic areas for the other lessons appear to be directly related to women, peace and security issues, such as Lesson 2.5, 'Protection of Civilians', Lesson 2.6, 'Conflict Related Sexual Violence', and Lesson 2.7, 'Child Protection' (UN Peacekeeping Resource Hub, 2017a). This suggests that the learning objectives for this module will not be truly satisfied unless each lesson is taught.

Looking at Lesson 2.4, the estimated one hour of in-class time is premised on spending one to two minutes on each slide, and using the rest of the time in the learning activities. For some of the slides, that might be accurate, but for many of the textual portions of the lesson plan to be covered to fully address the bullets on the slides, two minutes would be insufficient. Even the 'long' session option of 10 minutes for group work is likely a very optimistic estimate as to how much time would be required to cover the material in any depth. A more realistic estimate of the time needed to conduct Lesson 2.4 to the UN learning standard by even a skilled and practiced instructor is probably closer to two hours, meaning that the module in total should have two training days budgeted against it. This would likely be challenging in many pre-deployment training schedules.

Gender-focused specific military skills training – UN signals Academy Women's Outreach Course

The UN Signals Academy was established in 2015 at the UN Regional Service Centre in Entebbe as result of a partnership between the UN and supporting member states such as Japan. The goal of the Signals Academy is to increase both military and police signal capabilities on peacekeeping missions. It provides technical training on using UN communications equipment according to UN standard operating procedures. Thirty-eight students attended the 2017 Women's Outreach Course (UN Support Office in Somalia, 2017) held at the academy, and in May 2018, the Signals Academy held its fourth iteration of its course (UN Signals Academy, n.d.).

This course is specifically designed to train women military and police officers in using modern information and communications technology whilst also addressing issues of gender related to peacekeeping operations (UN DPKO, 2017a). Participants in the course are often very experienced in traditional signals work, such as by radio, but they have not necessarily had the chance to get hands-on training with advanced information and communications technology equipment (UN DPKO, 2017b). Given the importance of information and communications technology to military missions today, not only would courses of this nature provide national contingents with enhanced capabilities, they should also tend to enhance the opportunities for these female students to seek promotion to positions of higher rank and greater responsibility. This is particularly important given the generally male-normative nature of this technology sector today (*The Atlantic*, 2017).

Gender advisor courses

There are a number of institutions that have provided courses that have been specifically geared towards training personnel to be GENADs in an international security setting. For example, in 2014, the Italian Defence general staff conducted a two-week course that involved military officers and civil servants that featured lectures, discussion of case studies and exercises led by Italian Army and *Carabinieri* instructors (Ministry of Defence, 2014). Women In International Security (WIIS), a civil society organisation, has conducted one-week GENAD courses taught by its staff and members, complemented by a course on incorporating gender in planning of humanitarian assistance and stability operations and a gender military key leader seminar (WIIS, n.d.). These courses are offered on demand, and they are tailored to the client's requirements to cover issues from the strategic to the tactical levels (Chantal De Jonge Oudraat, 2018: personal communication). Sweden and Australia offer GENAD courses as well, and they merit studying together because they take two different approaches in course length, delivery and focus.

Sweden – 'Gender Advisor' course

Sweden has always ranked among the top five countries in terms of gender equality since the World Economic Forum's Global Gender Gap Report began measuring this in 2006 (Cann, 2017). Sweden was second only to Denmark in promulgating its national action plan, in 2006 (Egnell, Hojem and Berts, 2014: 50). Similarly, Sweden has been a pioneer in embedding a gender perspective in its military operations

(Egnell, Hojem and Berts, 2014: 50–57), and the Nordic Centre for Gender in Military Operations was established in 2012 in Kunsängen, outside Stockholm (Forsvarsmakten, n.d.: 'Welcome to NCGM'). The quality of the work done at the centre was recognised in 2013 when NATO designated the centre as its proponent for instruction and training on gender in operations (Forsvarsmakten, n.d.: 'Welcome to NCGM'), and again in 2018 when the Swedish government tasked it with providing increased support to the UN by providing greater training opportunities (Government Offices of Sweden, 2018).

The centre offers a number of courses on gender in operations that appear to complement each other well by providing instruction at different levels of leadership, staff experience and functionality. In 2018 for example, the centre offered short seminars for commanding officers and key leaders on gender, and two-week courses for training gender instructors and exercise developers, and for new GENADs (Forsvarsmakten, n.d.: 'Courses and Seminars'; n.d.: 'GToT'). Each of these courses has been approved for NATO personnel, which means that the instruction provided at the centre is likely to become an Alliance benchmark in terms of quality and content. Although the courses continue to evolve, their approach and content are significantly informed by the experiences of the Swedish Armed Forces in Afghanistan as part of ISAF from 2002 to 2012 (Egnell, Hojem and Berts, 2014: 77).

The centre's 'Gender Advisor' course provides comprehensive instruction on women, peace and security issues from the UN and international level, down to the national level in reviewing national action plans at the strategic level, and in implementing guidance and fostering inclusion at the operational level. Students do much of their work in syndicates based on presentations and discussions in plenary academic sessions, but there are individual assignments as well that include final assessment. Former and current GENADs from different NATO headquarters and countries complement the Swedish instructors. The students receive instruction on practical military staff tasks associated with staff planning, and how gender perspectives can be incorporated into the planning process. This includes advising command and staff audiences about the responsibility to prevent and respond to sexual and gender-based violence related to conflicts in their areas of operations, and how to translate UN mandates, ROE and directives set out in operational plans into gender-related actions and tasks that can actually be executed (NCGM, 2017).

Other blocks of instruction include how to incorporate gender in civil-military cooperation, comparing the duties of different GENADs, and how to conduct a gender analysis internal to a military organisation. Importantly, the course grapples with the reality that GENADs will often experience resistance from military personnel in working with gender. A full hour is devoted to explaining to students how they can handle this sort of resistance from those whom they advise (NCGM, 2017). This approach is likely well suited to developing staff skills in the gender context among more junior staff officers who might lack the experience necessary to function effectively as staff members in such a new area as gender in operations.

Australia – 'Operational Gender Advisor' course

In contrast to the Swedish course, the Australian Defence Force's 'Operational Gender Advisor' course (ADF, 2017) is only one week long and it is more directly focused on training GENADs as staff officers by having students generate specific staff products incorporating gender considerations. Consistent with the Australia's operational experiences, it covers the full range of operations from humanitarian and disaster relief to missions in which IHL is applied. The course includes presentations, panels and discussions on the roles and equities of different stakeholders in operations and the experiences of former GENADs. Importantly, it takes a scaffolding approach to having students develop the staff skills to generate the gender-related products that could be fed into the mission-planning process. One full afternoon is devoted to syndicate work in which students learn how to perform a gender analysis of a notional operational area, and then these products are briefed back to the whole class for critique and feedback (ADF, 2017).

After this, the students are familiarised with the Australian Defence Force's operations planning process as set out in doctrine, and given an operational scenario and specific staff tasks to complete. Working again in syndicates they devise the gender products that are fed into the planning process for this scenario. The students then engage in individual work completing an operations order with these inputs, which are then evaluated by the former and current gender advisors who are serving as course cadre. Students are also required to provide eight to ten-minute evaluated presentations on a range of topics to assess the students' ability to advocate for gender in challenging circumstances (ADF, 2017). The overarching theme of the course is to emphasise the benefits of considering gender through operational planning, execution and evaluation to achieve increased operational effectiveness (Stacey Porter, 2018: personal communication). Whilst the language of instruction in both the Swedish and Australian GENAD courses is English, the pace and focus of the Australian Defence Force course suggest that it is perhaps best suited for students with a high degree of English-language proficiency and previous staff experience.

Live exercises – Talisman Sabre

Talisman Sabre is a large-scale biennial live exercise that is conducted primarily by Australian and US forces, located for the most part in and around Australia. It involves as many as 30,000 participants, and it has been conducted continuously since 2005 (Australian Civil-Military Centre, 2015; Cronk, 2017). Other nations, such as Canada, Japan and New Zealand, have also participated (Australian Army, 2017) as have representatives from different Australian and US government agencies, the UN Office for the Coordination of Humanitarian Affairs and the International Committee of the Red Cross (Sheridan, 2015: 5).

Sparked by a small group of Australian civilian and military staff in 2014 (Sheridan, 2017), the 2015 iteration of *Talisman Sabre* was the first time that UNSCR 1325 and women, peace and security issues were fully incorporated into this multinational exercise (Sheridan, 2015: 5). Gender experts worked together with scenario managers to include gender narratives in the exercise, and storylines and supporting situational changes, or 'injects', were designed by civilian and military exercise planners to drive the exercise play. Importantly, presentations on women, peace and security were given at all major planning events, and the push to include gender considerations in the exercise enjoyed the support of the senior Australian and US military leaders (Sheridan, 2015: 5).

The most useful document to come out of *Talisman Sabre* 2015 for purposes of this chapter was a forthright and thorough article written by a senior US non-commissioned officer, Master Sergeant Vince Lowery, who served as an additional-duty GENAD in the US I Corps headquarters element that participated in the exercise. Lowery (2017) noted that without the emphasis of senior leadership on women, peace and security issues in operations, subordinate units that were unaccustomed to working with these issues were unlikely to effectively incorporate them into their planning. Similarly, he observed that in preparation for the exercise, many staff members were confused as to what gender considerations in operations were actually about, again as a result of unfamiliarity with the topic, and that this resulted in resistance to working with these concepts (39). Those staff members already familiar with the concept were inclined to include it in their staff work, and even ask for additional information on how to do so (40).

Once the exercise began, its fast pace left little time for the GENADs to provide any meaningful training to the exercise participants. This made it difficult to bring liaison officers and soldiers from major subordinate units up to speed with the significance of gender to the exercise, because many had unfortunately missed the introductory training that had been given on it for exercise participants. Further, many were not aware that working with gender considerations was actually a training objective. Although taking an online familiarisation course was required of all participants by the exercise order, there was no mechanism in place to ensure that this actually had occurred (40).

Lowery noted a significant difference in the two US Army divisions that participated in the exercise. One was an active duty unit, and in large part it did not incorporate gender considerations into its planning or execution of its exercise tasks. The other unit was a US National Guard division, a reserve unit. This division engaged with the exercise requirement to incorporate concerns regarding gender into its operations without resistance, and successfully used them to their advantage in the conduct of the exercise (40–41).

Lowery concluded that this difference was based on three primary factors. First, for the active duty division, this was just another exercise among the many in which it had participated, and it had already set its training objectives to meet the large mission set it was expected to be able to handle far in advance of the exercise. Issues regarding women, peace and security were not among its ordinary training objectives, and therefore it did not register with the commander or staff as a priority. Second, the reserve unit, being composed of officers and soldiers who spent most of their time as civilians in their communities, was only resourced for a limited amount of training in any given year. Thus, the assessment of their preparedness to conduct their missions was largely based on how well they did in the exercise, and that meant successfully meeting the specific exercise objectives, including their work with gender (40–41).

The third factor Lowery identified is particularly relevant for purposes of this book. Whilst the active duty division was accustomed to taking a conservative approach in conducting its training, reusing earlier training materials and methods that had worked before, the reserve unit had a different mind-set. Because its officers and soldiers came from civilian life, they had different life and professional experiences that made it easier for them to see things more holistically, and they understood how gender considerations could be important in their missions (41). This finding suggests that the incorporation of issues involving women, peace and security into educational and training curricula, buttressed by sound and relevant doctrine, is essential to moving large, multi-mission militaries to work effectively with gender (Wittwer, 2016).

Talisman Sabre also appears to have had a significant capacity-building effect in both the Australian and US forces. The Australian GENAD course described earlier was prompted in large part by the need to develop a sufficient number of trained GENADs to participate in *Talisman Sabre* 2017 (Gibbon, 2018). In 2017, the now-US Indo-Pacific Command, the US combatant command with overall responsibility for US forces in the Pacific and Indian Ocean areas, issued its first command instruction explaining the requirement to implement US domestic authorities regarding women, peace and security, and the responsibilities of commanders, GENADs and Gender Focal Points in this regard (US PACOM, 2017). Perhaps most significantly, in June 2018, the US held its first operational gender course in Hawaii. Designed by an Australian exchange officer, this week-long course more than doubled the number of trained GENADs in the US Department of Defense, from 17 to 41 (Bradley Smith, 2018: personal communication). Dedicating dollars and training time to building GENADs shows increasing understanding of the significance of gender considerations to operations within the US military establishment, at least in areas where interoperability with valued allies requires developing such capacities.

Planning – Allied Command Operations gender functional planning guidance

The use of planning guidance issued by higher headquarters is essential in any operational military organisation. It promotes uniformity across the planning activities of different subordinate units and their staffs, and it saves commanders time because they are pestered with fewer questions as to what their intent is and how they want it executed. There are a number of well-developed guidelines regarding the incorporation of gender in planning, including the UN's peacekeeping guidelines published in 2010 (UN DPKO and DFS, 2010), and the NATO planning guidance published in 2015. Because of detailed treatment of NATO strategy and doctrine provided in Chapter 4, 'NATO strategy, doctrine and gender', this section will focus on the NATO planning guidance.

Allied Command Operations (ACO, 2015) has published a planning guide that informs commanders and staffs as to how gender considerations are to be incorporated into the NATO planning process. It is a thoughtfully prepared document, constructed in conformance with command planning directives (4), and it is intended to serve as a guide to effectively operationalising gender considerations at the strategic, operational and tactical levels even in the absence of the GENAD (4; Annexes A–C). Importantly, it notes the relevance of conducting a gender analysis of an operational environment, and provides examples of how this analysis could generate inputs into the planning process (Annex A).

However, in terms of its overall approach, it is fair to say that although it emphasises that gender is important in kinetic as well as non-kinetic operations, it does not really incorporate gender considerations as part of the application of armed force by NATO (2, n. 1; Annex B, B-4). Further, gender considerations in the application of armed force are not mentioned in the strategic level annex to the guidance. In the operational-planning level annex, the planning guidance notes the importance of the GENAD participating in the planning dealing with the employment of forces by focusing on 'targeting, StratCom, Information Strategy and possible cooperation with National and International Actors' and ensuring that the J-3 (Operations) staff understands the gender-related aspects of the operational plan (B-6). Although 'targeting' is mentioned two other times, nowhere does the guidance explain how the GENAD is expected to functionally interact with this highly specialised area of operations.

Addressing this, one of Allied Command Operations' primary operational subordinate headquarters, Joint Force Command (JFC) Brunssum (NATO, 2018), has developed a standard operating procedure that sees the GENAD's role in targeting as largely nominating targets that should be prosecuted to eliminate risk to 'vulnerable' groups of civilians (JFC Brunssum, 2016: 14), rather than addressing the impacts to these groups were they to be caught up in a kinetic situation involving NATO weapons. This is an intriguing approach, and it appears to sidestep the question of whether a NATO commander's proportionality analysis in deciding to use armed force would need to consider the differentiated impact of armed conflict on women and girls.

Practically speaking, targeting actors attacking or imminently threatening to attack at-risk groups with lethal force could be allowed under potential NATO ROE if the at-risk groups were made Persons with Designated Special Status (PDSS) (NATO Military Committee, 2003: Appendix 1, Annex A, A-1–A-2). This authority, however, is intended to bridge the gap created by certain troop-contributing nations not recognising the concept of self-defence as extending to non-NATO forces (NSO, 2015: B-21). Proposing that certain at-risk groups be made Persons with Designated Special Status so that they can be defended if attacked is quite different from nominating as targets for attack those who might pose a threat to at-risk groups. Ordinarily, the individuals belonging to such groups would first need to be combatants under IHL before they could be engaged with kinetic means under the NATO attack ROE.

Returning to the 2015 ACO operations planning guidance, the substantive development of the operations plan gender annex at the operational level (as shown by its subordinate appendices) is primarily geared to preventing and investigating sexual and gender-based violence in the area of operations. Appendix 1 covers sexual and gender-based violence in general, Appendix 2 sets out monitoring and reporting measures and Appendix 3 repeats the NATO Standards of Behaviour (Annex B, Appendices 1, 2 and 3). Similarly, the tactical level annex provides a very detailed list of gender considerations to be employed, but in terms of actual operations under the J-3's purview, it mentions only general gender awareness, the possibility of information operations and key leader engagements as a means of bringing gender into play and using mixed or single-gender engagement teams to interact with local populations on patrols (Appendix C, C-2–C-3). It is not likely that a deployed J-3 would be concerned with such details in any depth.

Perhaps the most important area of the guidance that could be prioritised for enhancement is its treatment of operational risk and gender. Annex E to the guidance sets out a gender analysis tool to be used by planners. This annex states that the aim of the gender analysis is crucial, and that it 'orientates the whole analysis work' (Annex E, E-1). The aim is 'always its relevance to the planned/ongoing NATO mission' (Annex E, E-1). Further, the guidance states that the product of this analysis could take different forms, such as tasks, recommendations or requests for further information. Importantly, it then notes that there 'may also be the need to develop a risk assessment associated with the implementation or non-implementation of that' product (Annex E, E-2).

The question of course is risk to whom of what? Johnson (2012) observes that the management of risk in a military environment is complicated by a number of factors not present in many civilian settings, such as resourceful and adaptive enemies, fatigue and compromised or incomplete communications. The risks to different operations presented by a similar threat will differ depending upon the circumstances. As Johnson points out, Improvised Explosive Devices (IEDs) have been demonstrated to pose a threat to ground forces. The risks posed by IEDs to mission accomplishment by UK forces in Afghanistan have been high. In Northern Ireland, not so much, because of the decreasing likelihood of their use over time (10). Similarly, consider the risks posed to mission accomplishment by gender inequality in an engagement area in a traditional force-on-force engagement between two industrially developed nations. Gender inequality might still pose the same threat of social and economic instability in general in that area, but that risk might only be operationally relevant if the military operation were civilian-centric or had a significant civilian-centric component (such as a stability operation), rather than being almost purely kinetic.

The Allied Command Operations guidance itself provides little in the way of assessing risk to the operation as a result of risks encountered by populations based on gender. At the strategic level, it directs the GENAD to identify risks of sexual and gender-based violence as part of fleshing out different actions that could be taken by the military, but does not take the next step of differentiating the risk of this potential threat to the mission as a whole (ACO, 2015: Annex A, A-2.). At the operational level, GENADs are directed to provide information on 'threats and risks' to the assessment of the operational environment and the Theatre Civil Assessment, but they are given no additional information on the way in which these inputs are to be developed in terms of their relevance (Annex B, B-1).

Further, the reporting requirements that are to be established in the operational area to provide data to gain visibility over gender issues do not ask for assessments of risks to the mission as a result of observed gender-differentiated risks, but instead just ask 'What risks similar and/or different do men, women, boys and girls face?' (Annex B, B-5). To more fully operationalise gender, the risk assessment conducted by the GENAD must take the analysis to a higher level – and the result of this might be an assessment that gender is not operationally relevant in certain missions or certain aspects of missions. If so, this would not be a symptom of gender blindness, but rather represent forthrightness in analysis, and likely contribute to the GENAD's credibility with the commander and the staff sections.

Operations

There appear to be two primary means by which gender considerations have been actively operationalised in missions through the use of specially trained and designated personnel since UNSCR 1325 was promulgated. The first is through the use of GENADs, as introduced in the earlier chapters on Australian, NATO and UK gender mainstreaming efforts. In large part, a GENAD's work tends to be internal to the military organisation. The second way in which gender has been operationalised through personnel is the use of gendered units, ranging from the two-woman Cultural Support Teams used by US Special Operations Forces to assist in counterinsurgency operations to the use of line light infantry battalions composed primarily of women by Israel. It is not accurate to say that all of these gendered units were intentionally designed with UNSCR 1325 in mind – some were perhaps, but many were created out of simple necessity. The work of these units is largely external to the military organisations, but the indirect impacts that the effective use of these units might have on the attitudes and behaviours within predominately male military organisations cannot be ignored.

Gender advisors

Sweden has been a pioneer in the use of GENADs in military operations, having fielded the first GENAD as part of a peacekeeping mission in the Democratic Republic of Congo in 2006 (Egnell, 2017: 138–59). The Swedish Armed Forces implemented a programme that sought to accelerate the incorporation of gender perspectives in its activities and operations by creating this new sort of advisor at different levels throughout the command structure. At the strategic level, a full-time senior GENAD was appointed to serve directly under the Chief of Joint Operations, rather than locating this position in a policy office or in human resources. This emphasised the operational nature of the GENAD's role, and reflected the decision by a small group of committed actors in the Swedish Armed Forces to purposefully not seek implementation of the Swedish National Action Plan's requirements as a human rights initiative (Egnell, Hojem and Berts, 2014: 6).

There was concern that if the GENAD's role was seen as primarily based in a human rights context, organisational resistance within the Swedish Armed Forces to this innovation could be significant. Whilst Swedish civil society strongly values gender equality, those implementing the requirements of the Swedish National Action Plan assessed that the conservative traditions of the Swedish military and its pronounced male-normative nature would likely present obstacles to their efforts (Egnell, Hojem and Berts, 2014: 70–72, 117–18). Mindful of this, they created a system that sought to embed gender mainstreaming work at the lower levels of the Swedish Armed Forces, such as the establishment of Gender Field Advisors to work in operations abroad, and Gender Focal Points in subordinate units who took on gender-related work as an additional duty (Egnell, 2017: 147).

Afghanistan was in many ways the laboratory in which the Swedish gender advisor system was developed and tested. For example, by 2008, Swedish units rotating into ISAF were required to have Gender Focal Points at the platoon level and in each staff section (Egnell, 2017: 147). The NATO GENAD structure and duties described in Chapter 4, 'NATO strategy, doctrine and gender', have been heavily influenced by the Swedish model. In building on the Swedish model, and in keeping with its Comprehensive Approach, NATO expanded the reach of its GENADs in Afghanistan by placing them in different Afghan ministries to help them build gender awareness and capacity within the Afghan National Security Forces.

Within the NATO countries in general, as of 2016 74 percent (22 out of 28) had trained GENADs in their ranks, for a total of 553, 85 of whom had been deployed (NATO, 2016: 33–34). The country with the greatest number of trained GENADS was Spain, with 230, whilst the US only reported two trained GENADs at that time. Further, ten NATO member nations reported that they were using Gender Focal Points in their subordinate units (NATO, 2016: 34). The Swedish gender advisor model has clearly found a place within the Alliance.

Among the non-NATO nations, Australia in particular has adopted important points of the Swedish model and has dedicated significant resources to embed GENADs throughout the breadth of its activities and operations. As of 2018, the Australian Defence Force Headquarters had two GENAD positions, an O-6 (colonel or captain) advisor to the Chief of Defence Force, and an O-5 (lieutenant colonel or commander) at the Peacekeeping Operations Training Centre. The Australian Defence Force Joint Operational Headquarters had three GENAD positions, the O-6 advisor to the Commander, Joint Operations, her O-5 deputy, and an O-4 (major or lieutenant commander) serving as a Gender Planner. Each of the service components has two GENADs at headquarters levels, and the Australian

Defence Force deploys the Senior GENAD for the Resolute Support headquarters in Afghanistan (the follow-on mission to ISAF since 2014), a GENAD to Combined Joint Task Force Inherent Resolve in Iraq, one to the UN Mission in South Sudan and a staff officer to the UN Women office in New York (Stacey Porter, 2018: personal communication). This level of commitment, if sustained, will likely lead to the development of a significant capacity to build and retain gender expertise within the Australian Defence Force.

Challenges faced by GENADs

Informal interviews with a small number of active duty or retired field grade officers who have served as deployed GENADS show consistent positive anecdotal support for the work they had done. They could point to concrete results in organising ministry-level working groups in host nations, in interacting with host nation service women who were dedicated and professional but who needed mentoring to progress in their careers, and the satisfaction of organising a GENAD programme and then recruiting quality officers to fill it. Other positive results included using staff skills developed in managing the staffing of home country military units to effectively modify host nation planning documents to create slots for policewomen, thereby creating a demand for their recruitment, and in taking their GENAD experiences and incorporating them into scientific research that could be used to justify removing barriers to women's participation and advancement in their respective militaries (Interviews, 2018).[1]

Most telling was that each identified the same challenge to effectively doing their jobs – the lack of genuine interest or understanding among those with whom they worked of the operational reasons why incorporating a gender perspective into operations was useful. Interestingly, whilst most of those who did not believe that gender was worthwhile to include in their units' work were men, there were a number of women as well (Interviews, 2018). Further, Scanlon (2015) has noted that in Afghanistan for example, both Afghan officials and their US military counterparts would both play the culture card. From the Afghans, 'You don't understand our culture'; and from the US personnel, 'You can't push gender integration on this culture. They're not ready for change' (4107).

Culture, and perceptions of culture, are powerful forces. Certain female military writers have in fact expressed concern that the emphasis on a 'Women, Peace and Security Agenda' actually works to the detriment of both women seeking a level playing field in terms of career opportunities and in trying to achieve greater protection for civilian women in conflict situations (Fryer, 2016: 5–11). All of this suggests that without investments in training and assigning personnel to gender-related operational positions (Scanlon, 2015: 4111), doctrine that pragmatically explains the operational significance of gender, and the development of curricula in military education and training institutions based on this doctrine, the inclusion of gender perspectives into military activities and operations consistent with UNSCR 1325's requirements could be delayed and its delivery sub-optimal.

Gendered teams and units

Female engagement teams

The use of Female Engagement Teams was noted earlier in Chapter 6, 'US strategy, doctrine and gender', in the discussion of the measures to be taken by the US military to fulfil its assignments under the 2016 US national action plan. At the height of the renewed US commitment of ground forces to Afghanistan that began in 2009, all US Army manoeuvre battalions and provincial reconstruction teams were required to include trained Female Engagement Teams in their formations (Walker, 2011). The US has not been alone in its use of these all-women units, and Australia (Department of Defence, n.d.: 'Gender Perspective in Military Operations'), Canada (National Defence, 2017) and the UK (Cacciottolo, 2012) employed them in Afghanistan as well. Sweden used both all-female and mixed-gender engagement teams in its operations in northern Afghanistan (Egnell, 2017: 147).

By many accounts, Female Engagement Teams or women on mixed teams proved to be useful in gathering actionable intelligence (Babin, 2014a; Bumiller, 2010). For example, female members of one mixed Swedish team were able to learn from local Afghan women that a Taliban commander was unlawfully collecting taxes from people in the area. The male members of the team worked with the Afghan National Police to arrest him (Egnell, 2017: 149). They did experience a number of challenges, however, that highlight the need for such detachments to be not only well resourced and well trained, but to have their missions clarified and made common knowledge among the military units they assisted. For example, the lack of fluency in any of the Afghan languages among the team members made it difficult to interact meaningfully with local women in many cases (Watson, 2011: 23; Ricks, 2016). Commanders (generally men) often did not understand the purpose of the Female Engagement Teams and their potential value, and therefore did not utilise them in many cases (Ricks, 2016). Because Female Engagement Teams have not been integrated into US Army doctrine and personnel management practices (Haring, 2012: 7; Babin, 2014b), for example, such unfamiliarity with their capabilities and uses is not likely to be resolved for US Army officers in the future.

These problems were not unique to the US Army. At different times, according to after action review materials generated by different British Army assessment activities, UK Female Engagement Teams also experienced problems with a lack of available translators (Army Secretariat, 2017b: 1), too shortened a time frame for conducting required skills training (Army Secretariat, 2017a: 4), a lack of doctrine applicable to Female Engagement Team operations for the training of the early teams (Army Secretariat, 2017b: 6), and the use of ad hoc teams of women soldiers rather than trained Female Engagement Teams by small unit commanders because there were an insufficient number of Female Engagement Teams available (Army Secretariat, 2017b: 17). Other issues included the lack of a standard description of Female Engagement Teams' purpose and capabilities that could be provided to commanders (Army Secretariat, 2017a: 1–2), and a perception among some commanders that the purpose of Female Engagement Teams was to engage only with women and girls (Army Secretariat, 2017a: 2).

Special operations forces teams – US

US special operations forces first deployed Cultural Support Teams generally composed of two women service members to support field operations in 2011. They have been used to conduct 'medical outreach programs, civil-military operations, key leader engagements, and searches and seizures' (US Special Operations Command, n.d.). For the US Army, special operations forces include Special Forces, US Army Rangers, Information Operations, and Civil Affairs units (US Army Special Operations Command, n.d.). Cultural Support Team candidates underwent a gruelling selection process, and the successful candidates received more in-depth instruction about working within different cultures than US female engagement teams did in general (Maurer, 2011; Katt, 2014).

As had Female Engagement Teams, Cultural Support Teams were able to gather intelligence about insurgents in Afghanistan that would not have been available to male soldiers given gender norms in that country, and to search women who sometimes were found to be carrying materiel for improvised explosive devices (Lemmon, 2015; Jessica Sepp, 2017: personal communication). Katt (2014) suggests that a major difference between the use of Female Engagement Teams and Cultural Support Teams was that whilst the former functioned 'to soften coalition force's footprint as they moved through an area', Cultural Support Teams 'were designed to provide persistent presence and engagement' with local populations, consistent with special operating forces' population-centric operations (109). This might have been true in general, but in keeping with their practical nature US Special Operations Forces in Afghanistan do not appear to have been picky about using women military personnel from other units on an ad hoc basis as their mission required (Eric Ahlness, 2013: personal communication; Ricks, 2016).

In general, the contributions of the Cultural Support Teams to these operations appear to have been well regarded by the Special Operations Forces soldiers with whom they served, particularly by those units that had already been briefed on what Cultural Support Teams could do operationally prior to the teams' arrivals in theatre (C-SPAN, 2015). This was no small achievement, and it brings into question the usefulness of the term, 'hyper-masculinity', first mentioned in Chapter 8, 'International humanitarian law and gender'. One definition of hyper-masculinity is the expression 'of extreme, exaggerated, or stereotypic masculine attributes or behaviors' (Rosen, Knudsen and Fancher, 2003: 326).

Arguably, US Special Operations Forces service members such as Special Forces, Rangers and SEALS would be likely candidates for this label. Their education and training, rigorous physical conditioning, and the risks and nature of the missions to which they are assigned promote the development of strongly male-normative, insular communities within the larger US military establishment (Simmons, 1998: 118–21; Zazanis et al., 1999: 23–26; Sarkesian, 2002). If these individuals, even today almost exclusively male, can learn to accept and value women soldiers whose expertise, endurance and courage approaches or matches their own and consider them part of their teams, then sociologists have a bit more work to do in explaining why the traits exhibited by these men are necessarily a negative thing, and whether, from a perspective of masculinities, they even see themselves this way.

Special operations forces – Norway

Norway has allowed women in combat positions since 1985 (Braw, 2016), and since 2015 it has had mandatory conscription for both women and men (Forsvaret, 2016). Norway recognised from its experiences in Afghanistan as part of ISAF that there was an operational requirement for highly trained women soldiers in the field who could collect intelligence and interact with women and children in ways deemed socially acceptable by host nation societies. In 2014, the Norwegian forces established the *Jegertroppen*, an all-female special operations forces unit designed to meet this requirement (Ponniah, 2017). Although it is not clear that the *Jegertroppen* has been used in the field yet (Ponniah, 2017), a number of its soldiers have used service in it to their careers' advantage, moving on to serving in the elite Telemark Battalion and the Intelligence Battalion, and attending the Defence Intelligence College (Korpela, 2016).

Policewomen units

Police units that are composed completely of women have become significant in peacekeeping operations, and police functions are often an area of multinational focus in stability operations as well. The different skill sets women police officers bring to post-conflict operations have been favourably recognised (Brown, 2015), and over the last few years, UN peacekeeping missions have averaged over 1,000 policewomen being deployed at any time (UN Peacekeeping, 2015, 2018). Further, the quality of their performance of service has also been noted. For example, an all-female Indian police unit that deployed to Liberia was found to have significantly lower rates of misconduct complaints than all-male units, significantly fewer instances of force being used improperly and significantly fewer instances of weapons not being used properly (Anderholt, 2012: 14–15). Bangladeshi units deployed to Haiti and the Democratic Republic of Congo have likewise proven successful in the quality of their interactions with host nation people (UN Peacekeeping, n.d.).

Military line units

Israel has taken a unique approach to gendered units by creating combat-capable light infantry battalions assigned to border defence that are composed approximately of two-thirds women and one-third men, such as the Caracal Battalion (Israel Defense Forces, n.d.). Today, approximately 2,700 women serve in the Israel Defense Forces in combat roles, about 6 percent of the total female strength of the force (Eran-Jona and Padan, 2018: 98, 103). The use of these units is controversial within Israeli society, particularly among more conservative Jewish sects, which are concerned that a combat role for women is itself irreligious, and that it breaks down their religious tenets regarding the separation of the sexes (Eran-Jona and Padan, 2018: 100–02; Gross, 2018).

Publically available assessments of unit effectiveness as compared to all-male infantry battalions tend to be anecdotal. For example, the commander of the Lions of the Jordan Valley Battalion reported that even religious civilians in the unit's area of operations were looking forward to their return after the unit deployed on training exercises (Gross, 2018). However, Finestone et al.'s (2014) medical study of the Caracal Battalion has judged it a success (8). This research indicates that women in these units, despite being more prone to stress fractures and other overuse injuries, are actually more likely to make it through their entire three-year term of military service as combat soldiers than their male colleagues (4).

Finestone and his co-authors speculated that this might be due to women self-selecting for what is considered by them to be an elite unit, whereas the men in these battalions were assigned to them out of the general pool of recruits. Regular Israel Defense Forces infantry soldiers undergo more rigorous training than the members of these battalions, but their attrition rates are lower than those of the Caracal soldiers (female or male) (5). Regardless, as these researchers pointed out, the women on average entered basic training in a disadvantageous position compared to the men. They were shorter, weighed less and were unable to complete as many repetitions of strength exercises such as push-ups and sit-ups as the men, but they were required to use the same equipment as the men throughout their training (5). The empirical rigour of the Finestone study stands out in an area where such data and analysis is generally lacking.

UN Interim Force In Lebanon – transformation whilst deployed

As recognised by the UN Interim Force In Lebanon (UNIFIL) (2014), a UN study conducted in 2008 that found that personnel from different countries contributing troops to UN missions who had received pre-deployment training on gender assessed it as being irrelevant to their missions. A later study also found that the quality of pre-deployment training on gender was uneven and varied from mission to mission. This likely reflected the limited capacity and resources that many of the troop-contributing nations have to effectively conduct the gender training modules created by the UN (7). Since 1978, with only a few interruptions, Defence Forces Ireland have contributed a significant number of personnel to the UNIFIL mission, most recently as part of a joint Finnish-Irish battalion (Defence Forces Ireland, n.d.: 'Ireland's Involvement with the UN'). Conscious of the training deficit on gender issues, Defence Forces Ireland provided strong support for a pilot project designed to promote concentrated engagement with operational gender considerations throughout the mission in Lebanon over a six-week period in 2014 (UNIFIL, 2014: Preface).

The project had two parts that were geared towards increasing the effective understanding of the role of gender in operations in missions such as UNIFIL. First, the project team interviewed the chiefs of branches and staff officers working at UNIFIL headquarters to discuss with them the significance of gender in operations and how gender considerations might be better integrated into military tasks. Importantly, they also solicited from those interviewed their recommendations to improve the gender training materials the mission was currently using. On the basis of the interviews and other information that the project team had gathered, they assessed that the current training materials were not sufficiently related to the realities of the mission and the challenges that the soldiers faced in operations. Using this assessment, the project team developed a new gender-training package that used the principles of adult learning and focused on the functional application of gender in the mission, and recruited personnel who understood this to be part of subsequent training sessions (UNIFIL, 2014: 9).

The project team then validated the revised curriculum using members of the Finnish-Irish battalion before launching the programme with the other units and personnel in the mission. Training sessions were then held with the Gender Focal Points in the headquarters and in one of the mission sectors. The training covered core concepts of gender considerations in a military context, and importantly, it emphasised the use of active learning techniques such as scenario-based exercises to trigger discussion of topics and experiences that might otherwise not be discussed (UNIFIL, 2014: 10).

The second part of the project was focused on engagement by the project team with specific headquarters staff sections and mission elements to explore how to tailor gender considerations into their work. For example, the J-1 (Personnel) was encouraged to influence the force generation process by highlighting 'the strategic level issue of women peacekeepers in certain roles to ensure that female candidates' were included (11). The J-3 (Operations) was informed of the need to issue operational reporting guidance that would assist units in developing and analysing sex-disaggregated data (11). Similarly, the Joint Operations Centre was informed of the importance of actually using sex-disaggregated data in its reporting to the command (UNIFIL, 2014:12).

Importantly, the project team conducted an assessment of their efforts in the pilot project. Their findings recognised that the 'language used in gender training must be tailored and pragmatic, avoiding any form of jargon and made specific to the needs of the military in a peacekeeping context' (14). The project team also found that within the mission there was some confusion between the incorporation of gender issues into military activities and the efforts to ensure gender equality and opportunities for women in the military. Further, among their other findings, the project team identified that coverage of gender issues in operations was often lacking in the pre-deployment training the soldiers from different countries had received (UNIFIL, 2014: 14).

This pilot project stands out for three main reasons. First, once the deficit in gender understanding in the mission was identified, the project team did not seek to impose some sort of top-driven corrective training across the headquarters and different units that composed UNIFIL. Instead, the project team met with individual commanders and staff officers in one-on-one interviews to identify the mission-specific nature of the problem (9). Second, on the basis of the actual data that it had collected, the project team developed a training package that was calibrated to appeal to an audience trained to solve problems, and to use adult learning techniques such as scenario play (UNIFIL, 2014: 10). Third, the project team undertook one of the most challenging training scenarios imaginable – reorienting attitudes and procedures in theatre whilst operations continued their always hectic pace. This was likely only possible because of the labour-intensive, tailored approach the project team had taken in working with the different headquarters staff sections and the units. As a result, a significant portion of the local civilian population whose perspectives until that point in time had not really registered with the mission now became part of the information flow about events and needs in the UNIFIL area of operations (Clare Hutchinson, 2018: personal communication). UNIFIL's approach should be seriously considered by any military organisation seeking to implement gender considerations in its activities and operations.

Summary

At different speeds and in different ways, many military organisations today are moving forward with programmes and initiatives to operationalise the principles and goals of UNSCR 1325 and national action plans, or at least to address specific issues regarding gender in operations. Whether it is in the areas of education, training, planning or execution, a survey of current practices and products across the globe shows broad international engagement with the concepts of providing greater protection for women and girls in areas afflicted by armed conflict, and greater inclusion of women in the decision-making and conflict resolution processes that play out in all phases of armed conflict situations. From one perspective, a survey of these efforts could represent a catalogue of potential best practices, in some cases even off-the-shelf products that could be easily modified and incorporated into new gender mainstreaming efforts with less investment in development and greater assurance of quality.

From a different perspective, all of this great work is not necessarily organised to achieve optimal effects in the military school houses, the training areas, the staff planning offices and the deployed headquarters. This is likely due in part to the general lack of information regarding women, peace and security in military doctrine at this point in time. Further, where high-level doctrine does address these topics, there is generally a lack of continuity between it and the lower-level operational and tactical doctrine that commanders and staffs on the ground will actually work with and which will help shape their formative experiences as young service members. Finally, without a solid doctrinal footing, it can be much harder to win the fight for resources and personnel within military organisations.

The international security environment continues to evolve in a time of global climate change. Operationalising gender is not just a question of increasing the holism of military efforts in civilian-centric operations by incorporating gender considerations such that we are finally listening to women's perspectives on their security needs in civilian-centric operations. Their social, physical and economic security is bound up now with environmental security, and the most tangible manifestations of this for many people in conflict-afflicted zones are likely environmental degradation, lack of effective ways to communicate with deploying forces about their needs and challenges, and food insecurity. Let's turn to some examples that show how these things could worked on together to achieve better solutions in military missions.

Note

1 Interview composed of three questions that were sent via email to five former and current female GENADs of field grade rank from Australia, the UK and US whom the author had previously met in-person. The three questions were: (1) What one thing in your experience as a GENAD do you consider to be the most successful or most meaningful project or initiative? (2) What one thing in your experience as a GENAD did you find to be the greatest challenge in conducting your work? (3) What one thing in your experience as a GENAD surprised you most?

References

ACO (Allied Command Operations) (2015) 'Gender Functional Planning Guide'. Supreme Headquarters, Allied Powers Europe. Available at: https://pksoi.armywarcollege.edu/conferences/psotew/documents/wg2/07-ACO-Gender-Functional-Planning-Guide.pdf (Accessed 23 September 2018).

Anderholt, C (2012) 'Female Participation in Formed Police Units: A Report on the Integration of Women in Formed Police Units of Peacekeeping Operations'. US Army Peacekeeping and Stability Operations Institute, *Peacekeeping and Stability Operations Papers*. Available at: http://publications.armywarcollege.edu/pubs/1270.pdf (Accessed 17 May 2018).

Army Secretariat (UK) (2017a) Letter to Unknown Researcher, Response to Freedom of Information Act Request FOI2017/00953/8259, After Action Review Materials Dated 2011. Army Secretariat (15 February). Available at: https://assets.publishing.service.gov.uk/government/uploads/system/uploads/attachment_data/file/600315/Female_engagement_in_the_Army.pdf (Accessed 3 February 2018).

Army Secretariat (UK) (2017b) Letter to Unknown Researcher, Response to Freedom of Information Act Request FOI2017/00953/8259, After Action Review Materials Dated 2011, 2012, and 2015. Army Secretariat (15 February). Available at: https://assets.publishing.service.gov.uk/government/uploads/system/uploads/attachment_data/file/600315/Female_engagement_in_the_Army.pdf (Accessed 3 February 2018).

The Atlantic Staff (2017) 'How Did Tech Become So Male-Dominated?' *The Atlantic* (14 March). Available at: www.theatlantic.com/video/index/519426/how-did-tech-become-so-male-dominated/ (Accessed 2 March 2018).

Australian Army (2017) 'Exercise Talisman Sabre'. Australian Army (15 December). Available at: www.army.gov.au/our-work/operations-and-exercises/major-exercises/exercise-talisman-sabre (Accessed 17 June 2018).

Australian Civil-Military Centre (2015) 'Exercise Talisman Sabre – An Overview'. *ACMC News* (August/September). Available at: www.acmc.gov.au/wp-content/uploads/2015/10/ACMC-Aug-Sep-Newsletter.pdf (Accessed 10 September 2017).

Australian Defence Force (2017) 'Operational GENAD Course' [syllabus] (5 May) (copy on file with author).
Babin, LB (2014a) 'Fact Sheet – U.S. Army Female Engagement Teams: Integration and Employment'. US Army Research Institute for the Behavioral and Social Sciences (April) (Copy on file with author).
Babin, LB (2014b) 'Fact Sheet – U.S. Army Female Engagement Teams: Overview'. US Army Research Institute for the Behavioral and Social Sciences (March) (Copy on file with author).
Barry, A (2017) 'Defence Forces "facing challenges" in Fulfilling its Gender Targets'. *The Journal* (22 December). Available at: www.thejournal.ie/gender-defence-forces-ireland-3720170-Dec2017/ (Accessed 3 May 2018).
Brandeis University (n.d.) 'HS 252F – Women, Peacemaking, and Peacebuilding'. Brandeis University. Available at: http://heller.brandeis.edu/courses/course?acad_year=2018&crse_id=014134&strm=1181&class_section=1 (Accessed 11 June 2018).
Braw, E (2016) 'Norway's "Hunter Troop"'. *Foreign Affairs* (8 February). Available at: www.foreignaffairs.com/articles/norway/2016-02-08/norways-hunter-troop (Accessed 23 May 2018).
Brown, RL (2015) 'Would More Female Soldiers Improve UN Peacekeeping Missions?' *The Christian Science Monitor* (21 September). Available at: www.csmonitor.com/World/Africa/2015/0921/Would-more-female-soldiers-improve-UN-peacekeeping-missions (Accessed 17 May 2018).
Bumiller, E (2010) 'In Camouflage or Veil, a Fragile Bond'. *New York Times* (29 May). Available at: www.nytimes.com/2010/05/30/world/asia/30marines.html (Accessed 30 March 2015).
Cacciottolo, M (2012) 'UK Women Soldiers Aim to in "hearts and minds"'. *BBC* (26 October). Available at: www.bbc.com/news/uk-20078269 (30 March 2015).
Cann, O (2017) 'Moving Backwards: Ten Years of Progress on Global Gender Parity Stalls in 2017'. World Economic Forum [Press release] (2 November). Available at: http://reports.weforum.org/global-gender-gap-report-2017/press-release/ (Accessed 17 June 2018).
CNN Staff (2013) 'By the Numbers: Women in the U.S. Military'. *CNN* (24 January). Available at: www.cnn.com/2013/01/24/us/military-women-glance/ (Accessed 30 March 2015).
Cronk, TM (2017) 'Exercise Talisman Saber 17 Commences in Australia'. *Department of Defense News* (29 June). Available at: www.defense.gov/News/Article/Article/1233218/exercise-talisman-saber-17-commences-in-australia/ (Accessed 29 May 2018).
C-SPAN (2015) 'The Ground Truth: Firsthand Accounts of Women in Combat'. *C-SPAN* [video] (27 April). Available at: www.c-span.org/video/?325610-4/discussion-women-combat (Part of the 'Women in Combat: Where they Stand' Event Held at the Carnegie Endowment for International Peace, Washington, DC) (Accessed 28 June 2018).
DCAF (Geneva Centre for Democratic Control of the Armed Forces) and the Partnership for Peace Consortium of Defence Academies and Security Studies Institutes (2016) *Teaching Gender in the Military – A Handbook*. Geneva: DCAF.
Defence Forces Ireland (n.d.) 'Ireland's Involvement with the UN'. Available at: www.military.ie/overseas/organisation/irelands-involvement-with-the-un/ (Accessed 28 May 2018).
Defence Forces Ireland (n.d.) 'United Nations Interim Force Lebanon'. Available at: www.military.ie/overseas/current-missions/unifil/ (Accessed 28 May 2018).
Defence Forces Ireland (2013) 'Irish Defence Forces Action Plan on the Implementation of UNSCR 1325'. Defence Forces Ireland (copy on file with author).
Defence Forces Ireland (2016) 'Irish Defence Forces Second Action Plan on Women, Peace and Security – 2016–2018'. Defence Forces Ireland (18 November) (Copy on file with author).

Defence Forces Ireland (2018) 'DF Implementation of National Action Plan on 1325, Women, Peace and Security'. Human Resources Branch (J-1) (5 March) (Copy on file with author).
Department of Defence (Australia) (n.d.) 'Gender Perspective in Military Operations'. Department of Defence. Available at: www.defence.gov.au/Women/NAP/Gender Perspective.asp (Accessed 28 May 2018).
Department of the Prime Minister and Cabinet (2017) *2016 Progress Report on the Australian National Action Plan on Women, Peace and Security 2012–2018*. Australian Government. Available at: www.pmc.gov.au/sites/default/files/publications/progress-report-nap-women-peace-security.pdf (Accessed 17 June 2018).
The Economist Staff (2014) 'Japanese Women and Work: Holding Back Half the Nation'. *The Economist* (29 May). Available at: www.economist.com/news/briefing/21599763-womens-lowly-status-japanese-workplace-has-barely-improved-decades-and-country (Accessed 30 March 2015).
Egnell, R (2017) Introducing Gender Perspectives in Operations: Afghanistan as a Catalyst. In: Holmberg, A and Hallenberg, J (eds.) *The Swedish Presence in Afghanistan: Security and Defense Transformation*. Abingdon: Routledge, pp. 138–59.
Egnell, R; Hojem, P and Berts, H (2014) *Gender, Military Effectiveness, and Organizational Change: The Swedish Model*. New York: Palgrave MacMillan.
Eran-Jona, M and Padan, C (2018) 'Women's Combat Service in the IDF: The Stalled Revolution'. *Strategic Assessment* 20(4): pp. 95–107. Available at: www.inss.org.il/wp-content/uploads/2018/01/adkan21-1ENG_3-97-109.pdf (Accessed 3 June 2018).
Finestone, AS; Milgrom, C; Yanovich, R; Evans, R; Constantini, N and Moran, DS (2014) 'Evaluation of the Performance of Females as Light Infantry Soldiers'. *BioMed Research International* 2014: pp. 1–8. DOI 10.1155/2014/572953.
Forsvaret (2016) 'Universal Conscription'. *Norwegian Armed Forces* [Press release] (28 June). Available at: https://forsvaret.no/en/newsroom/news-stories/female-conscription (Accessed 17 June 2018).
Forsvarsmakten (n.d.) 'Courses and Seminars at NCGM. Nordic Centre for Gender in Military Operations'. Available at: www.forsvarsmakten.se/en/swedint/nordic-centre-for-gender-in-military-operations/courses-at-ncgm-and-how-to-apply2/ (Accessed 28 May 2018).
Forsvarsmakten (n.d.) 'GToT: Nordic Centre for Gender in Military Operations'. Available at: www.forsvarsmakten.se/siteassets/english/swedint/engelska/course-curricula/gtot.pdf (Accessed 28 May 2018).
Forsvarsmakten (n.d.) 'Welcome To NCGM. Nordic Centre for Gender in Military Operations'. Available at: www.forsvarsmakten.se/en/swedint/nordic-centre-for-gender-in-military-operations/co-ncgm-start/ (Accessed 28 May 2018).
Fosher, K; Mackenzie, L; Tarzi, E; Post, K and Gauldin, E (2018) *Culture Guidebook for Military Professionals*. Quantico, VA: Center for Advanced Operational Culture Learning. Available at: www.usmcu.edu/sites/default/files/CAOCL/Guidebook26Jan2018c.pdf (Accessed 19 May 2018).
Frühstück, S (2014) 'The Modern Girl as Militarist: Female Soldiers In and Beyond Japan's Self-Defense Forces'. *The Asia-Pacific Journal: Japan Focus* 12(45): pp. 1–15. Available at: https://apjjf.org/-Sabine-Fr – hst – ck/4212/article.pdf (Accessed 30 March 2015).
Fryer, D (2016) 'Women, Peace and Security: The Agenda Is Not Women and it Won't Achieve Peace or Security'. *Australian Defence Force Journal* 200: pp. 4–14. Available at: www.defence.gov.au/adc/adfj/Documents/issue_200/Fryer_Nov_2016.pdf (Accessed 28 June 2018).
Georgetown University (2018) 'Class Detail: SEST-703–01, Women, Peace, and Security'. Georgetown University (Spring). Available at: https://gufaculty360.georgetown.edu/s/

course-catalog/a1o36000001MnNxAAK/SEST-703-01?id=00336000014SmxOAAS (Accessed 3 June 2018).

Gibbon, D (2018) 'WPS 2018: Navigating the Operational Gender Agenda'. *The Strategist* [online] (23 March). Available at: www.aspistrategist.org.au/wps-2018-navigating-operational-gender-agenda/ (Accessed 29 May 2018).

Government of Ireland (2015) *Ireland's second National Action Plan on Women, Peace and Security – 2015–2018*. Dublin: Department of Foreign Affairs and Trade (Available at: www.dfa.ie/media/dfa/alldfawebsitemedia/ourrolesandpolicies/ourwork/empowering-women-peaceandsecurity/Irelands-second-National-Action-Plan-on-Women-Peace-and-Security.pdf (Accessed 29 April 2018).

Government Offices of Sweden (2018) 'The Government of Sweden Tasks the Armed Forces to Support the United Nations With Gender Competence'. Government Offices of Sweden [Press release] (25 January). Available at: www.government.se/press-releases/2018/01/the-government-of-sweden-tasks-the-armed-forces-to-support-the-united-nations-with-gender-competence/ (Accessed 17 June 2018).

Gray, A (2017) 'The World's 10 Biggest Economies in 2017'. World Economic Forum (9 March). Available at: www.weforum.org/agenda/2017/03/worlds-biggest-economies-in-2017/ (Accessed 12 May 2018).

Gross, JA (2018) 'Arguments Rage Over IDF's Inclusion of Women, But Co-ed "Lions" Have a Job To Do'. *The Times of Israel* (11 March). Available at: www.timesofisrael.com/arguments-rage-over-idfs-inclusion-of-women-but-co-ed-lions-have-a-job-to-do/ (Accessed 13 May 2018).

Haring, E (2012) 'Female Engagement Teams: An Enduring Requirement with a Rocky Start'. *Peace & Stability Operations Journal Online* 3: pp. 7–12. Available at: http://pksoi.army.mil/PKM/publications/journal/pubsreview.cfm?ID=27.

Headquarters Supreme Allied Command Transformation Office of the Gender Advisor (n.d.) 'Gender Education and Training Package for Nations'. Allied Command Transformation ['hosted locally' version]. Available at: www.act.nato.int/gender-advisor (Accessed 7 July 2018).

International Peace Support Training Centre (n.d.) 'About Us: International Peace Support Training Centre'. International Peace Support Training Centre. Available at: www.ipstc.org/about-us.aspx (Accessed 28 May 2018).

International Peace Support Training Centre (n.d.) 'IPSTC Partners'. International Peace Support Training Center. Available at: www.ipstc.org/patnership.aspx (Accessed 28 May 2018).

International Peace Support Training Centre (2018) 'Calendar 2018: International Peace Support Training Centre'. Available at: www.ipstc.org/media/documents/coursecalender 2018.pdf (Accessed 28 May 2018).

Interviews, Prescott, JM; Fielding, A (4 May 2018); Grimes, R (29 April 2018); Porter, S (1 May 2018); Scanlon, S (8 May 2018) and Wittwer, J (7 May 2018) (email responses to questions posed at n.1).

Israel Defense Forces (n.d.) 'Caracal Battalion'. Israel Defense Forces. Available at: www.idf.il/en/minisites/caracal-battalion/ (Accessed 28 May 2018).

Johnson, CW (2012) *Military Risk Assessment: From Conventional Warfare to Counter Insurgency Operations*. Glasgow: University of Glasgow Press.

Johnson-Freese, J (2017) 'Women, Peace and Security in Professional Military Education'. *Small Wars Journal* (blog) (26 July). Available at: http://smallwarsjournal.com/jrnl/art/women-peace-and-security-in-professional-military-education (Accessed 29 May 2018).

Joint Force Command Brunssum (2016) 'Joint Headquarters Standard Operating Procedure 106, Gender Advisor's Functions in JFC & JTF Headquarters'. Joint Force Command Brunssum (March). Available at: www.forsvarsmakten.se/siteassets/english/swedint/

engelska/swedint/courses/genad/09-sop-106-gender-advisor-functions.pdf (Accessed 29 May 2018).

Lemmon, GT (2015) 'The Army's All-women Special OPS Teams Show us How We'll Win Tomorrow's Wars'. *The Washington Post* (19 May). Available at: www.washingtonpost.com/posteverything/wp/2015/05/19/the-armys-all-women-special-ops-teams-show-us-how-well-win-tomorrows-wars/?noredirect=on&utm_term=.baa50cde344e (Accessed 29 May 2018).

Lowery, V (2017) 'Coping with Noncombatant Women in the Battlespace: Incorporating United Nations Security Council Resolution 1325 into the Operational Environment'. *Military Review* (May/June): pp. 35–43. Available at: www.armyupress.army.mil/Journals/Military-Review/English-Edition-Archives/May-June-2017/Coping-with-Noncombatant-Women-in-the-Battlespace/ (Accessed 28 June 2018).

Katt, M (2014) 'Blurred Lines: Cultural Support Teams in Afghanistan'. *Joint Forces Quarterly* 75(4), pp. 106–14. Available at: http://ndupress.ndu.edu/Portals/68/Documents/jfq/jfq-75/jfq-75_106-113_Katt.pdf (Accessed 29 May 2018).

Korpela, A (2016) 'Jegertroppen: Norway's All-Female Special Forces Unit'. NATO Association of Canada (19 February). Available at: http://natoassociation.ca/jegertroppen-norways-all-female-special-forces-unit/ (Accessed 30 March 2018).

Maurer, K (2011) 'In a New Elite Army Unit, Women Serve Alongside Special Forces, But First They Must Make the Cut'. *The Washington Post* (27 October). Available at: www.washingtonpost.com/lifestyle/in-new-elite-army-unit-women-serve-alongside-special-forces-but-first-they-must-make-the-cut/2011/10/06/gIQAZWOSMM_story.html (Accessed 29 August 2012).

Ministry of Defence (Italy) (2014) 'Italian Defence: the first Gender Advisor Course kicks off today'. Ministry of Defence [Press release] (9 June). Available at: www.difesa.it/EN/Primo_Piano/Pagine/20140609_ItalianDefencethefirstGenderAdvisorCoursekicksofftoday.aspx (Accessed 17 June 2018).

Ministry of Defense (Japan) (2014a) *Defense of Japan 2014*. Ministry of Defense. Available at: www.mod.go.jp/e/publ/w_paper/2014.html (Accessed 30 March 2015).

Ministry of Defence (Japan) (2014b) 'Women Making a Difference at the MOD – Interview with a SDF Female Officer'. *Japan Defense Focus* 56. Available at: www.mod.go.jp/e/jdf/no56/topics.html (Accessed 30 March 2015).

Ministry of Defense (Japan) (2017a) *Defense of Japan 2017, Part III: Initiatives to Protect the Lives and Property of the People as Well as Securing the Territorial Land, Water and Airspace*. Ministry of Defense. Available at: www.mod.go.jp/e/publ/w_paper/pdf/2017/DOJ2017_3-3-2_web.pdf (Accessed 3 May 2018).

Ministry of Defense (Japan) (2017b) 'Defense of Japan 2017 Special Feature 3, Active Participation'. In: *Defense of Japan 2017*, Ministry of Defense. Available at: www.mod.go.jp/e/publ/w_paper/pdf/2017/DOJ2017_feature3_web.pdf (Accessed 3 May 2018).

Ministry of Defense (Japan) (2017c) 'Special Feature: JSDF Female Personnel Empowerment Initiative'. *Japan Defense Focus* 88, Available at: www.mod.go.jp/e/jdf/no88/specialfeature.html (Accessed 3 May 2018).

Ministry of Foreign Affairs (Japan) (2013) 'Japan's Initiative Regarding Women's Employment and Gender Equality (Toward a Society in Which all Women Shine)'. Ministry of Foreign Affairs of Japan. Available at: www.mofa.go.jp/fp/pe/page23e_000181.html (Accessed 30 March 2015).

Ministry of Foreign Affairs (Japan) (2014a) 'Dispatch of a Female Self-Defense Force Personnel to NATO Headquarters. Ministry of Foreign Affairs of Japan' [Press release] (4 November). Available at: www.mofa.go.jp/press/release/press4e_000488.html (Accessed 17 June 2018).

Ministry of Foreign Affairs (Japan) (2014b) 'Japan's Contribution to United Nations Peacekeeping Operations'. Ministry of Foreign Affairs of Japan (March). Available at: www.mofa.go.jp/policy/un/pko/pdfs/contribution.pdf (Accessed 28 May 2018).

National Defence (Canada) (2017) 'Integrating Gender Perspectives Makes Military Operations More Effective: Gender Advisor'. National Defence (8 March). Available at: www.forces.gc.ca/en/news/article.page?doc=integrating-gender-perspectives-makes-military-operations-more-effective-gender-advisor/izkjrba6 (Accessed 3 May 2018).

Nordic Centre for Gender in Military Operations (NCGM) (2017) 'Gender Advisor (GENAD) Course II' [syllabus] (4 October) (copy on file with author).

NATO (2016) *Summary of the National Reports of NATO Member and Partner Nations to the NATO Committee on Gender Perspectives*. Brussels: NATO. Available at: www.nato.int/nato_static_fl2014/assets/pdf/pdf_2018_01/1801-2016-Summary-NR-to-NCGP.pdf (Accessed 28 June 2018).

NATO (2018) 'Allied Command Operations (ACO)'. NATO (14 June). Available at: www.nato.int/cps/ua/natohq/topics_52091.htm (Accessed 17 June 2018).

NATO Military Committee (2003) *MC 362/1, NATO Rules of Engagement*. Brussels: NATO. Available at: file:///C:/Users/jmprescott/Downloads/MC_362-1_NATO_ROE.pdf (Accessed 28 June 2018).

NATO Standardization Office (NSO) (2015) *ATrainP-4, Training in Rules of Engagement*, Ed. A, Ver. 1 (5 May). Brussels: NATO. Available at: file:///C:/Users/jmprescott/Downloads/ATrainP-4_EDA_V1_FD_E.pdf (Accessed 28 June 2018).

Office of the Gender Advisor (2017) 'Gender Education and Training Package for NATO Allies, Partners and Beyond'. Allied Command Transformation. Available for download at: www.act.nato.int/gender-advisor (Accessed 7 July 2018).

Peace Operations Training Institute (n.d.) 'POTI Online Classroom. Peace Operations Training Institute'. Peace Operations Training Institute. Available at: www.peaceopstraining.org/users/shopping-cart/courses/ (Accessed 28 May 2018).

Ponniah, K (2017) 'Meet the Hunter Troop: Norway's Tough-as-nails Female Soldiers'. *BBC* (31 March). Available at: www.bbc.com/news/world-europe-39434655 (Accessed 13 May 2018).

Prescott, JM, Iwata, E and Pincus, B (2015) 'Gender, Law and Policy: Japan's National Action Plan on Women, Peace and Security'. *Asian-Pacific Law & Policy Journal* 17(1): pp. 1–45. Available at: http://blog.hawaii.edu/aplpj/files/2016/04/APLPJ_Prescott-Iwata-Pincus-FINAL.pdf (Accessed 7 July 2018).

Prime Minister of Japan and His Cabinet (1946) *The Constitution of Japan* (3 November). Available at: https://japan.kantei.go.jp/constitution_and_government_of_japan/constitution_e.html (Accessed 30 March 2015).

Razakamaharavo, VT; Ryan, L and Sherwood, L (2018) 'Improving Gender Training in UN Peacekeeping Operations'. *Women in International Security Policy Brief*, Women in International Security (February). Available at: www.wiisglobal.org (Accessed 3 March 2018).

Reuters Staff (2018) 'Japan's Navy Appoints First Woman to Command Warship Squadron'. *Reuters News* (8 March). Available at: www.reuters.com/article/us-japan-defence-female-commander/japans-navy-appoints-first-woman-to-command-warship-squadron-idUSKCN1GI0VC (Accessed 3 May 2018).

Ricks, TE (2016) 'Thinking About Afghan Operations: How a Female Engagement Team Failed'. *Foreign Policy* [blog] (14 January). Available at: http://foreignpolicy.com/2016/01/14/thinking-about-afghan-operations-how-a-female-engagement-team-failed/ (Accessed 17 June 2018).

Rosen, LN; Knudsen, KH and Fancher, P (2003) 'Cohesion and the Culture of Hyper-masculinity in U.S. Army Units'. *Armed Forces & Society* 29(3): pp. 325–51. DOI: 10.1177/0095327X0302900302.

Sarkesian, SC (2002) 'The New Protracted Conflict – The U.S. Army Special Forces Then and Now'. *Orbis* pp. 247–58. DOI: 10.1016/S0030-4387(02)00106-0.

Sato, F (2012) 'A Camouflaged Military: Japan's Self-Defense Forces and Globalized Gender Mainstreaming'. *Asia-Pacific Journal: Japan Focus* 10(36): pp. 1–23. Available at: https://apjjf.org/-Fumika-Sato/3820/article.pdf (Accessed 30 March 2015).

Scanlon, S (2015) 'We Don't Know What We Don't Know – But We Can Learn: Lessons Learned from Afghanistan on Women, Peace and Security'. *Procedia Manufacturing* 3: pp. 4106–14. DOI: 10.1016.j.promfg.2015.07.983.

Sheridan, A (2015) 'Exercise Talisman Sabre – An Overview'. *ACMC News* (August/September). Available at: www.acmc.gov.au/wp-content/uploads/2015/10/ACMC-Aug-Sep-Newsletter.pdf (Accessed 17 June 2018).

Sheridan, A (2017) 'WPS and Talisman Sabre: Learning From the Past, Looking to the Future'. *The Strategist* [online] (9 March). Available at: www.aspistrategist.org.au/wps-talisman-sabre-learning-past-looking-future/ (Accessed 27 May 2018).

Shoemaker, M (2017) *Five Eyes Plus Gender Conference – Proceedings Report*. Kingston, ON: Canadian Defence Academy.

Simmons, A (1998) 'How Ambiguity Results in Excellence: The Role of Hierarchy and Reputation in U.S. Army Special Forces'. *Human Organization* 57(1): pp. 117–23. DOI: 10.17730/humo.57.1.r3087137m1516871.

Simpson, J (2011) 'JSW Talks: Prof. Sabine Früstück'. *New Pacific Institute: Japan Security Watch* (30 July). Available at: http://jsw.newpacificinstitute.org/?p=7467 (Accessed 30 March 2015).

Spitzer, K (2013) 'Japanese Women Take Command, Finally'. *Time* (22 March). Available at: http://nation.time.com/2013/03/22/japanese-women-take-command-finally/ (Accessed 30 March 2015).

Tian, N; Fleurant, A; Wezeman, PD and Wezeman, ST (2017) 'SIPRI Fact Sheet: Trends in World Military Expenditure, 2016'. Stockholm International Peace Research Institute. Available at: www.sipri.org/sites/default/files/Trends-world-military-expenditure-2016.pdf (Accessed 23 April 2018).

Tufts University (n.d.) 'Courses/Gender, Culture and Conflict in Complex Humanitarian Emergencies'. Tufts University. Available at: https://nutrition.tufts.edu/academics/course/nutr-222 (Accessed 28 May 2018).

UN Department of Peacekeeping Operations (UN DPKO) (2017a) 'Women Signal Officers: Vital Assets for Field Missions'. UN Permanent Missions [Press release] (8 May). Available at: www.un.int/news/women-signal-officers-vital-assets-field-missions (Accessed 17 June 2018).

UN Department of Peacekeeping Operations (2017b) 'Women Signal Officers: A Portrait of Aigie Rose Okungbowa'. UN Permanent Missions [Press release] (9 May). Available at: www.un.int/news/women-signal-officers-portrait-aigie-rose-okungbowa (Accessed 17 June 2018).

UN Department of Peacekeeping Operations and Department of Field Support (UN DPKO and DFS) (2010) *DPKO/DFS Guidelines – Integrating A Gender Perspective Into The Work Of The United Nations Military In Peacekeeping Operations*. New York: UN.

UN Department of Peacekeeping Operations and Department of Field Support (2015) 'Increasing Awareness on Prevention and Response to Conflict-related Sexual Violence' (Entebbe, Uganda). UN (23 June). Available at: http://dag.un.org/bitstream/handle/11176/90523/ToT%20CRSV.pdf?sequence=1&isAllowed=y (Accessed 17 June 2018).

UN Department of Peacekeeping Operations and Department of Field Support (2017a) 'Core Pre-deployment Training Materials for United Nations Peacekeeping Operations'. UN. Available at: http://dag.un.org/bitstream/handle/11176/400592/FINAL%20CPTM%202017%20Introduction%20160517.pdf?sequence=7&isAllowed=y (Accessed 3 February 2018).

UN Department of Peacekeeping Operations and Department of Field Support (2017b) 'Instructor Guidance'. UN. Available at: http://dag.un.org/bitstream/handle/11176/400593/FINAL%20Guidance%20for%20Instructors%20180517.pdf?sequence=29&isAllowed=y (Accessed 7 July 2018).

UN Interim Force In Lebanon (UNIFIL) (2014) 'Women, Peace and Identifying Security: Piloting Military Gender Guidelines in UNIFIL'. UNIFIL. Available at: https://peacekeeping.un.org/sites/default/files/unfil_pilot_report_web_flat.pdf (Accessed 23 March 2018).

UN Kenya (n.d.) 'Bolstering Female Military Numbers: 8th UN Female Military Officers Course (FMOC)'. UN Kenya. Available at: http://daogewe.org/index.php/blogs-k2/item/1290-bolstering-female-military-numbers-8th-un-female-military-officers-course-fmoc (Accessed 17 June 2018).

UN Peacekeeping (n.d.) 'Women in Peacekeeping'. UN Peacekeeping. Available at: www.un.org/en/peacekeeping/issues/women/womeninpk.shtml (Accessed 17 June 2018)

UN Peacekeeping (2015) 'Gender Statistics by Mission, For the month of December 2015'. UN Peacekeeping. Available at: https://peacekeeping.un.org/sites/default/files/dec15.pdf (Accessed 7 July 2018).

UN Peacekeeping (2018) 'Summary of Troop Contributions to UN Peacekeeping Operations by Mission, Post and Gender, 31/05/2018'. UN Peacekeeping. Available at: https://peacekeeping.un.org/sites/default/files/7_gender_report_2.pdf (Accessed 7 July 2018).

UN Peacekeeping Resource Hub (n.d.) 'Peacekeeping Training'. UN. Available at: http://research.un.org/en/peacekeeping-community/training (Accessed 17 June 2018).

UN Peacekeeping Resource Hub (2017a) 'DPKO-DFS Core Pre-deployment Training Materials (CPTM 2017) for United Nations Peacekeeping Operations, Module 2'. UN Peacekeeping Resource Hub. Available at: http://research.un.org/revisedcptm2017/Module2 (Accessed 7 July 2018).

UN Peacekeeping Resource Hub (2017b) 'Lesson 2.4, Women, Peace and Security'. UN Peacekeeping Resource Hub. Available at: http://dag.un.org/bitstream/handle/11176/400595/FINAL%20Lesson%202.4%20160517.pdf?sequence=49&isAllowed=y (Accessed 7 July 2018).

UN Signals Academy (n.d.) 'Connecting the Classroom to the Field. Partnership for Technology in Peacekeeping'. UN. Available at: www.un.org/en/fieldsupport/events/UNPTP/UNSA/index.html (Accessed 17 June 2018).

UN Support Office in Somalia (2017) 'Women in Peacekeeping Undergo Specialised Training in Information and Communications Technology' [Press release] (11 March). Available at: https://unsos.unmissions.org/women-peacekeeping-undergo-specialised-training-information-and-communications-technology (Accessed 17 June 2018).

UN Women (2012) 'Gender Training for Peacekeepers in Argentina'. UN Women [Press release] (2 May). Available at: www.unwomen.org/en/news/stories/2012/5/gender-training-for-peacekeepers-in-argentina (Accessed 17 June 2018).

UN Women (2016) 'Training in Kenya Boosts Deployment of Female Military Officers for Peacekeeping'. UN Women [Press release] (14 December). Available at: www.unwomen.org/en/news/stories/2016/12/training-in-kenya-boosts-deployment-of-female-military-officers-for-peacekeeping (Accessed 17 June 2018).

UN Women Training Centre (n.d.) 'Trainings'. UN Women Training Centre. Available at: https://trainingcentre.unwomen.org/portal/ (Accessed 28 May 2018).

US Army Special Operations Command (n.d.) 'Headquarters Fact Sheet, About the U.S. Army Special Operations Command'. US Army Special Operations Command. Available at: www.soc.mil/USASOCHQ/USASOCHQFactSheet.html (Accessed 28 May 2018).

US Indo-Pacific Command (2017) 'United States Pacific Command Instruction 2200.1, Women, Peace, and Security'. US Indo-Pacific Command (17 August) (copy on file with author).

US Marine Corps University (n.d.) 'Center for Advanced Operational Culture Learning (CAOCL)'. US Marine Corps University. Available at: www.usmcu.edu/caocl (Accessed 28 May 2018).

US Marine Corps University (2016) 'Marine Corps Language, Regional Expertise & Culture (LREC) Strategy: 2016–2020'. US Marine Corps University. Available at: www.usmcu.edu/sites/default/files/CAOCL/2016-20_USMC_LREC_Strategy_MCCDC_22Dec15.pdf.

US Special Operations Command (n.d.) 'About the Cultural Support Program'. US Special Operations Command. Available at: www.soc.mil/SWCS/organization.html (Accessed 28 May 2018).

Walker, DR (2011) 'ATN Announces COIN, Female Engagement Team Training'. *Fort Leavenworth Lamp* (14 July). Available at: www.ftleavenworthlamp.com/article/20110714/NEWS/307149881/0/ (Accessed 30 March 2015).

Watson, J (2011) 'Female Engagement Teams: The Case for More Female Civil Affairs Marines'. *Marine Corps Gazette* 95(7), pp. 20–24. Available at: www.mea.marines.org/gazette/article/female-engagement-teams-case-more-female-civil-affairs-marines (Accessed 30 March 2015).

Wittwer, J (2016) 'Eight Reflections on "mainstreaming" Gender and Women, Peace and Security into the Military'. *The Strategist* [online] (25 November). Available at: www.aspistrategist.org.au/eight-reflections-mainstreaming-gender-women-peace-security-military/ (Accessed 3 February 2018).

Women in International Security (n.d.) 'Women, Peace and Security Leadership Program'. Women in International Security. Available at: www.wiisglobal.org/programs/women-peace-and-security-leadership-program/ (Accessed 28 May 2018).

Worldometers (2018a) 'Ireland Population'. Worldometers. Available at: www.worldometers.info/world-population/ireland-population/ (Accessed 28 May 2018).

Worldometers (2018b) 'Japan Population'. Worldometers. Available at: www.worldometers.info/world-population/japan-population/ (Accessed 28 May 2018).

Zazanis, MM; Kilcullen, RN; Sanders, MG and Crocker, DA (1999) 'Special Forces Selection and Training: Meeting the Needs of the Force in 2020'. *Special Warfare* (Summer) 12(3): pp. 22–31.

CONCLUSION

Conclusion

With the goal of greater protection for women and girls in situations involving armed conflict, gender considerations in military activities and operations must be operationalised in pragmatic and actionable ways. War in general has different and more severe effects on women and girls than it does on men and boys. At the same time, climates in conflict-afflicted regions are changing, with a general trend towards greater warmth and more variable and extreme weather events. Climate change in general also has different and more severe effects on women and girls then it does on men and boys. To be most effective in civilian-centric operations, military organisations must factor the potential compounding effect that the impacts of climate change might have on at-risk population groups such as women and girls into their planning and operations.

Armed conflict and gender

Whether they are civilians caught in the crossfire of armed conflict, refugees fleeing from it or combatants conducting the actual fighting, women and girls experience and endure a range of negative effects related to the fighting, particularly in developing countries. Prominent among these are the increased risk of sexual and gender-based violence, the breakdown of supporting social and economic structures that enable livelihoods and ostracism from cultural life in communities because of stigmas associated with women and girls who served as fighters. A case study of women's and girls' experiences in the civil conflict involving the jihadi group Boko Haram in Nigeria shows how these effects might play out in a complex social, economic and cultural environment. Because of the mediating effects of these influences, not all women and girls will likely experience the same effects to the same degree, or even in the same way. For example, whilst many women and girls have been sexually assaulted and enslaved, some appear to have found their place with Boko Haram.

Gender and climate change

Just as armed conflict has a gender-differentiated impact, so too can climate change. Particularly in areas of the world where women and girls are dependent upon agriculture for their livelihoods, decreases in agricultural productivity directly impact food security. Women and girls can also experience different and more severe effects to their health, and those who perform tasks within the household such as domestic chores and serving as family caregivers are also burdened with other labour-intensive work such as drawing water and collecting fuel for cooking. Increased risk of suffering sexual or gender-based violence because of out-migration of male family members seeking employment is a threat, whether women and girls remain at their homes or become internally displaced persons or refugees themselves. The case study of South African municipalities in KwaZulu-Natal Province shows how the impacts of climate change are strongly mediated by society, economy and culture. How people in communities adapt to or mitigate climate change's negative effects on their lives will depend not just on their human and capital resources, but potentially on evolving notions of gender roles as well. Accurately tracking such changes would likely pose a significant challenge for many military operations, but understanding these changes could be very important in making sure that scarce military resources were being directed towards efforts that complement mitigation and adaptation methods and lifeways.

Climate change and armed conflict

Whilst there is little doubt that armed conflict can lead to significant environmental degradation and make environments more susceptible to climate change's effects, there is currently no scientific consensus that a direct reserve correlation is also true. From a research perspective, one challenge is finding appropriate and measurable proxies to stand in the place of climate change in analysis. Another challenge is the length of time over which changes have been occurring. There appears to be a trend in research studies that suggests that phenomena more directly related to human impacts of climate change, such as food insecurity, are associated with increases in conflict, even armed conflict. The case study of the civil war in Syria demonstrates just how difficult it can be to establish whether and to what degree climate change has played a causal role in armed conflict. From an operational perspective, however, the lack of scientific consensus as to causation is not a show-stopper – militaries are accustomed to dealing with uncertainty and risk as they formulate actions that must be taken decisively. Although there is no consensus in the scientific community that climate change causes armed conflict, the realistic threat of armed conflict and climate change having a compounding effect upon at-risk population cohorts such as women and girls means that gender considerations must be taken seriously to mitigate the risks they face, particularly in civilian-centric operations.

Strategy, military doctrine and gender

NATO

NATO provides an excellent example of a large, diverse multinational collective-defence organisation that has strongly moved forward to foster gender equality in its internal workings and gender considerations in its activities and operations. In terms of climate change, NATO recognises it at the strategic level as a threat (NATO, 2014: para. 110). In its high-level military doctrine, which is important for both interoperability purposes and national resource commitments in its member states, NATO has begun to include gender considerations as factors that must be considered in all its activities and operations. In its subordinate doctrine, progress in this regard has been uneven. Whilst the inclusion of gender considerations is increasingly better developed in certain guidance applicable to civilian-centric operations, gender is not included at all in the guidance dealing with kinetic operations. NATO military doctrine is largely and necessarily pitched at the joint level, and although its influence on the doctrine of its member states and its partners is significant, the degree of influence will vary from nation to nation. NATO's approach to operationalising gender is heavily influenced by the Swedish experience, and it is ably supported by the educational, training and doctrinal development expertise housed at the Nordic Centre for Gender in Military Operations.

UK

At the strategic level, the UK has moved forward in a thorough and deliberate manner to construct, evaluate and update its national action plan to implement UNSCR 1325, which includes tasks for the Ministry of Defence that are keyed to its overseas missions and the resource constraints these missions face. From the perspective of climate change, the UK recognises that climate change will have effects on its activities and operations, but that these effects are not likely to register significantly for perhaps a couple of decades. Reflecting the UK's strong role in helping develop NATO doctrine, the UK largely accepts NATO doctrine as its own, modifying it when necessary to address national interests and equities. UK joint doctrine shows distinct progress in factoring gender considerations into the planning and conduct of civilian-centric operations, but not kinetic operations. In relevant service doctrine at the tactical level, however, such as in British Army doctrine, gender is not yet fully operationalised, although some progress is being made.

US

The US is a North American member of both NATO and the Five Eyes Community, and the enormous size of the US military establishment and the large scope of its world-wide military commitments cause it to stand out as unique in certain ways. At the strategic level, the new US national security strategy no longer mentions climate change, and barely mentions gender. Although the US has continued to evaluate and update its national action plan, the pace of implementation across the US Department of Defense has been slow and uneven. As a matter of law, however, the US has passed the *Women, Peace and Security Act of 2017*, which imposes legal requirements for implementing and reporting on national action plan tasks to the Department of Defense. Although important progress has been made in US joint-level doctrine in operationalising gender, particularly doctrine applicable to certain civilian-centric operations, there are still significant areas in which it does not register at all. This unevenness is also found in US Army doctrine, which is the most likely service-level doctrine to be employed by the US in civilian-centric operations. Certain subordinate doctrine within US Army doctrine is actually better developed regarding gender considerations than is its hierarchical superior. In terms of practical measures to operationalise gender in the Department of Defense, there are distinct indicators of progress, such as the work being done with executive education at the Daniel K. Inouye Asia-Pacific Security Studies Center, and the new US GENAD course.

Australia

A Five Eyes Community member located in Oceania, Australia is closely associated with NATO, the UK and the US in military planning and operations at all levels. In part due to domestic policies driven by a recognised need to improve gender equality in the Australian Defence Force, the Australian defence establishment has made quick and substantial progress in implementing the tasks assigned to it under the national action plan to include gender in its work. This progress has outstripped the pace of inclusion of gender considerations in Australian joint-level doctrine, although a plan is in place to review all joint doctrine and operationalise gender where appropriate, and certain evidence supports the view that this revision is being done productively. In terms of its military engagements with its allies such as Japan and the US, Australia brings gender issues to the fore and heightens their visibility in the multinational context. At the land force level, the Australian Army military planning and decision-making process doctrine has apparently been revised to include gender considerations, but due to its classification it is not available to the public to review. Although older Australian Army doctrine generally lacks any consideration of gender issues, there has been recent important progress as shown by the new operations and the new CIMIC doctrine.

International humanitarian law and gender

Largely ignored in international work on gender in military activities and operations but for those parts of it that prohibit sexual and gender-based violence, IHL requires a thorough reassessment in terms of its long-standing male-normative approach to armed conflict, its emphasis on affording protection to women and girls as a function of their weaker constitutions and their socially presumed characteristics of chastity and modesty, and its failure to consider the operational gender factors that are relevant to the kinetic part of kinetic operations, that is, the application of lethal armed force. The legal doctrine of NATO, the UK, the US and Australia are all silent in this regard. For example, for purposes of conducting a proportionality analysis to weigh the relative values of military advantage and civilian casualties and damage to their property, one civilian is no different than another, even though through UNSCR 1325 the international community established that this was not so. Recognition of these issues is not new – Australian feminist authors such as Gardam, Charlesworth and Jarvis have literally been writing about them for decades. In sum, IHL quietly both reflects and reinforces the traditional male-normative understandings of the significance of gender in operations despite its stated emphasis on impartiality. Recent attempts by the International Committee of the Red Cross to re-contextualise IHL's explicit provisions that embed gender discrimination within it lack the gravitas of the commentaries on IHL treaty law written closer to the time fundamental treaties were negotiated and by authors involved in the negotiations.

Gender in military activities and operations

Despite some troublesome continuity breaks in operationalising gender between strategy, joint-level doctrine and operational and tactical level doctrines in different military organisations, a survey of military activities and missions that work to operationalise gender throughout the international security environment shows a broad array of efforts in the areas of organisational development, education, training, planning and operations. If not best practices, these efforts are at least the solid beginnings for the establishment of a more thorough and comprehensive operationalisation of gender in the different military organisations of the world. At the moment, however, these efforts are not all occurring in one place at the same time, suggesting that the resources that are being harnessed to deliver these efforts are not likely being used as efficiently as they could be. Further, since many are geared towards national concerns, they are not necessarily 'plug and play' in other national settings. Another challenge to increased implementation of UNSCR 1325 throughout militaries is likely the lack of empirical evidence demonstrating the positive effects of including gender considerations across the span of activities and operations. Finally, these efforts do not yet meaningfully address relationships between armed conflict, women and climate change together, and not doing so ignores what could be a significant threat in the evolving international security environment, particularly in civilian-centric operations.

Potential paths forward

The lack of treatment of the relationships between armed conflict, women and climate change in military activities and operations is not because they do not lend themselves to being made actionable in effective ways. For example, even though the most recent NATO environmental protection doctrine fails to make these connections, viewing this doctrine through a gender lens shows the ease with which these considerations could be included in doctrine related to the environment. Further, work done by the Centre of Excellence for Civil-Military Cooperation demonstrates how these linkages could be developed in CIMIC doctrine that provides a basis for understanding the relationships between armed conflict, women and climate change together and acting upon them. Finally, one of the proxies for climate change that has proven useful in finding positive correlations between climate change and armed conflict, food security, provides a practical platform for considering what gender-related tasks are both properly within military skill sets and resource constraints. Food security would also provide a practical platform for military units to work more closely and effectively with civilian government organisations, and if non-governmental organisations were willing, a topic of information exchange between military units and non-governmental actors that could lead to more complementary efforts on the part of both without them having to work closely together.

NATO environmental protection doctrine

Even though some of it is quite recent, the NATO environmental protection doctrine that forms the basis for the environmental protection standardisation agreements by which the NATO partners agree to implement this doctrine includes no discussion of the significance of environmental activities in theatres of operation with regard to women or to gender. There are two separate tracks of environmental doctrine regarding environmental protection published by the NATO Standardization Office, the Allied Joint Environmental Protection Publication (AJEPP) series for land-based operations, and the Allied Maritime Environmental Protection Publication series for sea-based operations. This section will only examine some of the doctrine set out in the AJEPP series, and particularly those portions of the documents that would lend themselves to the inclusion of a gender perspective when they are next reviewed by the appropriate subject matter experts for updating.

Within the AJEPP series itself, it will be sufficient to focus on *AJEPP-3, Environmental Management System In NATO Military Activities* (NSO, 2017), and *AJEPP-4, Joint NATO Environmental Protection Doctrine during NATO-led Military Activities* (NSO, 2018). The most useful place to begin this examination is with *AJEPP-4*, which provides the overarching NATO doctrine and implementation guidance in this area. *AJEPP-4* states that 'While meeting their military mission, NATO Forces should be committed to taking all reasonably achievable measures to protect the environment' (NSO, 2018: 1–1), and that this requires knowing the relationships between the military activities and the environment. Such knowledge requires environmental planning, and the use of an Environmental Management System, as set out in *AJEPP-3* (NSO, 2018: 1–1).

AJEPP-4, Joint NATO doctrine for environmental protection during NATO-led military activities

Consistent with the purpose of the document, *AJEPP-4* addresses environmental planning, environmental risk management, a commander's responsibilities regarding the environment and the objectives of environmental education for personnel who are tasked with environmental duties. Chapter 2, 'Environmental Planning', provides a checklist for commanders to help them identify 'the characteristics of the environment that may be impacted by or have an impact on NATO-led military activities', including the climate, water and air quality, and the presence of birds or bird migration routes' (NSO, 2018: 2-2). It is odd that the commander should be prompted to enquire about bird migration routes, but not whether environmental conditions are having a differentiated impact on at-risk population cohorts such as women and girls. Similarly, in a listing of the impacts that different types of pollution might have in the area of operations, the special protections accorded wetlands and biodiversity in general by the international community are noted, but there are no impacts identified based on gender (2-2–2-4).

Chapter 3, 'Environmental Risk Management', identifies four key elements of a risk management framework that commanders should use to comply with their obligations regarding environmental protection. First, commanders should issue 'clear guidance' on environmental protection as early as possible in the mission planning process (3-1). Second, an environmental plan should be developed, including 'a list of identified risks and prescribed mitigation measures' (3-1). 'Clear guidance' could easily include gender considerations, and this would then necessarily include identifying potential threats to at-risk groups such as women and girls and steps that could be taken to mitigate these risks.

The third key element is implementation, and *AJEPP-4* notes that in addition to personnel being trained on environmental issues, units are expected to 'work with local authorities and the community to identify and resolve problems' (3-1). Finally, risk management requires that 'Activities should be continuously monitored to ensure consistency with the commander's [Environmental Protection] objectives' (3-1). These elements presume that there will be periodic data collection and inspections, and that working with local communities is an aspect of this assessment process (3-1–3-2). If the commander's guidance were clear that this process should also include sex and gender-disaggregated data collection and analysis from a gender perspective, then the likelihood of discovering any differentiated and more severe pollution impact upon women and girls would be greatly increased.

Chapter 3 sets out in some detail a commander's responsibilities with regard to environmental protection. These duties include considering environmental impacts in decision-making, and enhancing 'relationships with host nations and neighbouring communities by addressing environmental issues and maintaining appropriate levels of coordination' (4-1). Environmental impacts should not be narrowly construed as just being those upon the physical environment, but on the human environment as well. Enhancing relationships with local communities regarding environmental protection cannot likely be fully achieved if the conversations are channelled towards talking about things such as bird migration routes rather than deploying force impacts on host nation women and girls.

Chapter 4 of *AJEPP-4* addresses the need for personnel to be appropriately trained and educated on environmental protection. Commanders are advised that although the 'nature and applicability of training will reflect the recipient's rank and specialisation' (5-2), education of personnel should be geared towards raising awareness of environmental policy, environmental protection and resource conservation (5-2–5-3). Each of these areas would be suitable in some fashion to include awareness of gender issues in environmental matters, but none of them lists this item as a specific educational goal. Instead, for example, under resource conservation, heritage protection and energy efficiency are listed, and under environmental protection, the focus areas of noise abatement and 'landscape quality protection' are listed (5-2). Surely gender considerations must be at least as important as litter control.

AJEPP-3, environmental management system in NATO military activities

Like *AJEPP-4*, *AJEPP-3* is of recent vintage, and like *AJEPP-4* it contains no mention of UNSCR 1325, *Bi-SCD 040–001*, women or gender. This document sets out for the environmental protection officer supporting the commander the Environmental Management System that *AJEPP-4* prescribes be used to meet the commander's environmental obligations. It is a thorough and useful piece of doctrine, and it leads the commander through the development of the Environmental Management System Plan, its execution, how to conduct an assessment of the system and how a commander should approach a review of the system (NSO, 2017: 2-1–5-1). It also contains two practical appendices, the first being a table that identifies possibly polluting activities, their vectors for pollution of the environment and how those pollution impacts might register in the environment (Appendix A, A-2). The second appendix is a template for drafting an Environmental Management System plan for military activities (Appendix B).

AJEPP-3 notes that the tasks assigned to the environmental protection officer will vary depending upon which phase of the Operations Planning Process the mission is in currently (1-3). For example, during Phase 1, Initial Situational Awareness of Potential or Actual Crisis, the environmental protection officer is expected to 'develop environmental intelligence products in conjunction with other subject matter experts' (1-3). In Phase 4, Operational Plan Development, the environmental protection officer is expected to liaise with host nation personnel and conduct environmental impact assessments (1-3). At the conclusion of this phase, the environmental protection officer is required to produce the environmental protection appendix to the Military Engineering Annex of the operations plan (1-3). Importantly, unless the environmental protection officer is aware of the need to enquire as to gender considerations regarding the impacts of activities on the physical environment, it is difficult to see how link-up with the GENAD occurs as part of subject matter expert consultation. Without this connection, I struggle to see how the environmental impact assessments will be as fully developed as they need to be regarding the impacts upon the human environment.

Further, this means that the risk assessment that is set out as an integral part of the planning process (2-2-2-4) could quite easily miss the differentiated and possibly more severe impacts that pollution from military activities would have on at-risk segments of the local communities, such as women and girls. Finally, execution of the Environmental Management Plan includes the use of an Environmental Management Board at the tactical level to manage the plan's implementation for the commander (3-1). This board is composed of regular members such as the organisation's senior environmental protection officer and representatives from J-3 (Operations), J-4 (Logistics), the military engineering staff and the medical staff. Special members as required include public affairs officers and the LEGAD, but there is no mention of the GENAD (3-1). As with *AJEPP-4*, small but important changes could be easily made in *AJEPP-3* to help ensure that the intent of *Bi-SCD 040–001* to achieve greater protection for women and girls in NATO military activities and operations is realised in the area of environmental protection.

The NATO-accredited Military Engineering Centre of Excellence (MILENG COE) located in Ingolstadt, Germany, is the proper institution to begin consideration of how gender perspectives could be best included in NATO environmental protection doctrine and practice. Through its Policies, Concept and Doctrine Development Branch, it supports the development of NATO engineering doctrine and practice, and its serves as the proponent of military engineering-related NATO doctrine and policy. Importantly, as part of its instructional work, it also teaches a week-long course on NATO Military Environmental Protection Practices and Procedures (MILENG COE, n.d.: 'About Us'). The combination of these important capacities should be leveraged to make NATO more responsive to an evolving international security environment increasingly influenced by climate change and marked by the occurrence of missions that are civilian-centric in nature, by revising its environmental protection doctrine to take account of gender.

Civil-military cooperation centre of excellence

The Civil-Military Cooperation Centre of Excellence (CCOE) is another NATO-accredited training, education and doctrinal development centre like the MILENG COE, and it is located in The Hague, The Netherlands (CCOE, n.d.: 'Meet Us'). It has published materials that are very relevant to NATO civil-military cooperation, because even though they are not considered official NATO doctrine, they incorporate lessons learned from many military operations, including those in Afghanistan. Although CCOE's field handbook on civil-military cooperation does not mention climate change specifically, it does note that crises are influenced by environmental factors and the importance of understanding the physical environment, climate and ecosystems and their impacts on the human environment

(CCOE, 2016: I-3-2, III-3-6, III-3-11). It also recognises that CIMIC officer liaison with engineers on environmental considerations can be effective in supporting the joint headquarters staff (III-1-5). The field handbook also includes a full chapter on gender awareness in operations, the operational significance of gender and practical pointers on how to integrate a gender perspective with operations (III-6-1–III-6-5). This chapter is complemented by an earlier chapter on achieving cross-cultural competence (III-5-1–III-5-8).

The field handbook is supplemented by two additional publications that help link gender and environmental considerations together from a CIMIC perspective. The first, by Groothedde (2013), is a guide on operationalising gender in the course of civil-military interactions. This guide provides a more detailed description of gender, explains what a gender perspective in operations is and how it might be practicably used in operations (2). It differentiates between sex and gender and explains the importance of mainstreaming a gender perspective in planning and operations as a way to ensure gender impartiality (15, 17). It also describes the role of GENADs in advising commanders and staff to incorporate a gender perspective in their planning and decision-making (24–27). Finally, it includes detailed guidance and case studies on how to incorporate a gender perspective into operations to improve operational efficiency (44–57).

The second publication, by Gallagher and Wit (2012), concerns the environment. It provides examples of how ecosystems are at risk because of climate change, a discussion of relevant international conventions on climate change and best practices to be employed by deployed units to avoid compromising local environmental conditions (27, 52–57). It also addresses the holistic approach required to implement sustainable resource use and management (75). Importantly, this publication also explicitly links women's roles in host nation societies to issues of resource use and ecosystem restoration, and notes how women's equities in these matters are often underestimated or overlooked (35–36). Further, this guide also directs readers to the gender guide to find useful methods for gathering information regarding women's access to 'resources such as land, water and wood' so that they are not marginalised by 'well intended measures' (36).

This field handbook and these two field guides show that it is not difficult to address the practical aspects of climate change and gender in situations involving armed conflict in CIMIC doctrine. It could serve a model for national CIMIC doctrine, using the field handbook as the general reference document for practitioners and the field guides as supplements that could be easily updated as new information and techniques developed. Although the model format of the Theatre Civil Assessment set out in the handbook contains a food and agriculture section that does not specify a gendered approach to evaluating the people involved in this aspect of the operational environment (CCOE, 2016: Annex 4, 4-9–4-10), I believe this approach might also lend itself to more fully develop the significance of food security and to set out practicable ways in which gender could be included in its operationalisation.

Food security

Many studies have looked at precipitation patterns in assessing the relationship between climate change and armed conflict. Although the findings are mixed, as described in Chapter 3, 'Climate change and armed conflict', more recent studies suggest that water availability in areas dependent on rain-fed agriculture is linked to food scarcity, and food scarcity is linked to increased violence, both at the local scale and at the interstate level (Devlin and Hendrix, 2014: 27–39; Wischnath and Buhaug, 2014: 6–15; Von Uexkull, 2014: 16–24). Gender is a cross-cutting issue in food security. When food is scarce in many developing countries, women and girls often eat after the men and boys (Gardam and Charlesworth, 2000: 155). Sexual violence against women can destabilise communities' economic productivity and cause food insecurity, because women might not be able to safely leave their homes and tend to crops or take their harvests to market to sell and obtain other necessary food items (Dharmapuri, 2011: 64). In certain countries, some crops are considered 'women's crops', planted and tended to by women (Aguilar, 2009: 22). Potentially, if climate variability impacts these crops more severely, this would have a disproportionate effect on these women's livelihoods. Two examples of the use of military assets that could have the ability to identify such information and either act on it directly, as in the case of the US National Guard's Agriculture Development Teams that were used in Afghanistan, or provide it to others once it had been obtained, as in the case of surveys conducted by deployed military units in Burundi, are worth considering because they make effective use of military skill sets and resources in ways that are sensitive to both gender and changes in the environment.

Agriculture development teams

Beginning in 2007, and continuing up to the conclusion of the ISAF mission in Afghanistan, US National Guard units called 'Agribusiness Development Teams' and then later 'Agriculture Development Teams' (ADTs) were deployed to different provinces to help Afghan farmers rebuild the country's once productive agricultural sector (Knight, 2012). Because these teams were state units, found neither in the active duty US Army nor the US Army Reserve, their structure was not constrained by doctrine. This provided the respective state National Guards some flexibility in determining team composition in terms of number of members and the different fields of civilian expertise that were represented on them. Importantly, ADTs were able to work with their respective state universities to receive additional agricultural training, and then as a reach-back resource once they deployed to Afghanistan (Center for Army Lessons Learned, 2009: 3–6).

Once in theatre, the ADTs enjoyed a significant degree of freedom of movement, because they brought with them their own transportation assets and security personnel. US agricultural experts attached to the US-run Provincial Reconstruction Teams often did not have these capabilities available to them and instead had to rely on manoeuvre units in their areas of operations for this sort of support. Further, because the different states committed to deploying replacement ADTs to the same areas, these teams enjoyed a level of continuity with local Afghan officials and farmers not shared by other US forces (Eric Ahlness, 2013: personal communication). They were generally assessed as having been successful in their missions (Leppert, 2010: 58; Fortune, 2012: 14), but there was some criticism that the quality of the ADTs was uneven, and highly dependent on the personnel comprising the teams (Carreau, 2010: 141–43).

As the international community began winding down its involvement in the ISAF mission, the ADTs began to focus more on sustainable projects that Afghans could maintain and carry forward with little or no outside support (Jeffrey Farrell, 2013: personal communication). Examples of this included demonstration farms and orchards that also served as training centres for local villagers (Arkansas National Guard, 2010), complementing existing agricultural practices with inexpensive and low-technology projects (Kentucky National Guard, 2011), basic animal husbandry skills (Cheryl Wachenheim, 2018: personal communication) and supporting greenhouse growers' associations so as to promote agricultural stock and price stability by extending vegetable growing periods (Jeffrey Farrell, 2013: personal communication).

The later rotations of ADTs into Afghanistan specifically included Female Engagement Teams, or a variation of the Female Engagement Team called a Women's Initiative Training Team, to engage with female Afghan officials and support projects seeking to teach Afghan women practical skills. This included training on how they could preserve food better (Grant, 2012: 13), how to raise small animals and poultry, how to keep bees (Holliday, 2012: 91) and how to grow crops for their own use and saplings for sale in reforestation projects in their backyard gardens (Prescott, 2014: 799). It was this last skill, learning how to grow saplings, which made the local Afghan women part of an ambitious long-term project that provides a useful example of how military efforts can support women becoming involved in practical climate change adaptation and mitigation measures.

The second ADT rotation from the US state of Georgia, working in Logar and Wardak Provinces, supported a project to restore seven watersheds in an area with a number of villages using local, low-technology methods. First, local water officials and village elders were included in the project and received training in water management skills. Later, in conjunction with US Department of Agriculture experts, the ADT helped conduct training for Afghan government officials to increase their appreciation of the need to take a holistic approach across government agencies to integrate water management into their different activities (Jeffrey Farrell, 2013: personal communication).

In the next step of the project, contractors were hired to identify areas that were best suited for watershed restoration, and the contractors were required to hire local men for the project so they would receive on-the-job training. Once suitable areas were identified, local men were again hired to undertake simple construction works, such as building stone dams and contour ditches to slow water flow and thereby reduce soil erosion and enhance water infiltration into the soil (Jeffrey Farrell, 2013: personal communication). These workers also received training from Afghan government officials and local teachers on various topics for two hours of their workday (Grant, 2012: 13). Finally, when the construction was completed, saplings grown by local women were purchased by the contactor and local men were hired to plant them in upland portions of the watershed restoration areas (Jeffrey Farrell, 2013: personal communication).

Respected non-governmental organisations have critiqued military development efforts as being disruptive, poorly focused and non-sustainable in many cases (Oxfam International, 2010). Conversely, those efforts that the military supports that seem more sustainable such as the watershed restoration project described earlier could take decades to mature – much longer than any military mission is likely to be around to check up on how the saplings are growing. Importantly though, initiatives such as the watershed restoration project provide an example of how military organisations could work in a gendered fashion to treat climate change as a process and not just a fact. In doing so, they could provide local people in conflict-afflicted areas the skills, the resources and potentially the opportunities to take steps to build resilience to climate change's negative effects in the at-risk ecosystems upon which they rely for their livelihoods.

Gender-sensitive surveys

Burundi has in recent history experienced significant periods of food scarcity and violent conflict (World Food Programme, 2005; Jones, 2016). Dr Richard Byrne, a senior lecturer at Harper Adams University in the UK, has conducted a great deal of research and field work on food security in the context of stabilisation operations. In a presentation he gave at the Security, Conflict and International Development Symposium at Leicester University in 2015, he talked about a pilot project in which he was involved in Burundi using military personnel to conduct agricultural surveys to develop the mission's understanding of food security. In at least one area, the ordinary metrics that might be tracked to gauge food security, such as availability of fertiliser and the safety of roads from the farms to the towns from bandits, all seemed adequate, yet the price of food in the marketplace still seemed artificially high (Byrne, 2015).

Fortunately, the surveys that the military personnel were conducting had a gender component. When the women of farm families were interviewed, they explained that often the soldiers who were patrolling the roads and keeping them safe from trouble makers were themselves often levying unofficial taxes on the women as they tried to bring their food to market. This discouraged the women from going to market, leading to higher prices for food items in the local markets (Byrne, 2015). Conceivably, this sort of situation could lead to imminent food security issues in the urban areas as townspeople ran out of affordable food, and longer-term food security problems in the countryside perhaps as farmers turned their focus towards subsistence crops and missed opportunities to earn cash.

This is a very subtle point in terms of intelligence collection, but very important. Many Western militaries devote a great deal of time to making sure that there are no conflicts between their own forces, so-called 'blue on blue' engagements, ironically also known as 'friendly fire'. They are also very concerned about 'blue on green' engagements, in which their forces mistakenly attack friendly host nation forces or civilians. In Afghanistan, because of insurgent infiltration of the Afghan National Security Forces, Western militaries have also become concerned about avoiding 'green on blue' incidents in which their personnel are attacked by insurgents who have infiltrated the Afghan ranks (Long, 2013: 167–82). Some metrics might assess 'green on green' incidents, such as how host nation forces are treating host nation civilians, but this could become very complicated and politically fraught depending on the status of the deploying forces in the host nation (Prescott and Male: 2018: 678–85).

Unfortunately, as the case study from Burundi shows, the ordinary indicators that militaries might traditionally look at to assess food security, such as availability of fertilisers and insurgent attacks on roads and highways, might simply not be sufficient to capture the gendered components of food security that could have very significant impacts on the overall security situation in an area. This is fixable, however, because military units possess important enablers that many non-governmental organisations often lack, such as their own transportation assets, their own weapons for self-defence, reliable communications in the event things go wrong and access to data and analysis. Conducting these surveys would make use of existing military skill sets and resources, and leverage the strengths of the larger military unit to which the survey administrators would belong (Byrne, 2015).

Importantly, the surveys that military personnel would be conducting would use basic demographic and economic questions to address the four dimensions of food security – availability of food, access or the ability to buy and produce food, utilisation of the food and market stability of food items over time (Byrne, 2015). These concepts could easily be included in a detailed CIMIC assessment because they lend themselves to both qualitative and quantitative evaluation, and they all include highly gendered aspects. Finally, not only could militaries share this sort of sex and gender-disaggregated data with governmental agencies and non-governmental organisations, but they could also be in a position to actually help solve the problem.

Suppose that the troops in this case study had not simply gone rogue, but instead were short of rations because of insufficient host nation logistical support. This information could be passed to a deployed military liaison working with the host nation logisticians to make sure the soldiers received their supplies. Conversely, suppose in fact the soldiers were well fed but had turned to petty thievery to acquire cash. In this case, a deployed military liaison working with the host nation legal advisors could inform them of this criminal activity and assist them with gathering evidence and conducting disciplinary actions. I believe that these sorts of actions would be consistent with Dr Byrne's recommendation that the militaries could usefully engage in these areas, but that they should leverage their inherent capabilities rather than seek to become full-scale development actors.

Summary

The threat posed by the failure to effectively operationalise gender must be recognised as having both operational and political components. As Duyvesteyn (2011) has cogently pointed out in assessing the role of intelligence in counterinsurgencies, the use of armed force by Western militaries in this sort of civilian-centric operation has become more difficult over time because 'of international law and the mores of democratic societies' (456). Commanders and staffs must come to grips with operationalising gender not just because of the risk posed to their missions by the failure to do so, but also because their political leaders and the civil societies they represent will expect this of them. My recommendation – no, my message actually – to commanders and staffs who are still hesitant about including gender considerations in their activities and operations is respectfully this: if you do not figure it out yourselves, my guess is that it will be figured out for you.

Based on my experiences and what I have learned writing this book, I would not favour the second option. No one knows your mission like you do, and hesitation on your part invites top-driven solutions that will likely be sub-optimal. Importantly though, figuring it out is not rocket science. By incorporating a gender perspective generally in education, training, planning and operations, we would allow military personnel to develop the skills to effectively apply it to an assessment of any given situation. Applying a gender perspective to a pure force-on-force kinetic operation in an area with no civilians, for example, might result in an assessment that gender is simply not relevant to it. Such a finding after an honest and objective mission analysis would promote the goals of UNSCR 1325 and the different national actions plans to achieve greater protection of women and girls in situations involving armed conflict, just as much as drawing the opposite conclusion in a very civilian-centric operation. In the end, as with so many things in military life, it is simply a question of professionalism and integrity.

References

Aguilar, LR (2009) *Training Manual on Gender and Climate Change*. Global Gender and Climate Alliance: San José, Costa Rica. Available at: https://portals.iucn.org/library/sites/library/files/documents/2009-012.pdf (Accessed 3 June 2018).

Arkansas National Guard (2010) 'Arkansas Troops Developing Farm as Agriculture Education Center'. Arkansas National Guard [press release] (19 November). Available at: www.arguard.org/publicaffairs/index.asp?id=news/2010/11/ADT_Farm.htm (Accessed 14 May 2014).

Byrne, R (2015) 'Food Security and Conflict: Stabilisation Forces & Agricultural Awareness'. In: *Security, Conflict and International Development (SCID) Symposium*, Leicester: University of Leicester (1 April). Available at: https://uolscid.wordpress.com/2015/04/01/2015-scid-symposium-dr-richard-byrne-food-security-and-conflict-stabilisation-forces-and-agricultural-awareness (Accessed 3 June 2018).

Carreau, B (2010) 'Lessons from USDA in Iraq and Afghanistan'. *PRISM* 1(3): pp. 139–50. Available at: http://cco.ndu.edu/Portals/96/Documents/prism/prism_1-3/Prism_139-150_Carreau.pdf (Accessed 17 June 2018).

CCOE (Civil-Military Cooperation Centre of Excellence) (n.d.) 'Meet Us'. Available at: www.cimic-coe.org/meet-us/ (Accessed 3 June 2018).

CCOE (Civil-Military Cooperation Centre of Excellence) (2016) *CIMIC Field Handbook*, 4th ed. The Hague: Civil-Military Centre of Excellence. Available at: https://www.cimic-coe.org/ccoe-news/new-version-of-cimic-field-handbook/ (Accessed 23 September 2018)

Center for Army Lessons Learned (2009) *Agribusiness Development Teams in Afghanistan: Tactics, Techniques, and Procedures, Handbook 10–10*. Fort Leavenworth, KS: US Army Training and Doctrine Command.

Devlin, C and Hendrix, CS (2014) 'Trends and Triggers Redux: Climate Change, Rainfall, and Interstate Conflict'. *Political Geography* 43: pp. 27–39.

Dharmapuri, S (2011) 'Just Add Women and Stir?' *Parameters* (Spring): pp. 56–70.

Duyvesteyn, I (2011) 'Hearts and Minds, Cultural Awareness and Good Intelligence: The Blueprint for Successful Counter-Insurgency?' *Intelligence and National Security* 26:4; pp. 445–59. DOI: 10.1080/02684527.2011.580598.

Fortune, MD (2012) 'The Real Key to Success in Afghanistan: Overlooked, Underrated, Forgotten, or Just Too Hard?' *Joint Force Quarterly* 65 (2d Quarter): pp. 10–16. Available at: http://ndupress.ndu.edu/portals/68/Documents/jfq/jfq-65.pdf (Accessed 10 June 2018).

Gallagher, N and Wit, P (2012) *Society Stabilization by Winning the Environment: Ecosystems Assessment Makes Sense . . . Full Situational Awareness in CIMIC*. Civil-Military Cooperation Centre of Excellence. Available at: www.cimic-coe.org/wp-content/uploads/2015/11/Ecosystems_Assessment_complete_MQ.pdf (Accessed 29 May 2018).

Gardam, J and Charlesworth, H (2000) 'The Need for New Directions in the Protection of Women in Armed Conflict'. *Human Rights Quarterly* 22(1): pp. 148–66.

Groothedde, S (2013) *Gender Makes Sense: A Way to Improve Your Mission*, 2nd Ed. Civil-Military Cooperation Centre of Excellence. Available at: www.cimic-coe.org/wp-content/uploads/2015/11/Gender-makes-sense-final-version.pdf (Accessed 29 May 2018).

Grant, M (2012) 'The Georgia Rhythm Section'. *COIN Common Sense* 3: pp. 3–14. Available at: https://ronna.apan.org/CAAT/Shared%20Documents/COIN%20Common%20Sense%20Vol%203%20Issue%202.pdf (Accessed 3 June 2013).

Holliday, JR (2012) 'Female Engagement Teams: The Need to Standardize Training and Employment'. *Military Review* (March/April): pp. 90–94. Available at: www.armyupress.army.mil/Portals/7/military-review/Archives/English/MilitaryReview_20120430_art014.pdf (Accessed 3 June 2013).

Jones, S (2016) 'Burundi Close to "major crisis" as Hunger and Disease Take Hold, Warns UNICEF'. *The Guardian* (18 February). Available at: www.theguardian.com/global-development/2016/feb/18/burundi-major-crisis-unicef-violence-malnutrition-malaria-cholera (Accessed 3 March 2017).

Kentucky National Guard (2011) 'Afghanistan: Kentucky Guard ADT 3 Pulls Out the "Wood Carpet" for Afghan Farmers'. Kentucky National Guard (22 November). Available at: www.arng.army.mil/News/Pages/AfghanistanKentuckyGuardADT3pullsoutthe%E2%80%98woodcarpet%E2%80%99forAfghanfarmers.aspx (Accessed 14 May 2014).

Knight, R (2012) 'Missouri's Agribusiness Development Team Ends Mission in Afghanistan'. Missouri National Guard (18 July). Available at: www.moguard.com/07-18-12-missouris-agribusiness-development-team-ends-mission-in-afghanistan.html (Accessed 22 May 2014).

Leppert, MA (2010) 'Agrarian Warriors: The Quiet Success of National Guard Agribusiness Development Teams in Afghanistan'. *Army Magazine* 60(6): pp. 54–58. Available at: http://ausar-web01.inetu.net/publications/armymagazine/archive/2010/6/Documents/Leppert_0610.pdf (Accessed 3 June 2018).

Long, A (2013) '"Green on Blue": Insider Attacks in Afghanistan'. *Survival: Global Politics and Strategy* 55(3): pp. 167–82. DOI: 10.1080/00396338.2013.802860.

MILENG COE (Military Engineering Centre of Excellence) (n.d.) 'About Us'. Military Engineering Centre of Excellence. Available at: http://milengcoe.org/milengcoe/Pages/default.aspx (Accessed 6 July 2018).

NATO (2014) 'Wales Summit Declaration'. NATO. Available at: www.nato.int/cps/en/natohq/official_texts_112964.htm (Accessed 6 July 2018).

NSO (NATO Standardization Office) (2017) *AJEPP-3, Environmental Management System in NATO Military Activities*, Ed. A, Ver. 1 (May). Brussels: NATO.

NSO (NATO Standardization Office) (2018) *AJEPP-4, Joint NATO Doctrine For Environmental Protection During NATO-led Military Activities*, Ed. B, Ver. 1 (March). Brussels: NATO.

Oxfam International (2010) 'Quick Impact, Quick Collapse: The Dangers of Militarized Aid in Afghanistan'. Oxfam International.org. Available at: www.oxfam.org/sites/www.oxfam.org/files/quick-impact-quick-collapse-jan-2010.pdf (Accessed 17 June 2018).

Prescott, JM (2014) 'Climate Change, Gender, and Rethinking Military Operations'. *Vermont Journal of Environmental Law* 15(4): pp. 767–802.

Prescott, JM and Male, JM (2018) 'Afghanistan'. In: Fleck, D (ed.) *The Handbook of the Law of Visiting Forces*, 2nd ed. Oxford: Oxford University Press, pp. 650–85.

Wischnath, G and Buhaug, H (2014) 'Rice or Riots: On Food Production and Conflict Severity across India'. *Political Geography* 43: pp. 6–15.

World Food Programme (2005) 'Food Shortages in Burundi Deepen as Country Embraces Peace' (World Food Programme.org (16 March). Available at: www.wfp.org/news/news-release/food-shortages-burundi-deepen-country-embraces-peace (Accessed 3 March 2017).

Von Uexkull, N (2014) 'Sustained Drought, Vulnerability and Civil Conflict in Sub-Saharan Africa'. *Political Geography* 43: pp. 16–26.

INDEX

2014 Progress Report Australian National Action Plan on Women, Peace and Security 2012–2018 161

Afghanistan: Australian efforts in 158, 168; CCOE lessons learned from 243; counterinsurgency in 143; deployment to 4–5; detention operations in 176; environmental damage claims in Zabul Province 6–7; environmental degradation due to armed conflict 59; 'green on blue' attacks 248; status of forces agreement with NATO 97, 99; Swedish efforts in 211, 218, 220; UK focus country 116, 119; US efforts to promote governance 131; use of gendered staff and units in 114, 133, 145, 148, 160, 219, 220–1; water rights and land titles 41; wheat 6
Africa, East 10, 80
Africa, South 10, 37, 50–1, 186, 236
agricultural production 38–9, 236, 45, 46, 65, 75
Agriculture Development Teams: composition of 245; critique of 246, 247; deployment in Afghanistan 245–6; use of Women's Initiative Training Teams 246; watershed restoration project 246–7
Akilu, F. 28
Alaska 3, 4
Al-Assad, B. 72, 73, 75, 76, 77, 81
Al-Assad, H. 72, 73, 81
Al Hasakah Governorate 73, 74, 79
Allied Command Operations: location and role 93; *see also* Bi-Strategic Command Directives*; Gender Functional Planning Guide
Allied Command Transformation: location and role 93, 207; *see also* Bi-Strategic Command Directives*;
Allied Joint Environmental Protection Publication 3, Environmental Management System in NATO Military Activities 240, 242–3
Allied Joint Environmental Protection Publication 4, Joint NATO Environmental Protection Doctrine during NATO-led Military Activities 240–3
armed conflict, casualties: legal doctrine 239; Libya 176; Nigeria 25–6; Syria 81
armed conflict, effects: on civilian women and girls 15–16, 17–19; on combatant women and girls 21–4; on refugee women and girls 19–20
armed conflict, environmental degradation 53, 58–9, 71, 82, 226, 236
armed conflict, women and climate change: compounding effects of 1, 10, 80, 82, 235; relationships between 9–10, 49–50, 82–4, 235–6
Army Techniques Publication 3–07.5, Stability Techniques 144, 146–7, 148, 150
Army Techniques Publication 3–07.6, Protection of Civilians 148, 149, 150
Australian Defence Force: gender equality efforts 157–8, 159; operationalisation of gender 159–68
Australian Public Service Commission 159

Ba'ath Party 72–3, 76
Babugura, A. 50–2
Bangladesh 9, 38, 43, 47, 48, 190, 222
biomass *see* energy resources
Bi-Strategic Command Directive 40–1 (2009) 93, 105
Bi-Strategic Command Directive 40–1 Rev. 1 (2012) 93, 107
Bi-Strategic Command Directive 040–001 (2017): education and training 98–9; generally 93, 94, 242–3; handling sexual and gender-based violence 101–2; role of GENAD assisting other staff members and sections 96–7, 99–100; Standards of Behaviour and Code of Conduct 96–9; use of Gender Focal Points 100–1
Boko Haram: favourable view of by certain local women 24, 29; kidnapping of women 25; Nigerian government crackdown 25; rehabilitation and de-radicalisation of women associated with 27–8; sexual assault and enslavement of women 25–6; women as suicide bombers for 28
Bosnia-Herzegovina 2, 4
Brody, A. 37, 38, 39, 45
Buhaug, H. 68, 69, 70, 71, 245
Burundi 245, 247, 248
Byrne, R. 247, 248

camp conditions for refugees and displaced persons: food scarcity 20; lack of medical services for women 20, 26, 43; lack of privacy for women in 44; NATO directive on 'gender sensitive camp design and management' 102; persistence of domestic chores 44; sexual violence in 20, 26, 122
Canada 21, 90, 206, 212, 220
Caracal Battalion *see* Israel, mixed gender infantry units
Center for Advanced Operational Culture Learning 204
Central Intelligence Agency 65
civilian shields, use of women and girls as 188, 141
civil-military cooperation (CIMIC) operations: bridge to civilian efforts 95; doctrinal lack of gender treatment 107, 121, 141, 147; and Gender Focal Points 100; gender-sensitive surveys 248; generally 105; improvements in doctrinal gender treatment 168; linking climate change and gender 240, 243–4
Civil Military Cooperation Centre of Excellence 243

climate change: and convenience bias 61; effects on sea level 60, 63, 64, 66; and ethnic fractionalisation 71; and government response 67, 69–70, 75; as risk multiplier 60, 62–3, 66–7, 82; severe weather events associated with 40, 41–2, 46, 48, 52–3, 63, 65–6
combat roles, women in: Australia 21, 160; Canada 21; generally 17–18, 21–3; as insurgents in doctrine 139; Israel 222–3; Japan 21; Kurdish 22; New Zealand 21; Norway 222; UK 21; US 21
convergence of norms, international humanitarian law and international human rights law 174–5
Cultural Support Teams 221
customary international humanitarian law and gender 188–9

damage, collateral 5, 103, 107, 145, 150, 151, 239
Daniel K. Inouye Asia-Pacific Center for Security Studies 132, 133, 238
Dar'a, Syria 72, 76, 78–9, 80
Defence Forces Ireland 199–201
Defence White Paper (2016) 159–60
Democratic Republic of Congo 17, 18, 22, 119, 217, 222
Development, Concepts and Doctrine Centre 122, 125
doctrinal hierarchies: Australia 163, 166–7; generally 91; NATO 102–3; UK 121, 122; US 131, 138, 143, 151, 238
doctrine: definition 89, 90, 102; environmental doctrine amenability to including gender 241–3; limitations of 3; nature and purpose of 2–4, 90, 91; relationship to training 2–3; UK and US approach to climate change 62–3
domestic chores: burden of 39, 43–4, 45, 48; impact on developing marketable skills for women and girls 20, 42; impact on education for women and girls 19, 23, 37, 40, 42; matrilineal assignment of 41
dynamic targeting 5, 7, 101, 102

education, training, planning, operations and doctrine, relationships between 2, 3, 10, 11, 89–91
Effects-based Approach to Operations 5–6
Eklund, L. 74, 78, 83
energy resources: availability and use 47; biomass collection 19, 40, 41, 42, 44–6, 51; building resilience in 53; climate change disruptions in 66–7; consumption

by militaries 59; health impacts of solid fuel use 40, 47; improvement in scholastic performance among girls 41; UK promotion of clean energy 63

Female Engagement Teams: Australia 161; generally 95, 96, 134; Sweden 220; UK 114, 220–1; US 132, 134, 145, 148, 220
female military officers courses 205, 209–10
feminist critique of international humanitarian law: customary IHL 189; failure to gain traction 189–90; generally 177–8
Field Manual 3–07, Stability 144–5, 147
Five Eyes Community 89, 90, 101, 169, 206, 237, 238
food security, gendered efforts to increase 39, 46–7, 53, 245, 247, 248
formed policewomen units 190, 222
Fröhlich, C. 10, 48, 49, 70, 80
Future Operating Environment 2035 63–4

Gardam, J. 16, 18–19, 20, 40, 177–8, 183–5, 188–9, 190–1, 239
gender advisor course: Australian Defence Force 212, 214; Italy 210; Nordic Centre for Gender in Military Operations 210–11; US 238; Women In International Security 210
gender advisors (GENADs): Australia 157–8, 160, 167, 168–9, 218–19; challenges faced by 219, 243; informal survey of 219, 226; Ireland 200; NATO 94, 97, 99–102, 107, 215–17, 218; role in headquarters 93, 94, 107, 244; role in operations 217; Sweden 217–18; UK 118; US 213, 218
gender analysis: in GENAD courses 211, 212; in general 11, 24, 29, 30, 52, 80, 117, 122; NATO definition of 94; in NATO functional planning guidance 214–16; operational 18, 37, 78, 96, 100–1, 123, 136, 169
gender and sex-disaggregated data 38, 49, 52, 95, 96, 146, 169, 224, 241
Gender Equality Advisory Board 158
Gender Focal Points 100–1, 118, 199, 200, 214, 218, 224
Gender Functional Planning Guide 94–5, 96, 97, 214–17
gender in military education, lack of academic courses 203–4
gender mainstreaming: in Australia 169; in IHL 191; in military activities and operations 200, 205, 207, 209, 217–18, 225; in NATO 94–5, 104; in US 134
Gender Makes Sense: A Way To Improve Your Mission 244
gender roles: changes in response to climate change 236; generally 8, 10, 21, 23, 40, 51, 52, 122, 123
gender statistics: on equality 36, 37; violence 102, 135
Geneva Convention I and Geneva Convention II: deficiencies in protection for women 183; protections for women 182
Geneva Convention III: deficiencies in protection for women POWs 183–5; protections for women POWs 183
Geneva Convention IV: deficiencies in protection for women civilians 185–6; protection for women civilians 185–6
Geneva Conventions of 1949: commentaries and gender bias 183, 184–5, 186, 191–2; Common Article 3 181–2
Geneva Conventions of 1949, Conference of Government Experts: First Commission 178–9; generally 178; Second Commission 179; Third Commission 179
Geneva Conventions of 1949, Diplomatic Conference: Committee I 179; Committee II 180; Committee III 180–1; generally 179
Global Strategic Trends – Out to 2045 64–5
Goh, E. 37, 44–9

humanitarian operations 64, 106, 122, 168, 201
Hurricane Katrina 36
Hurricane Mitch 43
hybrid conflicts 125

'I Know Gender 1–2–3' 205–6
impacts mapping, climate change and gender: Brody et al.'s approach 38–44; Goh's use of methodology 44–6
information operations 141, 164, 216, 221
International Committee of the Red Cross (ICRC): changes in commentaries to Geneva Conventions I and II 191–2, 239; generally 173, 212; position on gender under customary international humanitarian law 189; role in facilitating the Geneva Conventions negotiations 178–80

international humanitarian law (IHL): conservative nature of 11, 174, 181, 191, 193; feminist critique of 177–8; impartial protections irrespective of sex 7, 173, 181–9; overlap in prohibiting sexual or gender-based violence 177; relationship to international human rights law, 174–6
international human rights law 16, 137, 162, 174–5, 176, 177, 190
International Peace Support Training Centre 204–5
International Security Assistance Force (ISAF) 4–6, 97, 99, 133, 176, 211, 218, 222, 245–6
Israel, mixed gender infantry units: Caracal Battalion 222, 223; generally, 217, 222–3; Lions of Jordan 223

Japan Self-Defense Forces, Female Personnel Empowerment Initiative 202–3
Japan Self-Defense Forces, women in the 201–2
Joint Doctrine Note 4/13, Culture and Human Terrain 122, 204
Joint Doctrine Publication 3–52, Disaster Relief Operations Overseas: The Military Contribution 122–3
Joint Doctrine Publication 05, Shaping a Stable World: the Military Contribution 121–2
Joint Force Command Brunssum, Joint Headquarters Standard Operating Procedure 106; Gender Advisor's Functions in JFC & JTF Headquarters 215–16
Joint Publication 3–07, Stability 140–1
Joint Publication 3–07.3, Peace Operations 142–3
Joint Publication 3–57, Civil-Military Operations 141–2
Joint staff sections: J-1 (Personnel) 97, 121, 224; J-2 (Intelligence) 6, 96; J-3 (Operations) 101, 151, 215, 216, 224, 243; J-4 (Logistics) 243; J-5 (Plans) 94; J-7 (Training) 150, 200; J-9 (CIMIC) 95, 100, 121
Joint Warfare Centre 4

Kelley, CP 77, 78, 79
kinetic: definition of 5; IHL application in operations 8, 16, 94, 95; lack of IHL focus upon operations 100, 101, 103, 107, 125, 161; operations doctrine 107, 138–9, 144–7, 150; operations generally 104, 108, 161, 166, 215, 216, 237, 239, 249; weapon 7
Kurdish female combatants 22, 190
KwaZulu-Natal Province 50, 236

Land Warfare Doctrine 3–0, Operations 167
Land Warfare Doctrine 3–8–6, Civil-Military Cooperation 167–8
legal advisor (LEGAD) 4–6, 97, 99–100, 243

masculinities: and armed conflict 17; in *Bi-Strategic Command Directive 040–001* 95; military masculinities 95–6; and Special Operating Forces 222
matriarchy, military 190
medical needs, female soldiers 21, 185
megacity 9, 139
migration, related to climate change: in general 48, 63, 69, 82; impacts on women and girls 38, 46; from low-lying island states 64; male out-migration 41–2, 52; scepticism as to cause of armed conflict 69; Syria 77–8
militarisation: of climate change 59–60; of UNSCR 1325 17
Military Engineering Centre of Excellence 243
Mozambique 15, 47

Namibia 186
national action plan, Australia 160–3
national action plan, Ireland 199
national action plan, Japan 201
national action plan, Mozambique 15
national action plan, UK 115–20
national action plan, US 132–7
national caveats on rules of engagement 101
NATO, assessment of inclusion of gender perspectives in operations 107
NATO-Euro Atlantic Partnership Council 91–3
NATO *Military Guidelines* 93, 94, 97, 98
Nepal 41, 53
New Zealand 21, 64, 89, 90, 212
Nigeria: civilian vigilante groups 25; north-eastern generally 26–7; *see also* Boko Haram
non-convergence of international humanitarian law and international rights law in use of force 175–7
non-kinetic: definition of 4–5; ISAF operations 4, 5; in NATO strategy and doctrine 105, 107, 215; in UK

doctrine 125; in US doctrine 141–3; US operations 138–9, 141–4, 147–8
Nordic Centre for Gender in Military Operations 98–9, 207, 211
Norway, *Jegertroppen* 222

operations and environmental damage claims 2, 6–7
out-migration, female 36, 42
out-migration, male 42, 43, 48, 51, 52, 80, 236

peacekeeping operations: abuses during 17; Australia 166; gender mainstreaming in 16; Ireland 199; Japan 201; UK 115, 118, 119, 120; UN 137, 190, 208, 210, 222, 224; US 135, 136, 142, 143, 145
Pokot people, response to changing climatic conditions 17, 46, 79
proportionality analysis: in Five Eyes legal doctrine 239; in IHL 178; in NATO doctrine 104, 215; NATO operations 99; US operational doctrine 145
protection of civilians: in Australian doctrine 161, 163, 165–6, 167; Geneva Convention negotiations 179; UN training materials 208–9; in US doctrine 142, 143, 144, 148–50
Protocol Additional to the Geneva Conventions of 1949, I: deficiencies in protection of women 187–8; generally 175–6; negotiation context of 186
Protocol Additional to the Geneva Conventions of 1949, II 187, 188

Quinquepartite Combined Joint Warfare Conference 89, 163, 164

recruitment and retention, women service members: Australia 160; generally 219; Ireland 200–1; Israel 223; Japan 201–2; UK 114, 119; US 135, 200
reintegration difficulties: Boko Haram supporters 27–8, 29; Boko Haram victims 27; former female fighters 22, 23, 136, 140, 145–6
resources, gendered access to 19, 21, 37, 38, 40–1, 46–7, 53, 122, 139
risk: environmental management 241; operational 50, 52, 63, 71, 216
rules of engagement (ROE) 101, 166, 176, 211, 215–16

Salehyan, I. 68–9
Selby, J. 78–9

sexual and gender-based violence as risk multiplier 146
Sierra Leone 23
Smith, R. 8–9
Social Development Direct: first report 116–17; generally 116; second report 117; third report 117–18
social media and the internet: aspect of modern international security environment 9; as education platform 205; role in Syrian Civil War 77, 81
Society stabilization by winning the environment: Ecosystems assessment makes sense . . . Full situational awareness in CIMIC 244
stability operations 8, 71, 125, 139–47, 148, 210, 222
Strong and Secure: A Strategy for Australia's National Security 158–9
surveys, gender-sensitive food security 247–8
Sweden: in ISAF, 211, 220; use of gender advisors, 217–18; *see also* Nordic Centre for Gender in Military Operations
Syria: changes in precipitation 71–2, 74, 77; civil war 71–7; drought 72–5; ethnic and religious fractionalisation 75, 76, 77; internally displaced persons and refugees 75–6; irrigation 74, 79; mismanagement of water resources 74; population growth 75, 76, 77; reductions in government support to agricultural sector 74, 75, 77; repression of dissent by government 75, 76–7; wheat crop 72, 73–6

Talisman Sabre exercise: assessment of US units involved 213–14; gender considerations embedded within 212–13; generally 169, 212
Teaching Gender In The Military – A Handbook 206
Turkana people, responses to changing climatic conditions 17, 46, 79, 214

UK national security strategy: *Building Stability Overseas Strategy* 113–14; *National Security Strategy and Strategic Defence and Security Review* 114
UN Assistance Mission in Afghanistan 102
UN Core Pre-Deployment Training Material 2017 208–9

UN High Commissioner of Refugees 19, 20
UN Intergovernmental Panel on Climate Change: Fifth Assessment Report 49, 62; Fourth Assessment Report 61; generally 61
UN Interim Force In Lebanon: incorporating gender perspectives in operations 223–4; Irish participation 199, 223
UN Security Council Resolution 1325: Australian implementation of 160–2; critique of 16–17; generally 1, 16; militarisation of 17; NATO-EAPC implementation of 91–3; NATO strategic headquarters implementation of 93–102; UK implementation of 113, 114–16, 118–20; US implementation of 132–7
UN Security Council Resolution 1820 1, 162
UN Security Council Resolution 1960 1, 162
UN Signals Academy 209–10
UN Women Training Centre 205–6
urbanisation and urban operations 9, 40, 63, 64, 139
US Army Peacekeeping and Stability Operations Institute 144, 145
US Army values 96
US Department of Defense 59, 60, 65, 67–8, 102, 130, 131–6, 137–8, 150
US Marine Corps University 204
US *National Defense Strategy of the United States of America* 131
US National Intelligence Council 66
US *National Security Strategy* 133, 237
US *National Security Strategy of the United States of America* 131
US *Quadrennial Defense Review* (2014) 67, 68, 131, 133

violence, sexual and gender-based: Australian response to 161, 164, 166–8; effects of 18–19; forms of 18–19; generally 102, 104; Geneva Conventions negotiations concerning 178–83; ICRC commentary concerning 191–2; investigation of 97, 99, 106, 114, 140; men as victims 20, 192; NATO directives' response to 96–8, 99–102; NATO doctrine responding to 104, 106, 216; prohibited under IHL and international human rights law 16, 174–5, 189; in refugee camps 20, 26, 43–4; result of male out-migration 42, 48; training programmes to combat 204, 207, 208–9, 211; UK response to 114–15, 118, 119, 122, 123–5; US response to 135, 136, 140, 142–3, 146, 148–50; women as victims of Boko Haram (*see* Boko Haram)
Von Clausewitz, C. 6, 7, 8
Von Uexkull, N. 69, 245

'war amongst the people' 8, 9, 29, 64, 103
water: gendered access to 6, 7, 38, 244; gendered collection and use of 19, 40, 41, 45, 47, 236, 51, 52; management of 2, 7, 59, 74, 75, 79, 246; salinization of 9, 48; scarcity 40, 61, 63, 64, 65, 67, 79
watershed restoration 246–7
Women, Peace, and Security Act of 2017 60, 130, 137, 138, 139, 147, 150, 200
women farmers: economic disadvantages 38; excluded from decision-making 38; gender-specific crops 39, 245
Women's Initiative Training Teams *see* Agriculture Development Teams
World Health Organization 47

Yusuf, M. 24, 25

Zabul Province *see* Afghanistan
Zimbabwe 186

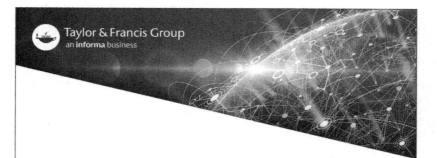